CHITIN AND BENZOYLPHENYL UREAS

SERIES ENTOMOLOGICA

EDITOR

K.A. SPENCER

VOLUME 38

CHITIN AND BENZOYLPHENYL UREAS

Edited by

JAMES E. WRIGHT and ARTHUR RETNAKARAN

1987 **DR W. JUNK PUBLISHERS**
a member of the KLUWER ACADEMIC PUBLISHERS GROUP
DORDRECHT / BOSTON / LANCASTER

Distributors

for the United States and Canada: Kluwer Academic Publishers, P.O. Box 358, Accord Station, Hingham, MA 02018-0358, USA
for the UK and Ireland: Kluwer Academic Publishers, MTP Press Limited, Falcon House, Queen Square, Lancaster LA1 1RN, UK
for all other countries: Kluwer Academic Publishers Group, Distribution Center, P.O. Box 322, 3300 AH Dordrecht, The Netherlands

Library of Congress Cataloging in Publication Data

```
Chitin and benzoylphenyl ureas.

   (Series entomologica ; 38)
   "The original contributions composed the symposium
'Chitin and benzoylphenyl urea' organized by the co-
editors at the International Congress of Entomology in
Hamburg, Germany, August 1984"--Pref.
   Bibliography: p.
   Includes index.
   1. Benzoylphenyl ureas--Congresses.  2. Chitin--
Congresses.  I. Wright, James E., 1940-      .
II. Retnakaran, Arthur, 1934-      .  III. International
Congress of Entomology (17th : 1984 : Hamburg, Germany)
IV. Series: Series entomologica ; v. 38.
SB952.B43C45  1986        632'.951        86-7215
```

ISBN-13: 978-94-010-8638-7   e-ISBN-13: 978-94-009-4824-2
DOI: 10.1007/978-94-009-4824-2

Copyright

Preface

The opportunity to explore a developing new technology in a single biological system, chitin, from the molecular basis and with the inter-relationship of the utilization of benzoylphenyl ureas in effective pest agroecosystem management strategies, represents a new evolution for integration of knowledge in this highly complex area.

The degree of great progress and interest in the understanding of the interaction of chitin ultrastructures, biochemistry, and the unique ben-zoylphenyl ureas attest to the timeliness of this effort. The purpose of the book that follows is to provide up-to-date and well illustrated details of current research knowledge including the latest of research results. The combination of the basic to the applied aspects rarely occurs specifically at the levels presented by the international contributors within.

The original contributions composed the symposium "Chitin and Ben-zoylphenyl Urea" organized by the co-editors at the International Congress of Entomology in Hamberg, Germany, August 1984. We extend our appreciation to everyone who made the conference an outstanding success and highlight of the Congress, as well as making this book possible. We thank the authors, especially for their excellent international presentations, discussions, and preparation of the manuscripts. The knowledge evolved from many areas in chitin research, as well as the intricate complexity of successful utilization of specific knowledge involving benzoylphenyl ureas in system management strategies for insects will continue to be in the forefront throughout the world's agroecosystems; and hopefully this book will enhance further research and development.

We sincerely appreciate the efforts of Ms. Thea Swavely for the preparation and help in putting this book together.

James E. Wright
Arthur Retnakaran
Co-editors

Contents

Contributors

Daniel C. Alder, Departments of Chemistry, Statistics and Zoology, Brigham Young University, Provo, UT 84602

Gary M. Booth, Departments of Chemistry, Statistics and Zoology, Brigham Young University, Provo, UT 84602

Melvin W. Carter, Departments of Chemistry, Statistics and Zoology, Brigham Young University, Provo, UT 84602

Ephraim Cohen, Department of Entomology, The Hebrew University of Jerusalem, Faculty of Agriculture, Rehovot 76-100, Israel

Jeffrey Granett, Department of Entomology, University of California, Davis, CA 95616

A. C. Grosscurt, DUPHAR B.V., Crop Protection Division, P.O. Box 4, 1243 ZG,' s-Graveland, The Netherlands

R. H. Hackman, Division of Entomology, C.S.I.R.O., G.P.O. Box 1700, Canberra, A.C.T., Australia 2601

T. Haga, Central Research Laboratories, Ishihara Sangyo Kaisha Limited, Kusatsu, Shiga 525, Japan

Isaac Ishaaya, Department of Entomology, ARO, The Volcani Center, Bet Dagan 50 250, Israel

B. Jongsma, DUPHAR B.V., Crop Protection Division, P.O. Box 4, 1243 ZG,'s Gravenland, The Netherlands

T. Koyanagi, Central Research Laboratories, Ishihara Sangyo Kaisha Limited, Kusatsu, Shiga 525, Japan

Milton G. Lee Gee, Departments of Chemistry, Statistics, and Zoology, Brigham Young University, Provo, UT 84602

Edwin P. Marks, Metabolism and Radiation Research Laboratory, Agricultural Research Service, U.S. Dept of Agriculture, State University Station, Fargo ND 58105

Bernard Mauchamp, Lab. Phytopharmacie INRA 78000, Versailles, France

R. Nishyiyama, Central Research Laboratories, Ishihara Sangyo Kaisha Limited, Kusatsu, Shiga 525, Japan

Odile Perrineau, Lab. Phytopharmacie INRA 78000, Versailles, France.

Arthur Retnakaran, Forest Pest Management Institute, Canadian Forestry Service, P.O. Box 490, Sault Ste, Marid, Ontario, Canada P6A 5M7

Robert E. Seegmiller, Departments of Chemistry, Statistics, and Zoology Brigham Young University Provo, UT 84602

T. Toki, Central Research Laboratories, Ishihara Sangyo Kaisha Limited, Kusatsu, Shiga 525, Japan

Gordon B. Ward, Jr. Department of Biochemistry, North Dakota State University, Fargo, ND 58105

Robert C. Whitmore, College of Agriculture and Forestry, West Virginia University, Morgantown, WV 26506

James E. Wright, National Program Staff, Agriculture Research Service, U.S. Department of Agriculture, Beltsville, Maryland 20705

Sara Yablonski, Department of Entomology, ARO, The Volcani Center, Bet Dagan 50 250, Israel

1. Chitin and the fine structure of cuticles

R. H. Hackman

1 INTRODUCTION

The arthropod integument consists of the cuticle and the underlying epidermis (Figures 1, 2) and the cuticle serves as both the outer covering of the animal and as the skeleton. The cuticle is not uniform over the entire body, its properties differ according to its function. Depending on the

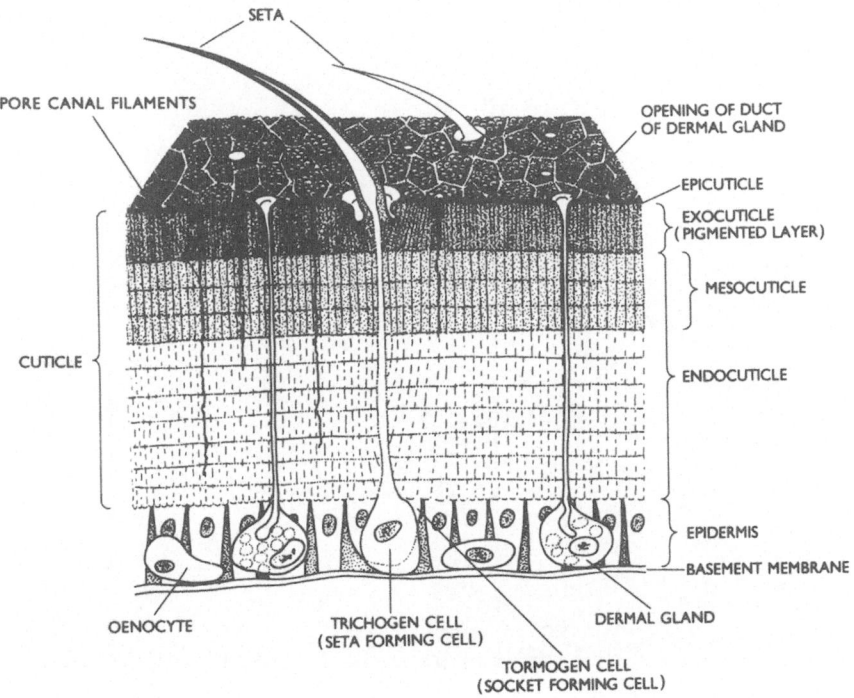

Figure 1. Diagrammatic representation of the insect integument (from Hackman, 1971).

Wright, J. E. and Retnakaran, A. (Eds), Chitin and Benzoylphenyl ureas. ISBN 978-94-010-8638-7.
© *1987, Dr W. Junk Publishers, Dordrecht.*

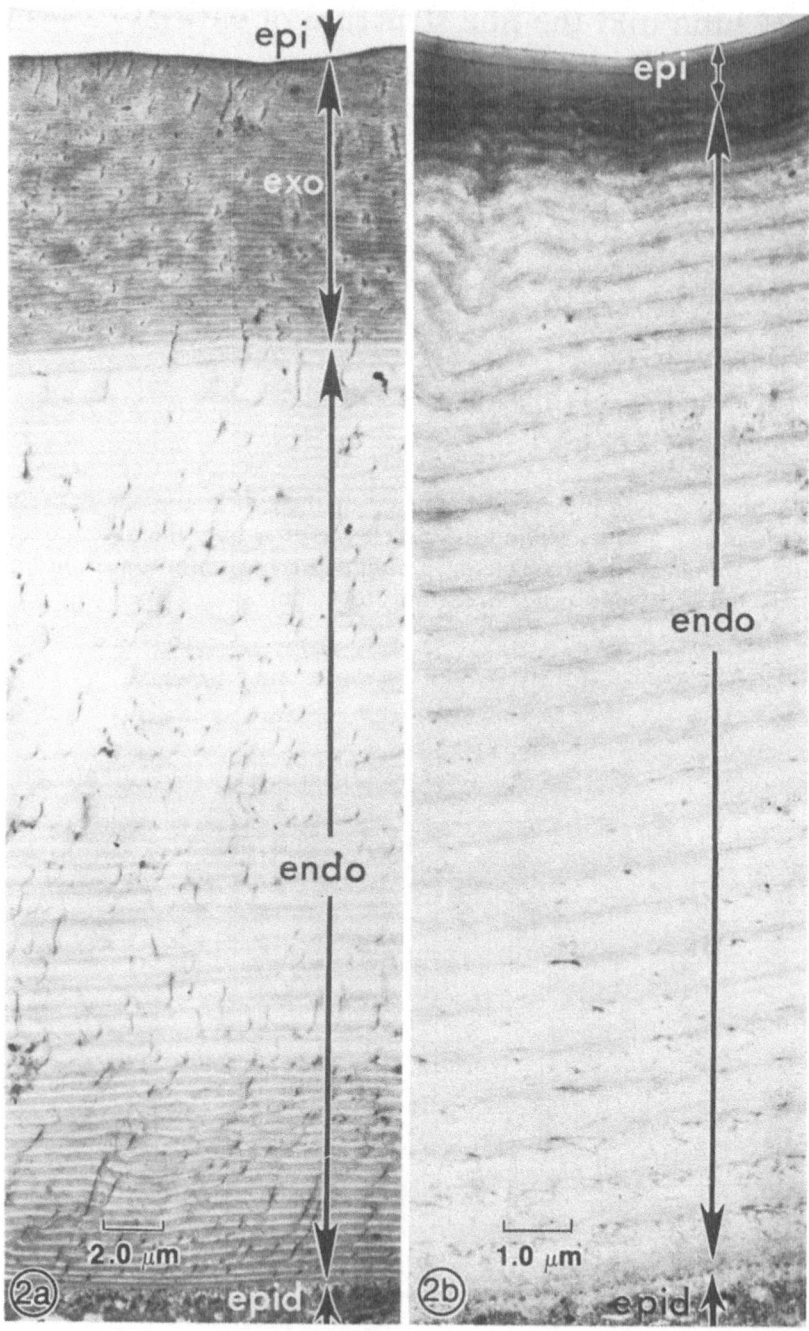

Figure 2. Electron micrographs of transverse sections of (a) the femur cuticle from the metathoracic leg of the locust *Locusta migratoria* and (b) larval cuticle of *Lucilia cuprina*. epi = epicuticle, exo = exocuticle, endo = endocuticle, epid = epidermis.

species, it may vary from a very thin, pliable membrane to a thick, horny armour. Cuticle also lines the foregut, hindgut, tracheae, tracheoles and cuticular glands. Apodemes are of cuticular origin. An arthropod's life cycle is made up of a number of stadia, each terminating in a moult when the cuticle is shed and a new one secreted. The moulting process also permits changes in form to take place.

Cuticle is an extracellular, multilayered, composite material secreted from the epidermis. It consists of a very thin epicuticle and a thicker, inner procuticle (Richards (1951) terminology). The outer layers of the latter (usually those secreted before ecdysis) may become sclerotized to form an exocuticle, the inner layers then form the endocuticle. In the absence of an exocuticle procuticle becomes the endocuticle. The outer part of the endocuticle may stain differently from the inner part. This outer region is then known as the mesocuticle. During secretion of the cuticle a deposition zone lies between the procuticle and epidermis. The epicuticle generally has an outer layer of lipid which may be covered by a cement layer. In some structures, e.g. the linings of tracheoles and of some glands, the procuticle is absent. Chitin and protein are the principle components of the procuticle. The epicuticle is reported not to contain chitin.

Most cuticles contain a network of ducts known as pore canals which vary in size and may be quite conspicuous. They originate as cytoplasmic extensions of the epidermal cells and extend up to the epicuticle where they connect with finer filaments which traverse the epicuticle. These canals and filaments appear to provide a link between the epidermis and the different layers of the cuticle but their function remains controversial. In regions where muscles are attached to the cuticle, pore canals are associated with the muscle attachment fibres (Caveney 1969). The ducts of dermal and sensory glands may also traverse the cuticle.

2 CHITIN

2.1 *Properties and structure*

Chitin is always associated with protein in cuticles and no sample of purified chitin has been found to be free of amino acids. The nature of the association has yet to be firmly established but from what is known about chitins and the cuticular proteins (e.g. Brine 1982; Hackman and Goldberg 1958, 1978; Rudall and Kenchington 1973) it is deduced that protein is bound non-covalently, and possibly also covalently, to the chitin and the link may be through aspartic acid. Blackwell and Weih (1980) have developed a model for the chitin-protein complex as it occurs in the ovipositors of the ichneumon flies *Megarhyssa lunator* and *M. atrata*. These ovipositors contain highly oriented arrangements of the chitin chains. The model shows a sheath of protein subunits arranged in a helix

(6 subunits per turn) around a chitin core. Not all cuticular proteins are bound to chitin, some form the protein matrix in which is embedded the chitin-protein complex. These chitin-protein complexes as ordered structures are discussed by Blackwell in chapter. 2.

Drastic chemical treatments are needed to free the chitin of other cuticular components and chitin is, by definition, the colourless residue remaining after a cuticle has been extracted exhaustively with hot, dilute alkali (Odier 1823). Chemical analysis has shown this insoluble material to be essentially a $(1 \rightarrow 4)$-β-linked linear polymer of 2-acetamido-2-deoxy-D-glucose (i.e. N-acetylglucosamine) residues (Figure 3 in which the structure of chitin is shown in two projections). When chitin is prepared in this way some amino groups may not be acetylated (Hackman and Goldberg 1974) and small amounts of other substances may be present, in particular the outer epicuticle and inorganic compounds (Hackman 1982). This definition of chitin, which is based on its insolubility in alkali, lacks precision, but in practice it has proved to be adequate. Cuticular chitins from all arthropod species are considered to have the same chemical structure but not necessarily the same crystalline form.

Chitin is insoluble in water, dilute alkalis, cold dilute mineral acids and all common organic solvents. It is soluble in concentrated salt solutions, e.g. LiSCN, from which it can be precipitated, apparently unchanged (Clark and Smith 1936), by the addition of water, ethanol or acetone. Some fluorinated compounds, e.g. hexafluoro-2-propanol and hexafluoro-acetone sesquihydrate, and a number of aqueous and anhydrous systems of acids modified with organic solvents dissolve chitin. Rutherford and Austin (1978) have reviewed solvents for chitin and report two new solvent systems based on tertiary amides, viz. 5% LiCl in N,N-dimethylacetamide and 5% LiCl in N-methyl-2-pyrrolidone, as the best inert solvents for chitin. The stability of chitin in these solvents has yet to be investigated. Rarely is a chitin preparation completely soluble and there may be substantial differences in the solubilities of chitins, even from a single arthropod species, depending on the method of isolation (Brine and Austin 1981). This is probably a reflection of the physical state of the chitin.

Chitin is a polydisperse, high molecular weight material, m.w. 1–2×10^6 but the value depends on which m.w. average is measured (Brine and Austin 1981; Hackman and Goldberg 1974; Strout et al. 1976). Therefore chitin has a chain length of 5000 to 10 000 N-acetylglucosamine residues. It is soluble in cold, concentrated mineral acids, but undergoes rapid and extensive degradation (Hackman 1962). Hot acids hydrolyze chitin to glucosamine and acetic acid, and the enzyme combination chitinase (EC 3.2.1.14) and β-N-acetylglucosaminidase (EC 3.2.1.30) hydrolyzes it to N-acetylglucosamine. Before chitin in a cuticle can be hydrolyzed enzymically it may be necessary to remove the associated protein (Bade and Stinson 1978, 1979; Jeuniaux 1963, 1965).

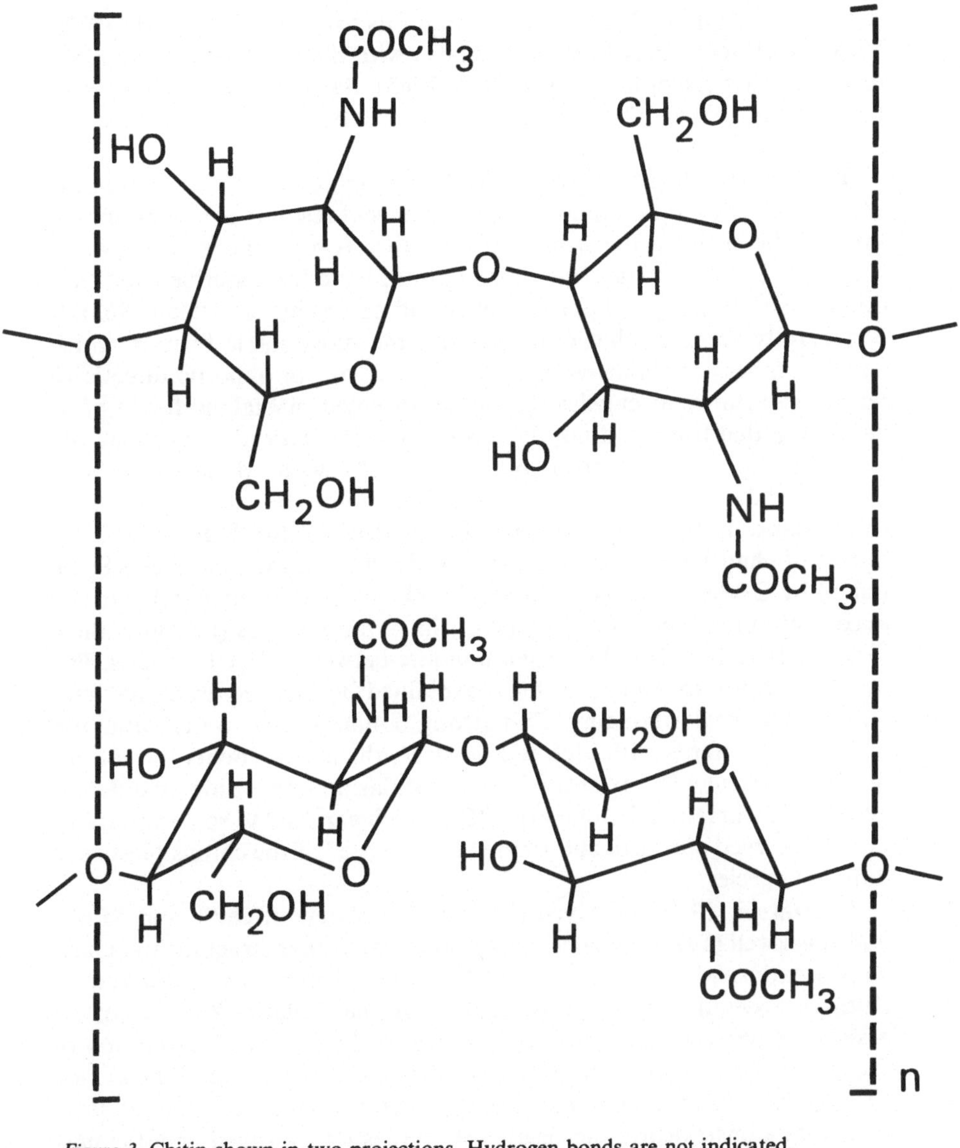

Figure 3. Chitin shown in two projections. Hydrogen bonds are not indicated.

Hot, concentrated alkali deacetylates chitin and degrades it to an ill-defined chitosan (Hackman 1982). Because of differences in molecular bonding some acetyl groups are more readily lost than others. Chitosan is by definition a linear polymer built up of glucosamine residues and is soluble in dilute acids. However, the name is also given to all partially deacetylated chitins which are soluble in dilute acids. During the preparation of chitosan there is a rapid shortening of the main chain as is shown by the rapid fall in the viscosities of its solutions (Muzzarelli 1977). Yields of chitosan may be as low as 50% with the chain length shortened to 20 residues (Horton and Lineback 1965). Sannan et al. (1976) report that 40% NaOH at 25° brings about deacetylation of chitin accompanied by a shortening of the chain.

The β-glucosidic link between the N-acetylglucosamine residues in the chitin chain causes the amino groups in adjacent residues to be on opposite sides of the chain so the structural unit, therefore, is the biose shown in Figure 3. Several crystalline forms of chitin have been ascribed and they differ in the packing and polarity of the adjacent chains (Rudall 1963). It is generally accepted that only α-chitin, the most stable form in which equal numbers of chains with 2_1 screw axes run in opposite directions, occurs in arthropod cuticles. In highly oriented crystalline forms of α-chitin, e.g. that from tendons, the biose units from two adjacent chains are arranged in an orthorhombic unit cell with dimensions $a = 0.474 \pm 0.001$ nm, $b = 1.886 \pm 0.002$ nm and $c = 1.032 \pm 0.002$ nm (fibre axis); the spacing group is $P2_1 2_1 2_1$ (Minke and Blackwell 1978). Sheets of chains are arranged in stacks along the a axis, the sheets being linked by $C=O \ldots H-N$ hydrogen bonds approximately parallel to the a axis. Within a sheet, in the plane of the pyranose rings (i.e. along the b axis), there is interchain hydrogen bonding between CH_2OH groups. The failure of chitin to swell in water is explained by this very extensive inter and intramolecular bonding. This strong bonding leads to the formation of long microfibrils with high tensile strengths and accounts for chitin's fibrous character. The extents of the molecular bondings differ in different cuticles (Hackman and Goldberg 1965) which can lead to some variations in the observed chemical and physical properties of the chitins present in different species.

Muzzarelli (1977) has given an extensive review of chitin and Minke and Blackwell (1978) the latest views on its molecular structure and chain conformation. Much of our knowledge of the properties and structure of chitin has been derived from studies on "purified" chitin. Removal of the associated protein, e.g. with alkali, increases the degree of orientation of the chitin chains, as shown by X-ray diffraction patterns, and the attraction between them is such that they aggregate to give a continuous sheet of chitin. Consequently the results of studies on "purified" chitins may have to be modified when considering chitin as it occurs, associated with proteins and other components, in cuticles.

2.2 Detection and estimation

Pearse (1985) and Richards (1951) discuss histochemical methods for the detection of chitin, e.g. the Schultz chor-zinc-iodine reagent, alkaline tetrazolium, alcian dyes, periodic acid-Schiff reagent and colloidal iron, and conclude that none is specific. The methods give positive results with a variety of compounds. Benjaminson (1969) has described fluorescent conjugates of the enzyme chitinase for use as stains for chitin (the affinity of an enzyme for its substrate). This technique does not appear to have been developed further, perhaps because of the need for a pure enzyme and because chitin in cuticles may not be accessible to the enzyme (see below). Chitin binds wheat germ agglutinin (WGA) and Mauchamp and Schrevel (1977) have incubated cuticle with the fluorescein-isothiocyanate-WGA complex and detected the fluorescent chitin by fluorescent microscopy. The usefulness of this method will depend on its specificity and on the accessibility of chitin in a cuticle to the complex. The van Wisselingh colour test for chitin is a test for chitosan, the chitin must first be deacetylated in 60% w/w KOH at 160° (see above). On a small scale this drastic treatment may completely destroy the sample or the sample may become transparent and be lost. Consequently the test is not necessarily reliable when negative (Hackman 1982). The only physical method by which chitin has been identified is X-ray diffraction but the sample must be free of contaminating materials and the method cannot be used on very small samples.

Removal of non-chitinous components by repeated alkaline extraction (1N NaOH at 100°) is the only method which gives absolute values for the chitin content of cuticles and as an analytical method it has been used on macro and semi-micro scales. The values, when corrected for the presence of epicuticular membranes and inorganic compounds, are accurate (Hackman 1982). The membranes, unlike chitin, are stable to acidic hydrolysis and inorganic compounds are assayed by ashing the material. Calcified cuticles are first decalcified with cold dilute acid. Microgram amounts of chitin can be estimated as soluble chitosan (i.e. deacetylated chitin) derivatives (Hackman and Goldberg 1981) and the method is not affected by the presence of non-chitinous materials. Hackman (1982) has reviewed methods for the detection and estimation of chitin in cuticles.

Hot acids, e.g. 6N HCl at 100°, hydrolyze chitin to glucosamine and acetic acid and both compounds have been assayed to estimate chitin (e.g. Brine 1978; Chen and Mayer 1981; Fischer and Nebel 1955; Holan et al. 1980). However, recoveries of glucosamine may be variable, the rate of hydrolysis depends on the source of the chitin and glucosamine may be destroyed when hydrolyzates are evaporated, especially in the presence of amino acids and neutral sugars (Fischer and Nebel 1955; Hackman 1962). Also compounds other than chitin yield glucosamine and acetic acid on hydrolysis. Consequently the method cannot be considered always reliable even though it gives reproducible results when experimental conditions are rigorously controlled. The method can only give relative values and needs to be standardized on known amounts of chitin.

Enzymic hydrolysis of chitin is not a suitable method for estimating chitin in cuticles because most of the chitin is not accessible to the enzyme (Bade and Stinson 1978; Jeuniaux 1963, 1965). The method has been used to measure the amounts of free and bound chitin in cuticles (Jeuniaux 1963, 1965). Bound chitin is hydrolyzed enzymically only after the cuticle has been extracted with hot alkali (to remove non-chitinous material) and it includes complexes such as glycoproteins. Highly crystalline forms of chitin resist enzymic digestion because the enzyme cannot penetrate the structure.

The chitin contents of cuticles (as per cent of dry weight) range from 1.4 per cent in third instar *Loxostege sticticalis* larvae (Pepper and Hastings 1943) to 60.3 per cent in adult *Glossina morsitans* (Hackman and Goldberg unpublished) though they commonly fall within the range 25 to 50 per cent (Table 1). Cuticle from different developmental stages of a species, the different types of cuticle from one insect or even the same type from male and female insects of one species may have different chitin contents. Richards (1947) did not detect chitin in the wing scales of some lepidopterans. Decalcified crustacean cuticles contain 18.1 to 72.1 per cent chitin (Brine and Austin 1981; Welinder 1974) and arachnid and myriapod cuticles 3.2 to 38.2 per cent (Hackman 1982; Hackman and Goldberg 1985; Shrivastave 1970).

The relative proportions of protein and chitin in cuticles change during cuticulogenesis. In *Locusta migratoria migratorioides* the cuticle deposited immediately after apolysis contains 90 to 95 per cent protein but as cuticle synthesis and deposition proceeds this falls to about 70 per cent and the chitin content rises to about 30 percent (Cassier et al. 1980). Similar changes take place in the femur of 5th instar *Schistocerca gregaria*, at ecdysis the cuticle contains 24 per cent chitin, the outer half of the endocuticle (laid down during the first day after ecdysis) 38 per cent and the inner half (laid down from day 2 to day 6) 55 per cent (Andersen 1973).

3 FINE STRUCTURE OF CUTICLE

The fine structure of the cuticle is discussed in terms of the deposition and organization of chitin microfibrils and their assembly into the lamellate and non-lamellate structures seen in electron micrographs of procuticle. Two of the five layers of the epicuticle are also considered, namely the outer and inner epicuticles. Both are relevant in a discussion on chitin in cuticles, the outer epicuticle because it occurs as a contaminant in isolated chitins and the inner epicuticle because, in at least one species, it contains microfibrils similar to those in procuticle. Descriptions of other aspects of cuticle such as pore canals and lipid distribution in procuticle and the entire epicuticle are given in reviews, e.g. Filshie (1982) and Neville (1975).

Table 1. Chitin content of some insect cuticles.

Species	Chitin % of dry weight
Blattodea	
Blatta orientalis abdomen sternite ♀, ♂	34.2, 36.0[a]
tergite ♀, ♂	31.7, 35.0[a]
Periplaneta americana abdominal and pronotal	33.8[b]
Phasmatodea	
Dixippus morosus abdomen dorsal	39.8[c]
ventral	42.5[c]
thorax, dorsal	36.5[c]
ventral	44.9[c]
Orthoptera	
Locusta migratoria abdominal	31.8[c]
abdominal intersegmental	42.2[d]
membrane	23.7[c]
elytra and wings	36.9[c]
femoral	
Hemiptera	
Rhodnius prolixus abdominal	11.2[e]
Coleoptera	
Agrianome spinicollis larval	37.2[b]
Dytiscus sp. elytra	37.4[c]
Tenebrio molitor larval	31.3[c]
Tribolium confusum larval	22.8[a]
Xylotrupes gideon pronotal	46.3[b]
Diptera	
Calliphora augur puparia	37.1[b]
Glossina morsitans adult abdomen ventral	60.3[d]
Lucilia cuprina larval	52.5[b]
puparia	41.1[b], 39.6[f]
pupal and moulting membranes	10.1[f]
Melophagus ovinus abdominal	34.5[d]
Phormia regina larval	48.6[a]
Phormia terranovae larval	59.5[c]
Lepidoptera	
Bombyx mori larval	45.4[b], 44.2[c]
Galleria mellonella larval	38.3[a]
Loxostege sticticalis 3rd instar larval	1.4[g]
4th instar larval	1.6[g]
5th instar larval	2.3[g]
Sphinx ligustri larval	50.3[c]

[a] Tsao & Richards (1952)
[b] Hackman & Goldberg (1971)
[c] Fraenkel & Rudall (1947)
[d] Hackman & Goldberg unpublished
[e] Hackman (1975)
[f] Gilby & McKellar (1970)
[g] Pepper & Hastings (1943)

3.1 *Procuticle*

Procuticle when viewed in transverse section (i.e. normal to the cuticle surface) usually appears to be built up of lamellae, which in electron micrographs alternate as dark and light bands (Figure 2) those in exocuticle being very narrow (Figure 2a). Mesocuticle shows less difference in electron density across the lamellae than does endocuticle but is otherwise similar to endocuticle (Fogal and Fraenkel 1970; Noble-Nesbit 1967). The lamellae are mostly parallel to the surface of the cuticle. However, in cuticles with textured surface patterns only the most distal lamellae follow the contours of the pattern. Figure 4 shows the lamellar structure beneath a tubercle on the cuticle of the tick *Argas robertsi*. A tangential section immediately beneath the tubercle would show the lamellae as a series of concentric rings. Electron micrographs of transverse sections of cuticle also show an array of small, electron-lucent (i.e. unstained) areas and electron-lucent filaments in a matrix of electron-dense (i.e. heavily stained) material. These small, electron-lucent areas may be packed hexagonally into groups (Figure 5) (see also Filshie 1982; Neville 1975; Rudall 1965). The filaments are considered to be microfibrils which when cut in cross section give the small, electron-lucent areas. The structure of the procuticle is determined by the nature and arrangement of these microfibrils.

3.2 *Microfibrils*

The chitinous nature of the microfibrils has been deduced from indirect evidence only, viz. microfibrils are found in chitin containing regions of cuticles (but see below and EPICUTICLE) and removal of protein from the cuticle reduces the heavily stained matrix. These facts led Rudall (1967) to conclude that cuticle is composed of microfibrils of chitin in a matrix of protein. Neville (1970), when examining the staining properties of structures composed of chitin and the protein resilin, came to the same conclusion. In addition estimates of the volume fraction of the microfibrils based on measurements taken from electron micrographs of these structures agreed with the known chitin content (Neville et al. 1976) and patterns presumed to be chitin disappear from locust cuticle deposited

Figure 4. Electron micrograph of a transverse section through a tubercle on the cuticle of *Argas robertsi* to show the lamellar pattern.

Figure 5. Electron micrograph of a transverse section of *Tenebrio molitor* sternite exocuticle showing the hexagonal arrangement of microfibrils (arrow heads).

Figure 6. Electron micrograph showing the arcuate pattern in lamellae of *Lucilia cuprina* larval cuticle sectioned obliquely.

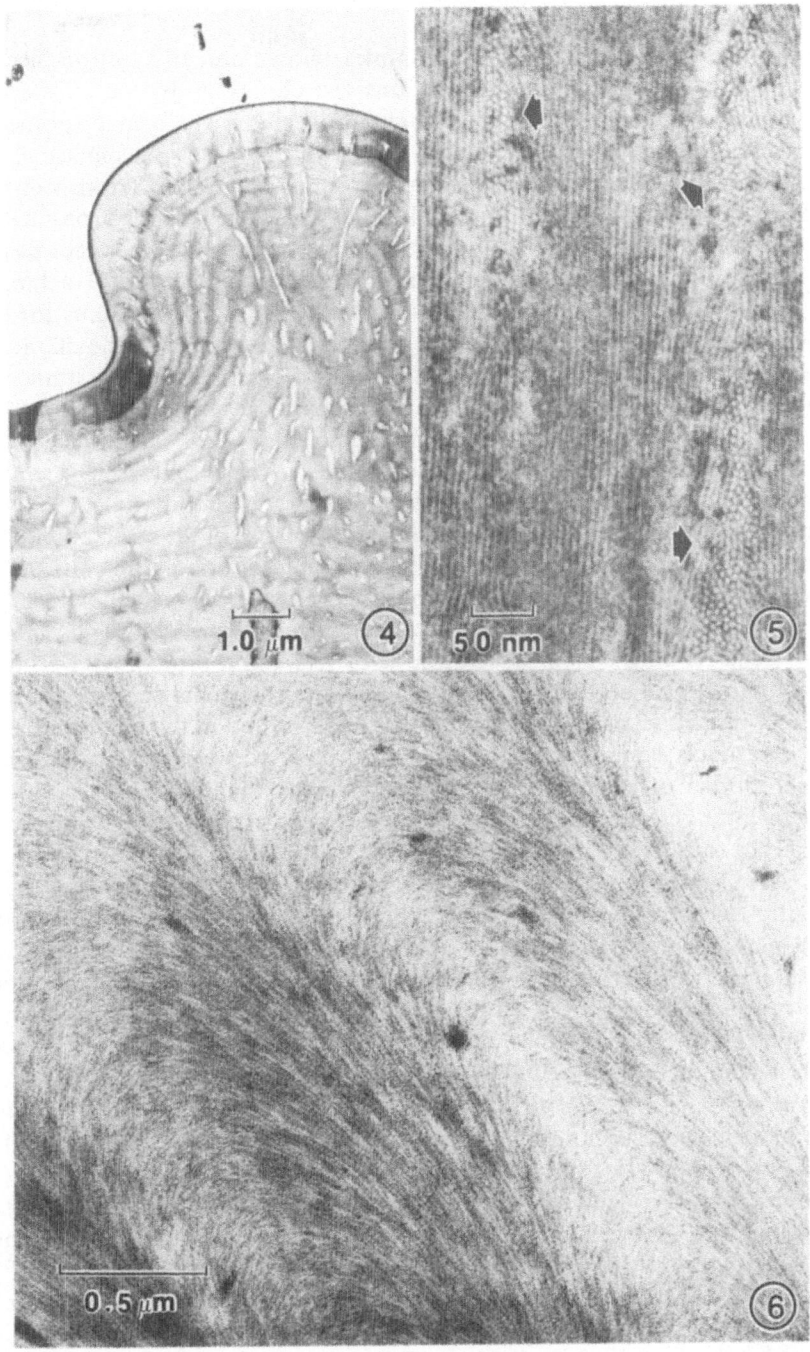

after the locusts have been treated with the chitin synthesis inhibitor diflubenzuron (Ker 1977, 1978).

However, as the following examples show, not all electron-lucent microfibrils in arthropod preparations are chitinous. In the colleterial region of the tortoise beetle, the protein fibrils which will form the oötheca do not take up heavy metal stains (uranyl acetate and phosphotungstic acid) but the protein matrix which surrounds them does (Atkins et al. 1966). The endocuticle of the tick *Boophilus* contains electron-lucent microfibrils which Filshie (1982) estimatede occupy 30–50 per cent of its volume, yet the chitin content of the cuticle is only 3.2 per cent (Hackman 1982). Most of these microfibrils must be protein which accounts for 94 per cent of the dry cuticle by weight. Electron micrographs of the silkmoth chorion are similar in appearance to those of cuticle and show a microfibrilar structure but the chorin does not contain chitin (Mazur et al. 1982).

High resolution micrographs of cuticular microfibrils show them to be poorly stained (Filshie 1982), not unstained as is usually stated, which indicates that either chitin takes up some stain (see Giraud-Guille 1984) or the presence of other material, e.g. protein. Because of the strong bonding between chitin chains it is assumed that the microfibrilar core consists of crystalline chitin and only the peripheral chains are associated with protein. However, chitin synthesis in *Sarcophaga bullata* and *Plodia interpunctella* requires concurrent synthesis of RNA and protein (Clever and Bultman 1972; Ferkovich et al. 1981) and so protein may also be associated with chitin chains within the core. Thus, although the microfibrils in many cuticle contain chitin, there is no convincing evidence that any microfibrilar core is pure chitin. Neville refers to the microfibrils as chitin crystallites but until their composition has been established they are best referred to as microfibrils.

The diameters of microfibrils (i.e. of electron-lucent rods) and their centre-to-centre spacings show large variations among groups of arthropods but their size and spacings appear to be uniform in any one cuticle. Diameters range from 2.5 nm in some insects and crustacea, with centre-to-centre spacings of 5 to 10 nm, to 25 nm in a crustacean copepod with a spacing of 50 nm (Table 2). Gardiner and Khan (1979) report microfibrils with a diameter of 300 nm in the jaws of the locust *Chortoicetes terminifera*. Neville et al. (1976) reported the mean diameter of microfibrils in 13 arthropod species as 2.8 nm and, by assuming them to be pure chitin and using the X-ray diffraction-determined dimensions of the unit cell of "pure" chitin, calculated that a microfibril contained 18 chitin chains arranged in 3 sheets each containing 6 chains. Previously Rudall and Kenchington (1973) estimated that in a highly flexible cuticle (blowfly larva) the microfibrils (diam. ca 2.5 nm) are composed of 3 sheets, each containing 7 chains of chitin whereas in the *Rhyssa* ovipositor (a rigid cuticle) the unit microfibril (diam. 5–6 nm) contained about 4 times as many chains. Recently Atkins (cited by Neville 1984) has calculated that the microfibrils in the cuticle of the *Rhyssa* ovipositor each contain 19 chains of chitin (2 rows of 6 and 1 of 7), and those in the locust tendon 25 chains (2 rows of 8 and 1 of 9).

Filshie (1982) has questioned the accuracy of published values of microfibril diameters derived from measurements taken from electron micrographs because of possible systematic and subjective errors. He considers the value 2.8 nm should be taken only as a lower limit. Consequently estimates of the number of chains in a microfibril calculated from such values and the volume fractions of microfibrils derived from electron micrographs are of unknown accuracy.

3.3 *Lamellae*

The lamellae seen in electron micrographs may have an arcuate pattern superimposed upon them. This pattern is very clear in Figure 6 which is of *Lucilia cuprina* cuticle sectioned obliquely. Drach (1953) considered that in the cuticle of the crustacean decapod *Homarus vulgaris* this arcuate pattern resulted from filaments which are continuous from one lamella to the next, the filaments being in parallel oblique planes. That is, the lamellae as seen in light and electron micrographs are real and are structural components of the cuticle. Scanning electron micrographs of the cuticle of the chelae of the decapod crustaceans *Homarus gammarus* and *Cancer pagurus* (Dalingwater 1975) appear to support this conclusion but Gubb (1975) regards the curved interlaminar filaments observed by Dalingwater as being displaced pore canal tubules, Dalingwater (1975a, b, 1977), Dennell (1973, 1974, 1975, 1976a, b, 1978), Ejike (1973), Mutvei (1974, 1977) and Rudall (1969) have favoured the Drach model when describing the structure of a number of arthropod cuticles. Locke (1961) modified the Drach model by having the curving filaments in parallel planes normal to the cuticle surface.

Bouligand (1965) took a different view and attributed the arcuate pattern seen in crustacean cuticle to an optical effect produced when sheets of parallel filaments are arranged so that successive sheets are rotated through a small angle about an axis perpendicualr to the sheets and the array is sectioned obliquely. The lamellae are not real and each results from 180° rotation of sheets. This arrangement, now known as the helicoidal model, has been likened to a pack of cards in which each card, representing a sheet of fibrils, is turned thorugh an angle slightly greater than that of the card immediately above it (Neville et al. 1969). Figure 7a depicts a number of sheets of fibrils and the arcuate pattern as it would be seen in a section which has been cut obliquely (45°) to the cuticle surface and viewed normal to the cut surface of the section. Only 5 sheets forming

Table 2. Dimensions of microfibrils.

Species	Diameter nm	Centre-to-centre spacing nm	Reference
Insecta			
Blattodea			
Blaberus discoidalis	3.04	7.2, 6.3	Neville et al. 1976
Isoptera			
Termite sp.	2.82	7.1, 7.2	Neville et al. 1976
Orthoptera			
Eutropidacris cristata	2.77	5.4, 5.3	Neville et al. 1976
Schistocerca gregaria	2.74	5.5, 5.3	Neville et al. 1976
Hemiptera			
Hydrocyrius columbiae	4.5	6.5	Neville & Luke 1969b
Coleoptera			
Tenebrio molitor	4	6.2	Delachambre 1973
	2.86	5.4, 5.3	Neville et al. 1976
	5.0		Neville 1970
Oryctes rhinoceros	2.71	5.9	Neville et al. 1976
Siphonaptera			
Xenopsylla cheopsis	3.08	9.1	Neville et al. 1976
Diptera			
Glossina austeni	2.41	5.7	Neville et al. 1976
Calliphora sp.	ca 2.5		Rudall 1967
Lepidoptera			
Aglais urticae	3.01	5.5, 5.9	Neville et al. 1976
Macroglossum stellarum	2.87	6.5, 6.1	Neville et al. 1976

Hymenoptera			
Apis mellifera	2.94	5.8, 5.6	Neville *et al.* 1976
Megarhyssa lunator	2.8 or 3.8	7.25	Blackwell & Weih 1980
Megarhyssa nortoni			
nortoni	4.5–6		Rudall 1967
Crustacea			
Copepoda			
Porcellidium fimbriatum	25	50	Gharagozlou-van Ginneken & Bouligand 1975
Decapoda			
Cancer maenas	3	4.6	Giraud-Guille 1984
Hemisquilla ensigera	2.50	4.4	Neville *et al.* 1976
Arachnida			
Araneida			
Cupiennius salei	3.5	5.2, 5.0	Barth 1973
Tegenaria agrestis	2.72		Neville *et al.* 1976
Scorpionida			
Hadrurus arizonensis	4–4.5		Filshie & Hadley 1979

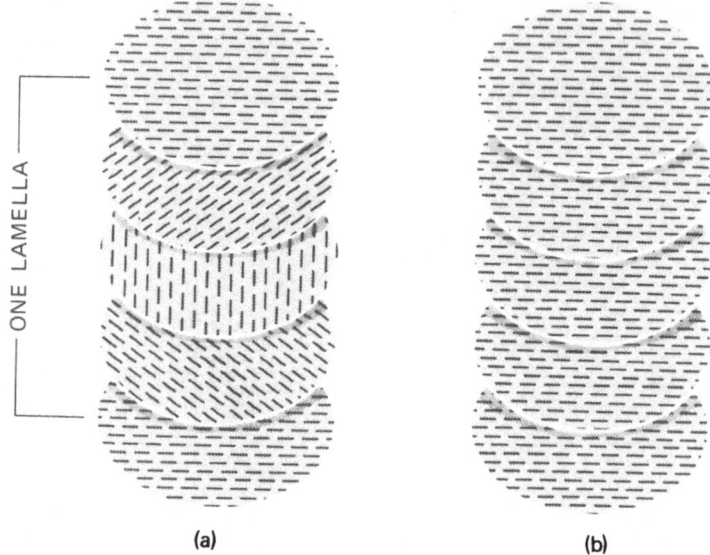

(a)   (b)

Figure 7. (a) Three-dimensional diagram of the rotation of circular sheets of microfibrils, arranged step-wise and viewed normal to the pitch of the steps, to show the origin of the arcuate pattern seen in Fig. 6. Fig. 7a represents the arrangement of microfibrils in a section of cuticle cut obliquely to the cuticle surface. For clarity the sheets are drawn for each 45° of rotation, in a cuticle a lamella is formed from many more sheets of microfibrils. The accurate pattern is formed from short lengths of straight microfibrils. (b) Unidirectional arrangement of microfibrils as would be present in non-lamellate cuticle. Although drawn as sheets of microfibrils the structure would be continuous.

a lamella have been drawn. Cuticular lamellae are formed from many more sheets and the arcuate patterns are formed from a large number of short lengths of straight fibrils. In the cuticle each sheet is built up from appropriately oriented fibrils as the cuticle is being secreted and laid down. The arcuate pattern does not appear in micrographs of sections cut normal to the sheets of fibrils, unless the section is tilted, and fibrils are seen in the plane of the section alternating with fibrils cut at various angles. The direction of rotation of the sheets of fibrils is anticlockwise down through the cuticle (Neville and Luke 1971). In insect cuticles microfibrils replace the macrofibrils seen in crustacean cuticles. Neville (1984) gives the pitch of the helicoid (i.e. 360° turn, 2 lamellae) in different cuticles as falling within the range 0.1 to 20 μm. (As calculated from the micrograph given by Filshie (1982, Figure 11) the pitch of the helicoid in *Lucilia cuprina* larval cuticle is approximately 1.5 μm).

Discussions continue on the validity of these two contrasting models, more so because published micrographs do not permit either continuous interlaminar fibres or patterns built up from short lengths of straight fibrils to be identified with certainty. However, in the majority of papers published on the fine structure of cuticles the Bouligand helicoidal model is favoured over Drach's continuous filament model, perhaps because of the apparent simplicity with which it explains the structural complexities observed. Objections to the helicoidal model have been explained as artefacts produced during the preparation of sections or as misinterpretations of micrographs (Bouligand 1972, 1978; Neville 1975). In the following references the helicoidal model is adopted to explain the structure of arthropod cuticles, Altner 1975; Barth 1973; Brück and Stockem 1972a, b; Credland 1978; Delachambre 1971, 1975; Filshie and Hadley 1979; Filshie and Smith 1980; Gharagozlou-van Ginneken 1974, 1976; Gharagozlou-van Ginneken and Bouligand 1973, 1975; Giraud-Guille 1984; Goudeau 1974; Green and Neff 1972; Gubb 1975; Hegdahl et al. 1977a, b; Livolant et al. 1978; Neville 1970; Neville and Berg 1971; Neville and Caveney 1969; Neville and Luke 1969a, b, 1971; Neville et al. 1969; Noirot and Noirot-Timotheé 1969; Pace 1972; Wigglesworth 1975; Zacharuk 1972; Zelazny and Neville 1972.

In the cuticles of a number of exopterygote insects (both nymphs and adults) and adult endopterygote insects, lamellate cuticle is laid down at night and non-lamellate cuticle during the day, the changes in orientation being controlled by a circadian clock (Neville 1975). Neville (1963a, b) first described this in locusts. The non-lamellate regions would differ from the lamellate regions in that the microfibrils are all oriented in the same direction, they do not rotate (Figure 7b). These daily growth layers have been used to determine the ages of insects (Neville 1983).

Neville and Luke (1969b) have extended the Bouligand model to give a two-system model of helicoidally oriented microfibrillar sheets and preferred oriented (i.e. unidirectional) microfibrils to describe the architecture of insect cuticles. The microfibrils may be helicoidally arranged throughout the cuticle as in some lepidopteran, dipteran and coleopteran larvae, follow a preferred orientation throughout as in orthopteran apodemes or both types of orientation may be present. When both types are present either may predominate and there may be a constant direction of all preferred orientations as in orthoptera, or a change in the direction of the regions with preferred orientations through a gradual rotation of sheets of microfibrils forming part of a helicoid, when passing from one major direction to the next. This latter arrangement gives the pseudo-orthogonal systems seen in some hymenopteran, coleopteran and odonatan cuticles.

The orientation of the microfibrils in the cuticles determines the architecture of pore canals which traverse the cuticle (see INTRODUCTION). In cross section pore canals are flattened and they form straight ribbons in regions with preferred orientation of microfibrils, but in helicoidally oriented sheets they take the form of twisted ribbons (Neville and Luke 1969b; Neville et al. 1969).

Some cuticles appear to have structures which deviate from those which the two-system model would give. Neville (1975) interprets such deviations as having been brought about by the stresses placed upon the cuticles after the microfibrils have been laid down. He instances the postecdysial expansion of the presumptive exocuticle prior to its sclerotization which would deform the cuticular pattern, as would the cuticular contraction which takes place during puparium formation in Diptera. Also the larval stages of endopterygotes grow, sometimes quite dramatically, during the inter-moult period and the expansion of the cuticle which accompanies this growth would bring about a reorganization of the lamellae, most notice-able in those laid down early in the instar (Figure 8). Again female *Boophilus microplus* adults, in common with some other arthropods, feed very rapidly which causes the abdominal cuticle to expand and its struc-ture to become disrupted (Figure 9) (Hackman 1975). These deviations are distortions brought about by a continuous deformation of the ideal model and they are functional.

A notable feature of many cuticles is the regularity of the lamellar patterns seen in micrographs and the absence of defects. Some defects have been reported, though, and these cause structures to depart from the ideal model, e.g. edge dislocations in *Locusta* cuticle (Bouligand 1972) where lamellae have been interrupted in places and disclinations in the metathoracic endocuticle of the beetle *Dynastes cerebrosis* (Zelazny and Neville 1972) which correct for rotational defects attributed to ducts traversing the lamellae. Defects, like distortions, appear to be functional.

However, the structure of some cuticles cannot be explained as post-depositional changes of the two-system model of Neville and Luke or by the presence of defects. Both the adult abdominal sternite exocuticle of *Tenebrio molitor* and the alloscutal cuticle of the adult female cattle tick *Boophilus micoplus* appear to contain lamellae composed of macrofibrils, rather than microfibrils. These lamellae are parallel to the cuticle surface and the macrofibrils in *Tenebrio* curve between verticle columns of micro-fibrils (Figure 10) and those in *Boophilus* curve between the large pore canals (Figure 11) (see Filshie 1982; Hackman and Filshie 1982). Both of these cuticles have the superficial appearance of being lamellate and helicoidal (*sensu* Bouligand). Delachambre (1971) had previously de-scribed the presence of numerous small cuticular rods in *Tenebrio* exocuti-cle. Vertical columns have also been reported in the cuticles of other beetles and of crustacea, millipedes and damsel flies (reference in Filshie 1982; Neville 1975) and in the larval cuticle of *Manduca sexta* (Wolfgang and Riddiford 1981). These last authors suggest that vertical cuticular columns may be a generalized structure in larvae which show large amounts of continual growth. Although the structures of the hyaline and inner exocuticles of the desert scorpion, *Hadrurus arizonensis*, are ex-plained by the helicoidal model, that of the endocuticle cannot be ex-

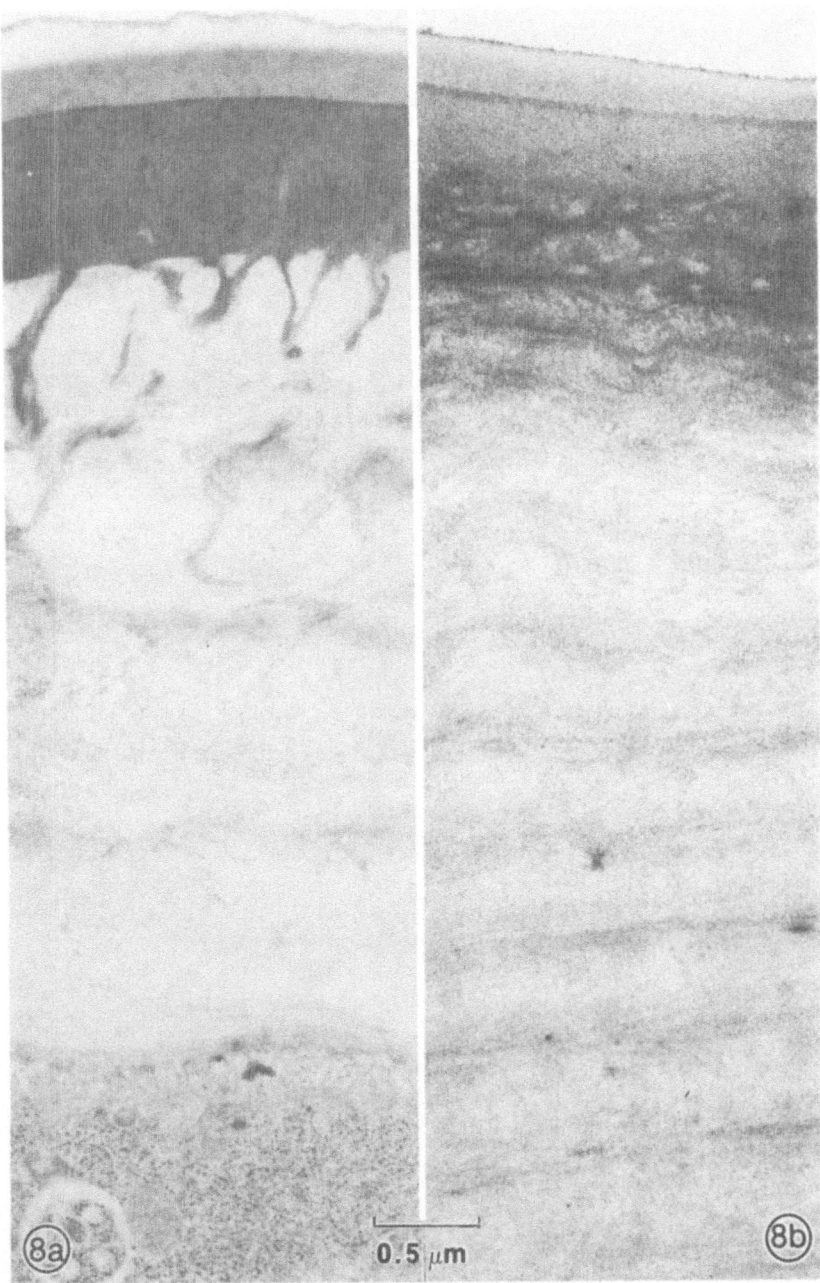

Figure 8. Electron micrographs of transverse sections of cuticle from (a) 6 h old and (b) 48 h old 3rd instar *Lucilia cuprina* larvae showing the reorganization of lamellae that takes place during larval growth.

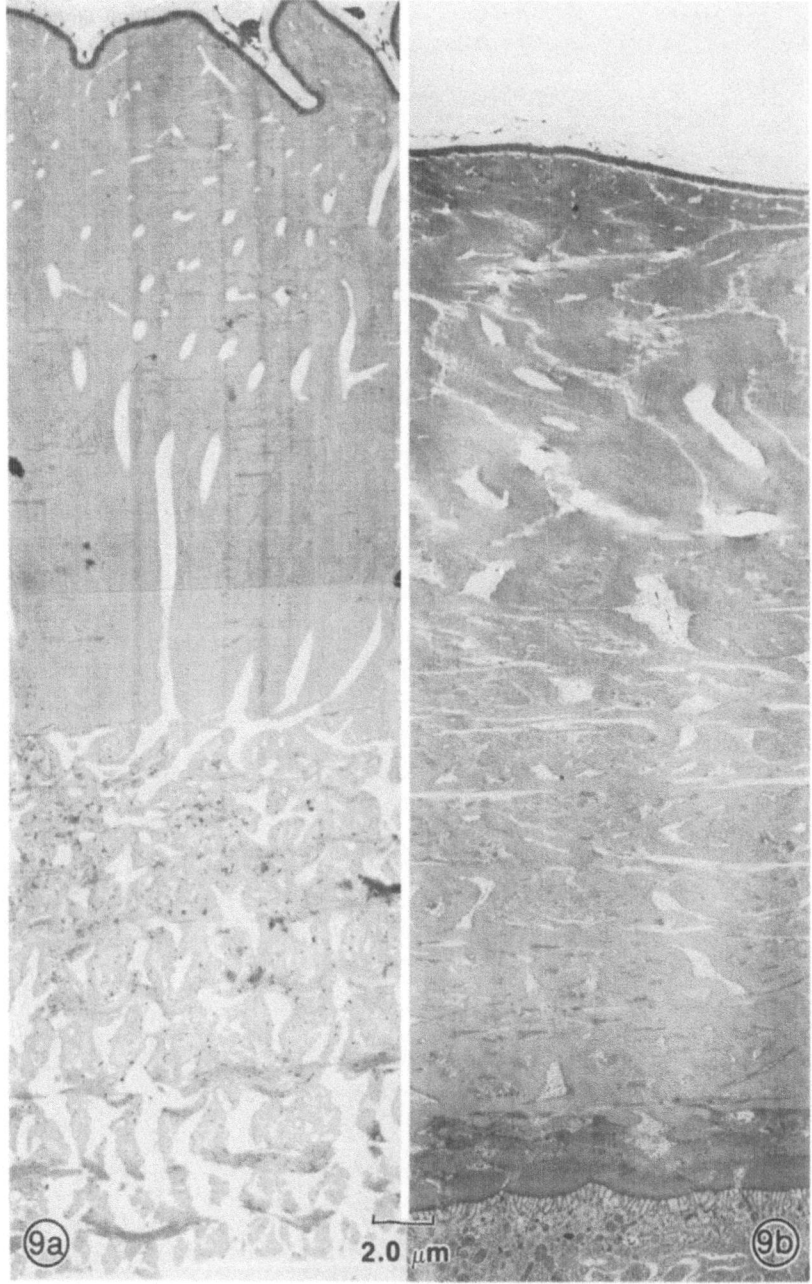

Figure 9. Electron micrographs of transverse sections of the alloscutal cuticle from adult female *Boophilus microplus* (a) before and (b) after rapid feeding. During feeding the cuticle expands and its structure becomes disrupted.

21

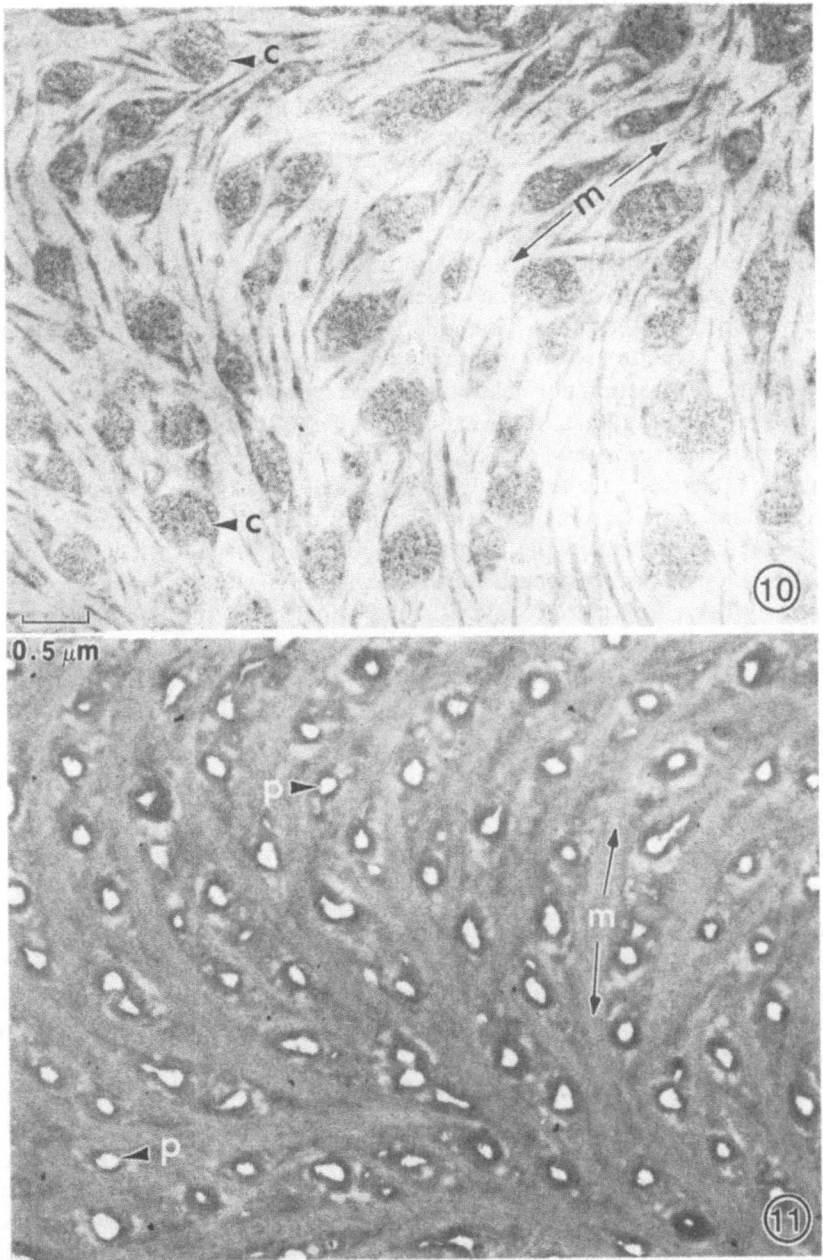

Figure 10. Electron micrograph of a tangential section of the sternite exocuticle from *Tenebrio molitor* showing macrofibrils (m) curving between vertical columns of microfibrils (c).

Figure 11. Electron micrograph of a tangential section of the alloscutal cuticle from adult female *Boophilus microplus* showing macrofibrils (m) curving between the large pore canals (p).

plained by either the Drach or the helicoidal model. It consists of bundles of microfibrils oriented horizontally and vertically (Filshie and Hadley 1979).

Chitin microfibrils, being composed of long molecular chains bound firmly one to another, contribute significantly to the mechanical strength of cuticles and their final orientation appears to be related to the stresses which will be placed upon the cuticle. The helicoidal model gives a plywood structure to the cuticle, each ply (or sheet) being made up of high modulus chitinous fibrils in a low modulus protein matrix. This would give a very strong structure, one which would resist bending forces in all directions. On the other hand microfibrils with a preferred orientation provide a structure strong in tension (when the strain is on the chitin) but weak in the direction normal to the chitin chains (when the strain is on the protein). For example, in the exocuticle of the locust femur the layers of microfibrils are closely packed and helicoidally arranged, while in the endocuticle there are axially oriented microfibrils between thin layers of helicoidal lamellae (Figure 2a). The femural cuticle is, therefore, well designed to withstand the longitudinal stresses to which it will be subjected. Newly fledged locusts fed benzoylphenyl ureas form cuticles with greatly decreased chitin contents, ones whose structural coherence and optical properties now depend primarily on the protein component (Ker 1977). The cuticle has only half the rigidity of normal cuticle and fails at about one-third the strain (Hunter and Vincent 1974).

3.4 *Liquid crystals and cuticle as a liquid crystal analogue*

Liquid crystals are highly organized geometric systems (Gray 1962). They are ordered liquids, being neither crystalline solids nor amorphous liquids. They consist of elongated molecules and, in the absence of bulk flow, show birefringence. Liquid crystals may exist in one of three basic states or mesomorphic phases, viz. smectic, nematic or cholesteric. Changes in temperature or in concentration may bring about a change in the phase adopted.

In the smectic phase (Figure 12a) the molecules are arranged in parallel layers, the heads and tails of all molecules being alligned, i.e. there is order in the direction of the molecular axes and in the position of the molecules. In the nematic phase (Figure 12b) the aligned molecules are arranged unidirectionally but there is no regular arrangement of the ends of the molecules. This represents a lower degree of order than that in the smectic phase. In the cholesteric phase (Figure 12c) the molecules are arranged in layers and within each layer there is a parallel alignment of molecules. Successive layers are displaced so that the molecular axes trace out a helix.

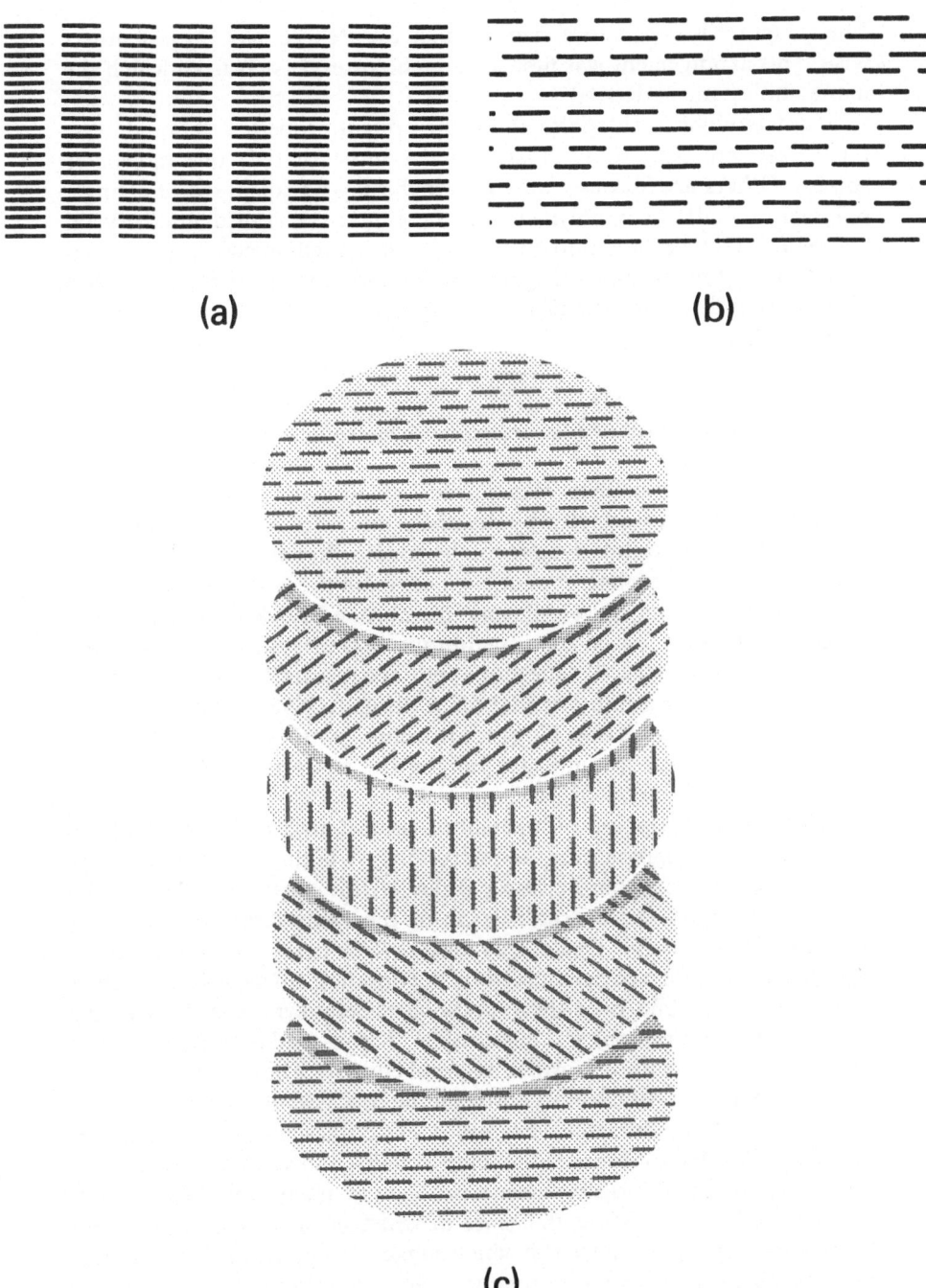

Figure 12. Idealized representations of liquid crystal structure showing the arrangement of molecules in (a) smectic, (b) nematic and (c) cholesteric mesophases.

The structure is continuous even though Figure 12c may suggest otherwise. This arrangement was first seen in cholesteryl esters from which the name cholesteric is derived.

Many biological materials exhibit the optical properties of liquid crystals. Some are true liquid crystals and consist of only one molecular species. Others are fibrous materials which may be composed of a single polymer or a mixture of polymers and these have been called liquid crystal analogues. Cuticle falls into this class, non-lamellate cuticle (preferred orientation of microfibrils) (Figure 7b) corresponds to the nematic phase (Fig. 12b) and the helicoidal model (Figure 7a) to the cholesteric phase (Figure 12c).

3.5 *Secretion of chitin and its assembly as microfibrils*

All cuticular components (except the cement layer which comes from epidermal glands) are secreted by the epidermal cells, but the components are assembled extracellularly to form the cuticle. Precursors for chitin, and also for the entire lamellate procuticle, pass from the cells through specialized plaques on the apical plasma membrane (Locke 1967a). Chitin appears to be synthesized by a membrane bound system located in the epidermis (Vardanis 1979). Questions yet to be resolved are how the chitin chains are assembled into oriented microfibrils to give the fibrous cuticle and what role the epidermis plays. Chitosomes synthesizing chitin microfibrils, such as occur in fungi (for references see Bartnicke-Garcia et al. 1978), have not been described from insects.

Procuticle is composed of microfibrils in a matrix of protein and the precision of the alignment of the microfibrils is high. They are generally arranged in sheets, more or less parallel to the surface of the cuticle. Within a sheet all the microfibrils lie in the same direction, but their direction in successive sheets may differ. The design and stability of organized structures are determined by the bonding characteristics of their component units and once these units have been synthesized, provided environmental conditions are favourable, they assemble themselves into the final structure without further external control. The principle of minimum free energy is sufficient to establish the self-assembling process. [The literature contains numerous examples of self-assembling systems (see Neville 1975)]. Microfibrils contain chitin chains and so the formation and orientation of microfibrils becomes a problem of assembling chitin chains.

In the helicoidal systems the sense of rotation of the sheets of microfibrils is bilaterally asymmetrical which suggests a crystallization assembly process. However, bilateral symmetry is shown both by the variations in the pitch of the helicoid and by the preferred (i.e. unidirectionally) oriented microfibrils and this suggests specific cellular control (Neville &

Luke 1969b). Two theories for a self-assembling system in cuticles have been proposed and both lack direct supporting evidence. One considers the cuticle to be a helicoidal system of stacked, discrete layers, the other that cuticle contains a continuous "screw carpet" helicoidal system.

Neville (1975), considering the sheets of microfibrils to be separate from each other, has developed a theory that cuticle is a self-assembling system, stabilized from a liquid crystalline deposition zone. In summary, cuticular components, immediately after secretion, would pass through a crystalline phase, either cholesteric crystals for helicoidally oriented sheets of micro-fibrils or nematic crystals for regions in which there is a preferred orientation of microfibrils. The epidermal cells control the environment within the cuticle and so would determine the type of cuticle by directing the type of crystal formed. A further command would be needed to control the direction of microfibrils in regions with preferred orientations and the suggestion is that this could involve cell polarity and gradient mechanisms. In locust cuticle the same cells secrete both types of orientation, helicoidal and preferred, one during the night and the other during the day. Neville & Luke (1969b) found no evidence to support Locke's (1967b) suggestion that the shape of the microvilli on the secretory surface of the epidermal cells determines whether microfibrils with a preferred or helicoidal orientaiton are secreted.

The "screw carpet" model described by Weis-Fogh (1970) requires the chitin microfibrils to originate on the external plasma membrane of the epidermal cells. The extreme insolubility of chitin suggests that its synthesis and crystallization may take place simultaneously and Weis-Fogh records (as others have, e.g. Arsenault et al. 1984; Filshie 1966; Riley and Banaja 1975) that microfibrils appear to begin at, and to radiate out from, the cell border. The chitin synthesis complex is located in the cell membrane (Vardanis 1979).

Microfibrils would be continuously secreted and a sheet of microfibrils, by curing around and changing its vertical position, would take part in more than one region of the helicoid. In this way a true continuous "screw carpet" would result rather than a stack of discrete sheets of microfibrils. The growing microfibrils would need to leave the cell surface at a very small angle so that they would be nearly parallel to the cuticle surface and the epidermal cells would need to control their rates of curving. To form microfibrils with a preferred orientation only the direction of the micro-fibrils would need to be controlled, they would not curve. In this model self-assembly would involve conventional crystallization rather rather than an intermediate liquid crystalline phase as required by Neville's model. Weis-Fogh has suggested that proteins may influence the assembly of the chitin chains. Here it should be recalled that chitin is always associated with proteins in cuticles and that chitins synthesis requires concurrent synthesis of protein.

4 EPICUTICLE

The fully formed epicuticle is extremely thin, usually 1 μm or less in thickness, and generally has five layers, an external cement layer and beneath this a lipid layer, the outer epicuticle, the cuticulin layer and the inner epicuticle. The inner epicuticle is contiguous with the procuticle, is the thickest of the layers and, because of its staining properties, is also known as the dense homogeneous layer or protein epicuticle. Although the epicuticle is described as non-chitinous the evidence is histochemical and the van Wisselingh test for chitin is unreliable when negative (Hackman 1982). In view of its extreme thinness there could only be a very small amount of chitin present and this might go undetected. Reports on the presence of chitin in tracheal epicuticle have been reviewed by Noirot & Noirot-Timothée (1982). Nevertheless the outer and inner epicuticles are relevant in a discussion on chitin in cuticles.

The outer epicuticle is insoluble in hot alkali and so is a contaminant of chitin isolated by alkaline extraction of cuticles. Chemically it appears to be built up from long chains of CH_2 groups (Hackman 1974). In many insects the outer epicuticle is no more than 10 nm thick and very fragile but, e.g., in calliphorid larvae it is much thicker. That of the last instar larva of *Lucilia cuprina* is 250 nm thick (Filshie 1970) and has been isolated as a "ghost" of the larval (or puparial) cuticle (Hackman 1974). In weight it represents 1.4 per cent of the dry larval cuticle and hence 2.7 per cent of chitin (as isolated), for puparia the values are 1.4 and 3.4 per cent respectively. The error introduced into an uncorrected chitin analysis increases as the chitin content of the cuticle decreases and in the alloscutal cuticle of the tick *Boophilus microplus* the outer epicuticle amounts to 12.5 per cent of the chitin content (Hackman 1982).

The inner epicuticle of the alloscutum of the adult tick *B. microplus* contains a close-packed array of electron-lucent microfibrils normal to the cuticle surface (Hackman and Filshie 1982, Figure 1.10). These microfibrils are about 3 nm in diameter and 6 nm apart and in electron micrographs resemble the chitin microfibrils seen in insect procuticles (Figure 5). There is no chemical evidence to suggest that these microfibrils are chitinous and similar microfibrils have not been reported in the epicuticles of other arthropods. The inner epicuticle is not secreted from the apical plasma membrane, as is chitin in the procuticle, but originates from the Golgi complexes within the cells and passes from the cells by a process of exocytosis (Locke 1969).

5 CONCLUSION

From this discussion a clear picture of chitin and its role in the insect cuticle is emerging. However, Filshie (1980, 1982) has drawn attention to the fact that electron micrographs provide only circumstantial evidence and that the interpretation of them is often equivocal. This uncertainty is brought about by the nature of the cuticle, its mechanical properties and chemical resistance, and to a lesser extent by the limitations of electron microscope techniques. It is not known what effects the harsh treatments needed to prepare sections have on the cuticle's molecular structure. Potassium permanganate, which is used for staining sections, is a powerful oxidizing agent and is known to leach protein from sections. Reference was made earlier to the fact that in cuticles chitin is always associated with protein and that in cuticles it could well have properties different from those of "purified" chitin. Study of cuticle structure should be considered a multidiscipline problem, because it is only in this way that the true model of an insect cuticle can be assembled. Chiting is possibly the most important cuticle component and interference with its synthesis or organization into microfibrils, or with the assembly of the microfibrils to form the fibrous cuticle would destroy the integrity of the cuticle and so destroy the insect.

6 ACKNOWLEDGEMENT

The author thanks the staff of the electron microscope laboratory, Division of Entomology, CSIRO, for the electron micrographs.

7 REFERENCES

Altner, H. 1975. The microfiber texture in a specialized plastic cuticle area within a sensillum field on cockroach maxillary palp as revealed by freeze fracturing. Cell & Tissue Res. 165: 79–88.

Andersen, S. O. 1973. Comparison between the sclerotization of adult and larval cuticle in *Schistocerca gregaria*. J. Insect Physiol. 19: 1603–1614.

Arsenault, A. L., Castell, J. D. and Ottensmeyer, F. P. 1984. The dynamics of exoskeletal-epidermal structure during molt in juvenile lobster by electron microscopy and electron spectroscopic imaging. Tissue & Cell 16: 93–106.

Atkins, E. D. T., Flower, N. E. and Kenchington, W. 1966. Studies on the oöthecal protein of the tortoise beetle, *Aspidomorpha*. J. R. Microsc. Soc. 86: 123–135.

Bade, M. L. and Stinson, A. 1978. Activation of old cuticle chitin as a substrate for chitinase in the molt of *Manduca*. Biochem. Biophys. Res. Commun. 84: 381–388.

Bade, M. L. and Stinson, A. 1979. Molting fluid chitinase: a homotropic allosteric enzyme. Biochem. Biophys. Res. Commun. 87: 349–353.

Barth, F. G. 1973. Microfiber reinforcement of an arthropod cuticle. Laminated composite material in biology. Z. Zellforsch. Mikrosk. Anat. 144: 409–433.

Bartnicki-Garcia, S., Bracker, C. E., Reyes, E. and Ruiz-Herrera, J. 1978. Isolation of chitosomes from taxonomically diverse fungi and synthesis of chitin microfibrils *in vitro*. Exp. Mycol. 2: 173–192.

Benjaminson, A. A. 1969. Conjugates of chitinase with fluorescein isothiocyanate or lissamine rhodamine as specific stains for chitin *in situ*. Stain Technol. 44: 27–31.

Blackwell, J. and Weih, M. A. 1980. Structure of chitin-protein complexes: ovipositor of the ichneumon fly *Megarhyssa*. J. Molec. Biol. 137: 49–60.

Bouligand, Y. 1965. Sur une architecture torsadée répandue dans de nombreuses cuticules d'Arthropodes. C.R. Hebd. Sceances Acad. Sci. Ser. D 261: 3665–3668.

28

Bouligand, Y. 1972. Twisted fibrous arrangements in biological materials and cholesteric mesophases. Tissue & Cell 4: 189–217.

Bouligand, Y. 1978. Liquid crystalline order in biological materials. In: Liquid crystalline order in polymers. Edited by A. Blumstein, pp. 261–297, Academic Press, New York.

Brine, C. J. 1978. Chitin content and variation with molt stage and carapace location in the blue crab, *Callinectes sapidus*. In: Proc. First Int. Conf. Chitin Chitosan, edited by R. A. A. Muzzarelli and E. R. Pariser, pp. 509–516.

Brine, C. J. 1982. Chitin-protein interactions. In: Proc. 2nd Int. Conf. Chitin Chitosan, edited by S. Hirano and S. Tokura, pp. 105–110.

Brine, C. J. and Austin, P. R. 1981. Chitin variability with species and method of preparation. Comp. Biochem. Physiol. B 69: 283–286.

Brück, E. and Stockem, W. 1972a. Morphologische Untersuchungen an der Cuticula von Insekten. I. Die Feinstruktur der larvalen Cuticula von *Blaberus trapezoideus* Burm. Z. Zellforsch. Mikrosk. Anat. 132: 403–416.

Brück, E. and Stockem, W. 1972b. Morphologische Untersuchungen an der Cuticula von Insekten. II. Die Feinstraktur der larvalen Cuticula von *Periplaneta americana* (L.). Z. Zellforsch. Mikrosk. Anat. 132: 417–430.

Cassier, P., Procheron, P., Papillon, M. and Lensky, Y. 1980. Contribution a l'étude des protéines cuticulaires du criquet migrateur, *Locusta migratoria migratorioides* (R. et F.) données quantitatives. Ann. Sci. Nat. Zool. Biol. Anim. Ser. 14 2: 51–65.

Caveney, S. 1969. Muscle attachment related to cuticle architecture in Apterygota. J. Cell Sci. 4: 541–559.

Chen, A. C. and Mayer, R. T. 1981. Fluorescamine as a tool for amino sugar analysis. J. Chromat. 207: 445–448.

Clark, G. L. and Smith, A. F. 1936. X-ray diffraction studies of chitin, chitosan and derivatives. J. Phys. Chem. 40: 863–879.

Clever, U. and Bultmann, H. 1972. Chromosomal control of foot pad development in *Sarcophaga bullata*. III. Requirement of RNA and protein synthesis for cuticle formation and tanning. Cell Differ. 1: 37–42.

Credland, P. F. 1978. An ultrastructural study of the larval integument of the midge, *Chironomus riparius* Meigen (Diptera: Chironomidae). Cell & Tissue Res. 186: 327–335.

Dalingwater, J. E. 1975a. The reality of arthropod cuticular laminae. Cell & Tissue Res. 163: 411–413.

Dalingwater, J. E. 1975b. SEM observations on the cuticles of some decapod crustaceans. Zool. J. Linn. Soc. 56: 327–330.

Dalingwater, J. E. 1977. Cuticular ultrastructure of a cretaceous decapod crustacean. Geol. J. 12: 25–32.

Delachambre, J. 1971. La formation des canaux cuticulaires chez l'adulte de *Tenebrio molitor* L. Étude ultrastructurale et remarques histochemiques. Tissue & Cell 3: 499–520.

Delachambre, J. 1975. Les variations de l'architecture dans la cuticle abdominale chez *Tenebrio molitor* L. (Ins. Col.). Tissue & Cell 7: 669–676.

Dennell, R. 1973. The structure of the cuticle of the shore-crab *Carcinus maenas* (L.). Zool. J. Linn. Soc. 52: 159–163.

Dennell, R. 1974.

Dennell, R. 1975.

Dennell, R. 1976a.

Dennell, R. 1976b.

Dennell, R. 1978.

Drach, P. 1953. Structure des lamelles cuticulaires chez les Crustacés. C.R. Hebd. Sceances Acad. Sci. 237: 1772–1774.

Ejike, C. 1973. Macrofibres in the cuticle of the crab *Callinectes gladiator* (Benedict). Zool. J. Linn. Soc. 53: 253–255.

Ferkovich, S. M., Oberlander, H. and Leach, C. E. 1981. Chitin synthesis in larval and puparial epidermis of the Indian meal moth, *Plodia interpunctella* (Hübner), and the greater wax moth *Gallaria mellonella* (L.) J. Insect Physiol. 27: 509–514.

Filshie, B. K. 1966. Structure and development of insect cuticle and associated structures. Ph.D. Thesis, Australian National University, Canberra.

Filshie, B. K. 1970. The fine structure and deposition of the larval cuticle of the sheep blowfly (*Lucilia cuprina*). Tissue & Cell 2: 479–498.

Filshie, B. K. 1980. Insect cuticle through the electron microscope — distinguishing fact from artifact. In: Insect Biology in the Future, VBW80, edited by M. Locke and D. S. Smith, pp. 59–77, Academic Press, New York.

Filshie, B. K. 1982. Fine structure of the cuticle of insects and other arthropods. In: Insect Ultrastructure, edited by R. C. King and H. Akai, vol. 1, pp. 281–312, Plenum. Pub. Corp., New York.

Filshie, B. K. and Hadley, N. F. 1979. Fine structure of the cuticle of the desert scorpion, *Hadrurus arizonensis*. Tissue & Cell 11: 249–262.

Filshie, B. K. and Smith, D. S. 1980. A proposed solution to a fine-structural puzzle: the organization of the gill cuticle in a crayfish (*Panulirus*). Tissue & Cell 12: 209–226.

Fischer, F. G. and Nebel, H. J. 1955. Nachweis and Bestimmung von Glucosamin und Galactosamin auf Papierchromatogrammen. Hoppe-Seyler's Z. Physiol. Chem. 302: 10–19.

Fogal, W. and Fraenkel, G. 1970. Histogenesis of the cuticle of the adult flies, *Sarcophaga bullata* and *S. argyrostoma*. J. Morph. 130: 137–149.

Fraenkel, G. and Rudall, K. M. 1947. The structure of insect cuticles. Proc. R. Soc. B. 134: 111–143.

Gardiner, B. G. and Khan, M. F. 1979. A new form of insect cuticle. Zool. J. Linn. Soc. 66: 91–94.

Gharagozlou-van Ginneken, I. D. 1974. Sur l'ultrastructure cuticulaire d'un crustacé copépode harpacticide: *Tisbe holothuriae* Humes. Arch. Zool. Exp. Gen. 115: 411–422.

Gharagozlou-van Ginneken, I. D. 1976. Particularités morphologiques du tégument des Peltidiidae (Crustacés copépodes). Arch. Zool. Exp. Gen. 117: 411–422.

Gharagozlou-van Ginneken, I. D. and Bouligand, Y. 1973. Ultrastructures tégumentaires chez un crustacé copepode *Cletocamptus retrogressus*. Tissue & Cell 5: 413–439.

Gharagozlou-van Ginneken, I. D. and Bouligand, Y. 1975. Studies on the fine structure of the cuticle of *Porcellidium*, Crustacea, Copepoda. Cell & Tissue Res. 159: 399–412.

Gilby, A. R. and McKellar, J. W. 1970. The composition of the empty puparia of a blowfly. J. Insect. Physiol. 16: 1517–1529.

Giraud-Guille, M. -M. 1984. Fine structure of the chitin-protein system in the crab cuticle. Tissue & Cell 16: 75–92.

Goudeau, M. 1974. Structures cuticulaires chez *Henioniscus balani* Buchholz, Isopode Epicaride. C.R. Hebd. Sceances Acad. Sci. Paris Ser. D 278, 3331–3334.

Gray, G. W. 1962. Molecular structure and the properties of liquid crystals. Academic Press, London.

Green, J. P. and Neff, M. R. 1972. A survey of the fine structure of the integument of the fiddler crab. Tissue & Cell 4: 137–171.

Gubb, D. 1975. A direct visualization of helicoidal architecture in *Carcinus maenas* and *Halocynthia papillosa* by scanning electron microscopy. Tissue & Cell 7: 19–32.

Hackman, R. H. 1962. Studies on chitin. V The action of mineral acids on chitin. Aust. J. Biol. Sci. 15: 526–537.

Hackman, R. H. 1971. The integument of arthropoda. In: Chemical Zoology edited by M. Florkin B. T. Scheer, Vol. 6, pp. 1–62. Academic Press, New York.

Hackman, R. H. 1974. Chemistry of the insect cuticle. In: The Physiology of Insecta, 2nd ed., edited by M. Rockstein, vol. 6, pp. 215–270. Acadmic Press, New York.

Hackman, R. H. 1975. Expanding abdominal cuticle in the bug *Rhodnius* and the tick *Boophilus*. J. Insect Physiol. 21: 1613–1623.

Hackman, R. H. 1982. The estimation of chitin in arthropod cuticles. In: Proc. 2nd Int. Conf. Chitin Chitosan, edited by S. Hirano and S. Tokura, pp. 5–9.

Hackman, R. H. and Filshie, B. K. 1982. The tick cuticle. In: Physiology of Ticks, edited by F. D. Obenchain and R. Galun, pp. 1–42. Pergamon Press, Oxford.

Hackman, R. H. and Goldberg, M. 1958. Proteins of the larval cuticle of *Agrianome spinocollis* (Coleoptera). J. Insect Physiol. 2: 221–231.

Hackman, R. H. and Goldberg, M. 1965. Studies on chitin. VI. The nature of α and β-chitins. Aust. J. Biol. Sci. 18: 935–946.

Hackman, R. H. and Goldberg, M. 1971. Studies on the hardening and darkening of insect cuticles. J. Insect Physiol. 17: 335–347.

Hackman, R. H. and Goldberg, M. 1974. Light-scattering and infrared- spectrophotometric studies of chitin and chitin derivatives. Carbohyd. Res. 38: 35–45.

Hackman, R. H. and Goldberg, M. 1978. The non-covalent binding of two insect cuticular proteins by a chitin. Insect Biochem. 8: 353–357.

Hackman, R. H. and Goldberg, M. 1981. A method for determinations of microgram amounts of chitin in arthropod cuticles. Anal. Biochem. 110: 277–280.

Hackman, R. H. and Goldberg, M. 1985. The expanding alloscutal cuticle in adults of the argasid tick *Argas* (*Persicargas*) *robertsi* (Acari: Ixodoidea). Int. J. Parasitol. 15: 249–254.

Hegdahl, T., Silness, J. and Gustavsen, F. 1977a. The structure and mineralization of the carapace of the crab. (*Cancer pagurus* L.) I. The endocuticle. Zool. Scr. 6: 89–99.

Hegdahl, T., Gustavsen, F. and Silness, J. 1977b. The structure and mineralization of the carapace of the crab (*Cancer pagurus* L.) 2. The exocuticle. Zool. Scr. 6: 101–105.

Holan, Z., Votruba, J. and Vlasakova, V. 1980. New method of chitin determination based on deacetylation and gas-liquid chromatographic assay of liberated acetic acid. J. Chromat. 190: 67–76.

Horton, D. and Lineback, D. R. 1965. N-Deacetylation. Chitosan from chitin. Meth. Carbohyd. Chem. 5: 401–406.

Hunter, E. and Vincent, J. F. V. 1974. The effects of a novel insecticide on insect cuticle. Experientia 30: 1432.

Jeuniaux, C. 1963. *Chitine et chitinolyse*, Masson, Paris.

Jeuniaux, C. 1965. Chitine et phylogenie: application d'une methode enzymatique de dosage de la chitine. Bull. Soc. Chim. Biol. 47: 2267–2278.

Ker, R. F. 1977. Investigation of locust cuticle using the insecticide diflubenzuron. J. Insect. Physiol. 23: 39–48.

Ker, R. F. 1978. The effects of diflubenzuron on the growth of insect cuticle. Pestic. Sci. 9: 259–265.

Livolant, F., Giraud, M. M. and Bouligand, Y. 1978. A goniometric effect observed in sections of twisted fibrous materials. Biol. Cellulaire. 31: 159–168.

Locke, M. 1961. Pore canals and related structures in insect cuticles. J. Biophys. Biochem. Cytol. 10: 598–618.

Locke, M. 1967a. The development of patterns in the integument of insects. Adv. Morphogen. 6: 33–88.

Locke, M. 1967b. What every epidermal cell knows. In: Insects and Physiology, edited by J. W. Beament and J. E. Treherne. pp. 68–92, Oliver & Boyd, Edinburgh.

Locke, M. 1969. The structure of an epidermal cell during the development of the protein epicuticle and the uptake of molting fluid in an insect. J. Morph. 127: 7–40.

Mauchamp, B. and Schrevel, J. 1977. Observations en microscopie a fluorescence de la cuticle des insectes: une methode faisant appel aux proprietes specifiques de la WGA vis-a-vis des glycoconjugues de la chitine. C.R. Hebd. Sceances Acad. Sci. Paris Ser. D 285: 1107–1110.

Mazur, G. D., Reiger, J. C. and Kafatos, F. C. 1982. Order and defects in the silkmoth chorion, a biological analogue of a cholesteric liquid crystal. In: Insect Ultrastructure, edited by R. C. King and H. Akai, Vol. 1, pp. 150–185, Plenum Pub. Corp., New York.

Minke, R. and Blackwell, J. 1978. The structure of α-chitin. J. Molec. Biol. 120: 167–181.

Mutvei, H. 1974. SEM studies on arthropod exoskeletons. Part 1: decapod crustaceans, *Homarus gammarus* L. and *Carcinus maenas* (L.). Bull. Geol. Inst. Univ. Upsala, N.S. 4: 73–80.

Mutvei, H. 1977. SEM studies on arthropod exoskeletons. 2. Horseshoe crab *Limulus polyphemus* (L.) in comparison with extinct eurypterids and recent scorpions. Zool. Scr. 6: 203–213.

Muzzarelli, R. A. A. 1977. Chitin, Pergamon Press, Oxford.

Neville, A. C. 1963a. Daily growth layers in locust rubber-like cuticle influenced by an external rhythm. J. Insect Physiol. 9: 177–186.

Neville, A. C. 1963b. Growth and deposition of resilin and chitin in locust rubber-like cuticle. J. Insect Physiol. 9: 265–278.

Neville, A. C. 1970. Cuticle ultrastructure in relation to the whole insect. Symp. R. Entomol. Soc. London 5: 17–39.

Neville, A. C., 1975. Biology of the Arthropod Cuticle, Springer-Verlag, Berlin.

Neville, A. C. 1983. Daily cuticular growth layers and the teneral stage in adult insects: A review. J. Insect. Physiol. 29: 211–219.

Neville, A. C. 1984. Cuticle: organization. In: Biology of the Integument, edited by J. Bereiter-Hahn, A. G. Matoltsy and K. S. Richards, Vol. 1, pp. 611–625, Springer-Verlag, Berlin.

Neville, A. C. and Berg, C. W. 1971. Cuticle ultrastructure of a Jurassic crustacean (*Eryma stricklandi*). Paleontology 14: 201–205.

Neville, A. C. and Caveney, S. 1969. Scarabaeid beetle exocuticle as an optical analogue of cholesteric liquid crystals. Biol. Rev. 44: 531–562.

Neville, A. C. and Luke, B. M. 1969a.

Neville, A. C. and Luke, B. M. 1969b. A two-system model for chitin-protein complexes in insect cuticles. Tissue & Cell 1: 689–707.

Neville, A. C. and Luke, B. M. 1971. Form optical activity in crustacean cuticle. J. Insect Physiol. 17: 519–526.

Neville, A. C., Thomas, M. G. and Zelazny, B. 1969. Pore canal shape related to molecular architecture of arthropod cuticle. Tissue & Cell 1: 183–200.

Neville, A. C., Parry, D. A. D. and Woodhead-Galloway, J. 1976. The chitin crystallite in arthropod cuticle. J. Cell Sci. 21: 73–82.

Noble-Nesbitt, J. 1967. Aspects of the structure, formation and function of some insect cuticles. In: Insects and Physiology, edited by J. W. L. Beament and J. E. Treherne, pp. 3–16, Oliver and Boyd, Edinburgh.

Noirot, C. and Noirot-Timothée, C. 1969. La cuticle proctodéale des insectes. I. Ultrastructure comparée. Z. Zellforsch. Mikrosk. Anat. 101: 477–509.

Noirot, C. and Noirol-Tomothée, C. 1982. The structure and development of the tracheal system. In: Insect Ultrastructure, edited by R. C. King and H. Akai, vol. 1, pp. 351–381, Plenum Pub. Corp., New York.

Odier, A. 1823. Mémoire sur la composition chimique des parties cornées des insectes. Mem. Soc. Hist. Nat. Paris 1: 29–42.

Pace, A. 1972. Cholesteric liquid crystal-like structure of the cuticle of *Plusiotis gloriosa*. Science 176: 678–680.

Pearse, A. G. E. 1985. Histochemistry Theoretical and Applied, 4th ed., Vol. 2, Churchill Livingstone, Edinburgh.

Pepper, J. H. and Hastings, E. 1943. Age variations in exoskeletal composition of the sugar beet webworm and their possible effect on membrane permeability. J. Econ. Ent. 36: 633–634.

Richards, A. G. 1947. Studies on arthropod cuticle 1. The distribution of chitin in lepidopterous scales and its bearing on the interpretation of arthropod cuticle. Ann. Ent. Soc. Am. 40: 227–240.

Richards, A. G. 1951. The Integument of Arthropods, University of Minnesota Press, Minneapolis.

Riley, J. and Banaja, A. A. 1975. Some ultrastructural observations on the integument of a pentastomid. Tissue & Cell 7: 33–50.

Rudall, K. M. 1963. The chitin/protein complexes of insect cuticles. Adv. Insect Physiol. 1: 257–313.

Rudall, K. M. 1965. Skeletal structure in insects. In: Aspects of Insect Biochemistry. Biochem. Soc. Symp. 25: 83–92.

Rudall, K. M. 1967. Conformation in chitin-protein complexes. In: Conformation of Biopolymers, edited by G. N. Ramachandran, vol. 2, pp. 751–765, Academic Press, London.

Rudall, K. M. 1969. Chitin and its association with other molecules. J. Polym. Sci. Part C 28: 83–102.

Rudall, K. M. and Kenchington, W. 1973. The chitin system. Biol. Rev. 48: 597–636.

Rutherford, F. A. and Austin, P. R. 1978. Marine chitin properties and solvents. In: Proc. First Int. Conf. Chitin Chitosan, edited by R. A. A. Muzzarelli and E. R. Pariser, pp. 182–192.

Sannan, T., Kurita, K. and Iwakura, Y. 1976. Studies on chitin. 2. Effect of deacetylation on solubility. Makromolec. Chem. 177: 3589–3600.

Shrivastava, S. C. 1970. Cuticular components of some common Indian arachnids and myriapods. Experientia 26: 1028–1029.

Strout, V., Lipke, H. and Geoghegan, T. 1976. Peptidochitodextrins of *Sarcophaga bullata*: molecular weight of chitin during pupariation. In: The Insect Integument, edited by H. R. Hepburn, pp. 43–61, Elsevier, Amsterdam.

32

Tsao, C. -H. and Richards, A. G. 1952. Studies on arthropod cuticle. IX. Quantitative effects of diet, age, temperature and humidity on the cuticles of five representative species of insects. Ann. Ent. Soc. Am. 45: 585–599.

Vardanis, A. 1979. Characteristics of the chitin synthesizing system of insect tissue. Biochem. Biophys. Acta 588: 142–147.

Weis-Fogh, T. 1970. Structure and formation of insect cuticle. Symp. R. Entomol. Soc. London 5: 165–185.

Welinder, B. S. 1974. The crustacean cuticle. I. Studies on the composition of the cuticle. Comp. Biochem. Physiol. 47A: 779–787.

Wigglesworth, V. B. 1975. Distribution of lipid in lamellate endocuticle of *Rhodnius prolixus* (Hemiptera). J. Cell. Sci. 19: 439–457.

Wolfgang, W. J. and Riddiford, L. M. 1981. Cuticular morphogenesis during continuous growth of the final instar larva of a moth. Tissue & Cell 13: 757–772.

Zacharuk, R. Y. 1972. Fine structure of the cuticle, epidermis and fat body of larval Elateridae (Coleoptera) and changes associated with molting. Can. J. Zool. 50: 1463–1487.

Zelazny, B. and Neville, A. C. 1972. Quantitative studies on fibril orientation in beetle endocuticle. J. Insect Physiol. 18: 2095–2121.

2. Regulation of chitin synthesis: mechanisms and methods

Edwin P. Marks and Gordon B. Ward, Jr.

1 INTRODUCTION

Chitin occurs in relation to two organs among insects, the body wall (epidermis) and the gut. The lining of the foregut and hindgut is continuous with the cuticle of the body wall. The midgut is lined by the peritrophic membrane (PM), a structure that contains roughly 7.5% chitin by dry weight (Richards and Richards 1977). We have found no evidence that in vivo production of the PM is linked to the molting cycle and therefore controlled by the titer of 20-hydroxyecdysone (20–OH–E) although Becker (1978) reports PM secretion is accelerated by 20–OH–E and inhibited by juvenile hormone in organ cultures. In addition he found that in the presence of the chitin synthesis inhibitor diflubenzuron (DFB), the rate of membrane production was decreased and the fine structure of the membrane was perturbed (Becker 1980).

The cuticle which is secreted by the epidermal tissues of the insect contains from 25 to 40% of chitin depending on the insect and portion of the epidermis examined (Richards 1978). The production of the cuticle and hence of the chitin that it contains is molt dependent and thus controlled by the titer of 20–OH–E both in vivo and in vitro (Marks & Leopold 1971, Riddiford et al. 1980.) In *Mamestra* the incorporation of glucosamine into the abdominal cuticle is highest immediately following molting and declines rapidly during the first half of the intermolt period. It then remains at a low level until the beginning of the next molt cycle (Mitsui et al. 1981). In the cuticle of isolated milkweed bug abdomens, chitin synthesis dropped rapidly after 10 hours postmolt (Hajjar and Casida 1979).

In cultured tissue, molting occurs only if the tissue has been exposed to high titers of 20–OH–E either before explantation or after this hormone is added to the culture (Marks 1972). The same pattern is evident for chitin synthesis (Figure 1, Table 1). From these findings, it appears that chitin synthesis in insects is stimulated by 20–OH–E. This coupling of chitin

Wright, J. E. and Retnakaran, A. (Eds), Chitin and Benzoylphenyl ureas. ISBN 978-94-010-8638-7.
© *1987, Dr W. Junk Publishers, Dordrecht.*

34

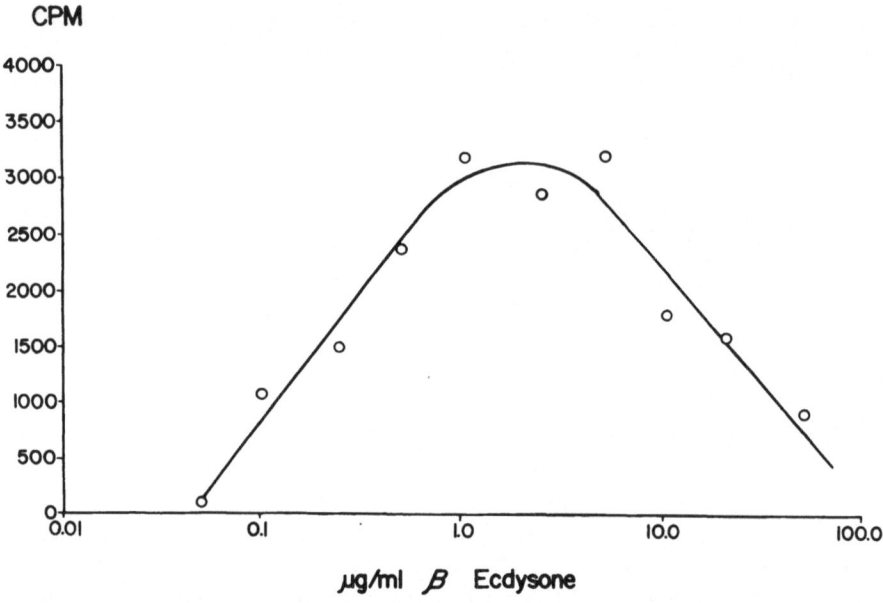

CPM

Figure 1. Dose curve for 20-hydroxyecdysone (20–OH–E) stimulation of chitin synthesis in cockroach leg regenerate cultures. Twenty six day regenerates were dosed with [^{14}C] N-acetylglucosamine and 20–OH–E for 48 hours and digested in Hyamine hydroxide. The remaining solids were collected on filter paper and counted by liquid scintillation (For details see Sowa and Marks 1975).

Table 1. The effect of 20-hydroxyecdysone (20–OH–E) on the incorporation of chitin precursors into cockroach tissue *in vitro*.

Tissue	[^{14}C] N-Acetylglucosamine (cpm) Control	20–OH–E (10^{-6}M)	[^{14}C] Glucose Control	20–OH–E (10^{-6}M)
Leg Regenerate	27	3054	3177	1516
Fat Body	48	46	1366	657

1 MM^3 of tissue was incubated 48 hrs at 27°C in 1 ml of M 2 medium. Tissue digested in Hyamine hydroxide, washed, and solids collected on a millipore filter and washed with 50 ml of water before counting by liquid scintillation (See Sowa and Marks 1975).

synthesis with the molting cycle is probably stronger in the epidermal tissue than in the gut where chitin synthesis appears to be more or less continuous.

2 CHITIN SYNTHESIS CONTROL

The evidence for juvenile hormone (J.H.) control of chitin synthesis is largely circumstantial. Becker (1978) found that juvenile hormone (J.H.)

reduced the rate of production of peritrophic membrane in cultured cardia of *Calliphora*. Leopold (personal communication) found that exogenous J.H. produced defective adult cuticles when applied early in the pupal stage of boll weevils. Application of J.H. at this time also greatly enhanced the effect of DFB in the later stages. In neither case was a direct effect of J.H. on chitin synthesis documented and repeated attempts to demonstrate an effect of J.H. on chitin synthesis in cockroach leg regenerates have failed (Marks unpublished data).

Another source of regulation of chitin synthesis in insects, is the availability of the immediate substrate UDP N-acetylglucosamine (UDPGlcNAc) and its precursors N-acetylglucosamine (GlcNAc), glucosamine (GA), and glucose. In many insect organ cultures, epidermal tissue is capable of synthesizing chitin from all three of these precursors (Fristrom 1968). Experiments comparing the incorporation of glucose and GlcNAc in cultured cockroach epidermal and fat body tissue showed that 20–OH–E (10^{-6}M) increased GlcNAc incorporation 100 fold in epidermal tissue (Table 1), but up to the time that 20–OH–E is applied much of the glucose incorporated by both tissues is in the form of glycogen since treatment with diastase removes from 50 to 70% of the labelled material (Marks unpublished data). The overall incorporation of glucose is decreased by roughly 50% when 20–OH–E is applied (Table 1). This decrease occurs primarily in the diastase labile portion. Both Fristrom (1968) and Surholt (1975) have suggested that the amination of fructose–6–P is a limiting step for the regulation of GA–6–P synthesis from glucose.

Sowa and Marks (1975) found that on an equimolar basis GlcNAc incorporation was an order of magnitude greater than GA incorporation in cockroach epidermal tissue. Similar results were obtained by Surholt (1975) with *Locusta* epidermis and by Ferkovich et al. (1980) with cultured *Plodia* wing disks. This suggests that either the acetylation of GA–6–P (Fristrom 1968, Turnbull and Howells 1982) or differences in uptake by the cells (Oberlander 1976) are control points in the synthesis of GlcNAc from glucosamine.

From these results it appears likely that substrate availability is a factor in controlling the rate of chitin synthesis. GlcNAc from the digestion of the old cuticle by chitinase and chitobiase (Muzzarelli 1977) becomes available to the insect prior to ecdysis probably through the swallowing of the exuvial fluid (Zacharuk 1976). This timely availability of the preferred substrate is accompanied by an increase in the rate of chitin synthesis and deposition in the new cuticle. Working with tissues of the migratory locust, Surholt (1975) found that prior to the time of ecdysis, the primary substrate for chitin synthesis was GlcNAc. The incorporation of glucose reaches a peak only after molting has occurred, the deposition of endocuticle in some insects continuing throughout the latter half of the instar. In *Bombyx* larvae the amount of glycogen present decreases rapidly during this period (Zaluska 1958). Under these conditions the substrates employed for chitin synthesis during the latter portion of the molting cycle are glucose derived from glycogen and acetate derived from lipids since the GlcNAc from the digestion of the old cuticle is no longer available.

3 ORGAN AND CELL CULTURE SYSTEMS

The complexity of the hormonal and nutritional melieu in living insects has made the study of chitin synthesis and its control difficult. As a result, work in this field is increasingly being carried out *in vitro*. *In vitro* studies have been made using target material at three different levels of complexity; organ cultures, cell free enzyme preparations, and, more recently, cultured insect cells. A comparison of the studies made at these three levels has shed some light on the control mechanisms in question.

Organ culture systems used to study chitin synthesis have ranged from the abdomens of milkweed bugs emptied of their contents and filled with the incubation medium (Hajjar and Casida 1979) to the imaginal disks of *Plodia* larvae (Oberlander and Leach 1974) and have included a wide variety of morphological entities and developmental stages. Among this diversity of available systems there are both similarities and differences that are of interest. The different systems appear to be able to utilize a variety of substrates, (with the exception of UDPGlcNAc), and to respond *in vitro* much like the insect itself. All appear to require activation of chitin synthesis by 20–OH–E although the form that this activation takes varies. When the tissue explant is taken from a developmental stage in which the titer of 20–OH–E is above the threshold for cuticle deposition, further stimulation *in vitro* is not required, having occurred before the tissue was explanted (Surholt 1975, Hajjar and Casida 1979, Turnbull and Howells 1982). However, if the explant is removed from a developmental stage in which the 20–OH–E titer is below the threshold for the tissue explanted, then chitin synthesis does not occur unless the hormone is added *in vitro* (Fig 1) (Fristrom 1968, Marks 1972). In either case, as in the intact insects, chitin synthesis *per se* is closely linked with the synthesis of cuticular proteins and the deposition of a new cuticle (Marks (1972), Riddiford et al. (1980), Ferkovich et al. (1980).

In cell free preparations, quite a different situation prevails. Here, since no cell membranes have to be crossed, UDPGlcNAc is the preferred substrate, and 20–OH–E is not required (Cohen and Casida 1980a, Mayer et al. 1980). The obvious implication of these findings is that the two control systems proposed depend upon the presence of intact cellular architecture and possibly the exchange of products between different kinds of cells (See Surholt 1975). In order to pursue this hypothesis further we proceeded to divise a cell culture system in which organized tissues are no longer present but the cellular architecture remains intact.

In 1977 while examining cell cultures which had been stored in liquid nitrogen for several years, we were startled to find a number of amber colored spheroids which were quite obviously the products of cells and which appeared to be cuticular in nature. Examination of a number of cultures turned up similar spheroids in two separate lines of cells, one derived from embryos of *Manduca sexta* (Figure 2) and the other from embryos of *Blattella germanica*. When samples were digested for 48 hours in hot Hyamine hydroxide they remained intact and, when examined by electron microscopy, they showed an ultrastructure typical of insect cuticle. Since both of these cell lines had been passed more than thirty times before they were frozen, we speculated that these continuous cell lines had retained the capacity for producing cuticle and the cuticle most likely contained chitin (Marks et al. 1984a). This hypothesis led us to begin a series of experiments to determine whether and under what conditions, continuous insect cell lines could synthesize chitin. Our initial studies showed that the CH34 line of *Manduca* embryo cells took up large amounts of [^{14}C] GlcNAc and retained more than 60% of it for 24 hours against a diffusion gradient. However the UMBGE-2 line of cockroach embryo cells took up almost none of the chitin precursor. When the contents of the cytosol of [^{14}C] labelled *Manduca* cells were examined most of the label was in the form of UDPGlcNAc (Figure 3). Apparently the cells cannot take up UDPGlcNAc but synthesize it in the cytosol using membrane impermeability to nucleotides and covalent modification of the substrate as means of retaining GlcNAc against the concentration gradient (Marks et al. 1984a). How this UDPGlcNAc in the cytosol is used by the cell to make the chitin which is deposited on the outer side of the cell membrane remains to be worked out.

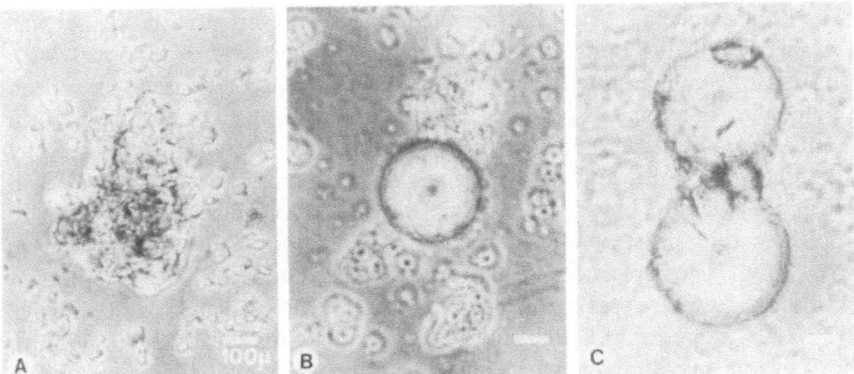

Figure 2. Cuticle spheres found in *Manduca* embryo cell line CH34 after freezing.

38

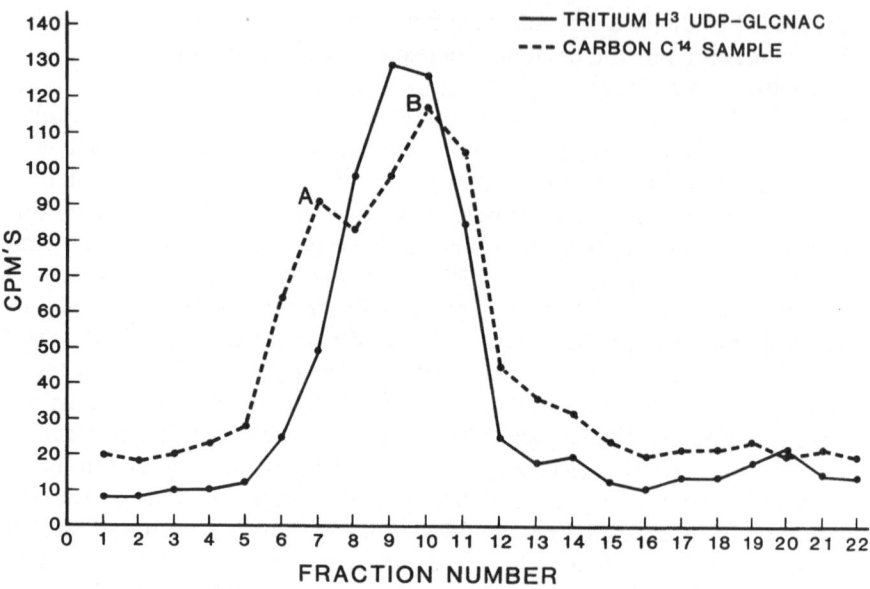

Figure 3. Column chromatograph of aqueous fraction of cytosol of CH34 cells. (---) Cells were incubated with [^{14}C] N-acetylglucosamine for 3 days, fractured by freezing and partitioned by Folch extraction. The water soluble fraction was applied to a P-4 (Bio-rad) column with 0.017 M NaCl solution and the fractions collected at 15 min intervals. (——) [^{3}H] labeled UDP-N-acetylglucosamine standard run under the same conditions.

4 BIOASSAY OF CHITIN SYNTHESIS

When we turned our attention to the water insoluble membrane fraction of these [^{14}C] labelled cells we found that approximately 1% of the label taken up by the cell remained in the cell pellet after digestion with Hyamine hydroxide for 48 hours. When this Hyamine treated pellet was washed to background levels it was then digested for 48 hours with chitinase. While a large portion of the chitin in the pellet must have been deacetylated by the Hyamine treatment, apparently enough raw chitin remained so that 15 to 30% of the remaining radioactivity was released by the chitinase digestion. To examine the nature of the material released from the pellet by chitinase, the supernate was lyophilized and silylated to produce the trimethyl silyl derivative and examined by GC mass spectroscopy. The presence of chitobiose, the characteristic product of chitinase digestion was confirmed (Figure 4). This in turn provide evidence that the *Manduca* cell line did synthesize chitin (Marks et al. 1984b). Following this discovery we examined a number of cell lines for the ability to synthesize chitin. Only one (UMBGE-4) of the four cell lines derived from *Blattella* possessed this ability. Of the eight cell lines derived from

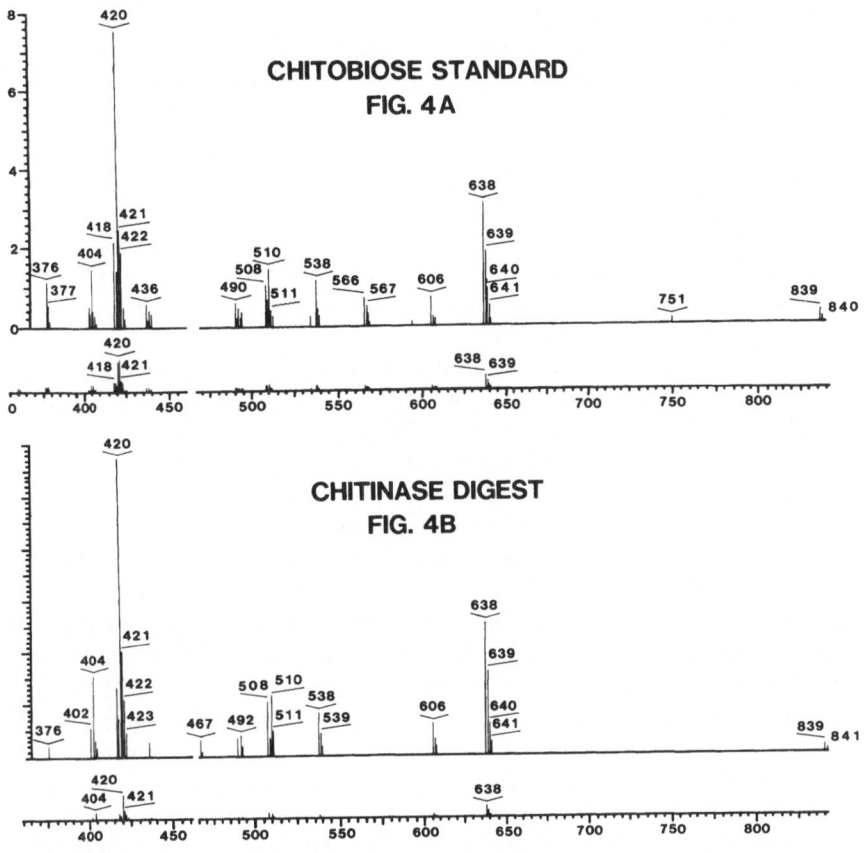

Figure 4. Mass spectra of trimethysilyl derivatives of (a) chitobiose standard, (b) main peak of GLC trace in Fig. 1b. GC-MS was performed using a modified Varian model 3700 gas chromatograph coupled to a Finnigan MAT 1125 mass spectrometer in the electron impact mode fitted with a Carter Cook jet separator. The GC glass column was 2.77 M × 0.635 cm o.d. and was packed with 3.5% OV–101R on Gas–Chrom QR, 100–120 mesh. Separations were performed by programming the GC from 200–335°C at 3°/min, flow rate 25 ml/min, and a MS scan was taken every 8 sec. Data storage and acquisition were through a PDP 11/34 computer using RSX–11m version 3.2 operating system dedicated to the mass spectrometer. Comparison of these mass spectra with those already in the literature leave no question that the GLC peaks in both the chitinase digest and the chitobiose standard represent chitobiose.

Lepidoptera seven synthesized chitin to at least some degree. Only the line TN derived from *Trichoplusia* was completely negative.

With this bioassay we could now examine the regulation of chitin synthesis at the level of the cell. Our initial studies showed that, in cell cultures as in organ cultures, GlcNAc was the favored substrate and UDPGlcNAc was not taken up. However, the response of the cells to treatment with 20–OH–E (10^{-6}M) decreased chitin synthesis by roughly

50% (Marks unpublished data). This was remarkably consistent for 10 of the 13 cell lines tested. In the other three cell lines, two derived from *Plodia*, and one derived from *Blattella*, the addition of 20–OH–E (10^{-6}M) produced increases in chitin synthesis of from 1 to 4 fold. Depression of chitin synthesis in lepidopteran cell lines is a more common response to 20–OH–E treatment than is increased chitin synthesis under the relatively high dosage levels and experimental conditions used in these preliminary experiments. However it is possible that the high concentrations of 20–OH–E used are inhibitory to cell lines that are particularly sensitive to this hormone and that lower doses will stimulate chitin synthesis. This possibility is being investigated.

The next question concerning the regulation of chitin biosynthesis that we asked of this new bioassay system was what is the response to the chitin synthesis inhibitor diflubenzuron (DFB). We were particularly interested because although organ culture systems were extremely sensitive to this compound (Marks et al. 1982) broken cell enzyme systems were not (DeLoach et al. 1981), (Cohen and Casida 1980b). We had postulated that the intact cell membrane was necessary for the action of this inhibitor (Marks et al. 1982). Our initial studies with cell lines derived from *Manduca* embryos showed that DFB (10^{-8}M) was highly effective in inhibiting chitin synthesis. However, there were differences in the level of inhibition among the other cell lines tested (Table 2). The difference between the mixed cell line CH33 from *Manduca* embryos and the CH34 line which is a clone from the CH33 line is particularly interesting and suggests that mixed lines contain different kinds of chitin producing cells possibly from both the epidermis and the gut which respond at different levels to the inhibitor while the clone CH34 contains only the cells more sensitive to DFB.

The finding that DFB inhibits chitin synthesis in cell lines supports the hypothesis that the action of DFB requires intact cell architecture. This finding in turn supports the concept that DFB, like many pesticides acts by partitioning into the cell membrane and disrupting the activities of the membrane bound enzymes (See Marks et al. 1982). Thus, by disrupting the lipoprotein lattice of the plasma membrane, DFB could either prevent the enzymatic activation of a chitin synthase proenzyme, or prevent the UDPGlcNAc precursor from reaching the site of polycondensation, or

Table 2. The effect of diflubenzuron (10^{-8}M) on chitin synthesis in lepidopteran cell lines.

Cell line	Origin of cells	Diflubenzuron inhibition of chitin synthesis
MDH	Malacosoma hemocyte	64%
CH34	Manduca embryo	95%
CH33	Manduca embryo (mixed line)	64%
PID	Plodia Imaginal Disks	67%

possibly both. This same disruption of the membrane lattice might well suffice to explain the effect of DFB on DNA synthesis reported by DeLoach et al. (1981).

5 SUMMARY

While the puzzle is far from complete, we are beginning to put together the pieces. We have evidence for at least three ways in which chitin synthesis may be regulated: direct regulation by the molting hormone; regulation by substrate availability; and regulation by xenobiotic compounds. The hormone 20–OH–E stimulates chitin synthesis in insects and cultured organs. It may stimulate or inhibit chitin synthesis in cell lines depending on their origin and the experimental conditions. The means by which 20–OH–E accomplishes this regulation is not yet clear. It has no detectable effect once the membrane system of the cell is destroyed. The form of the substrate is also critical and varies with the portion of the molting cycle. Chitin can be synthesized directly from glucose but GlcNAc is the favored substrate when it is released by digestion by the old cuticle. Labeled UDPGlcNAc accumulates in the cytosol of chitin producing cells. Although it cannot be taken in through the plasma membrane of intact cells, in broken cell preparations it is the immediate substrate of chitin synthase. The way in which UDPGLcNAc is transported from the cytosol to the outer surface of the plasma membrane remains obscure. Xenobiotics such as JH analogs and DFB affect chitin synthesis both directly and indirectly and, in the boll weevil, synergize to greatly elevate mortality at nominal dosage levels. DFB acts at the level of the cell membrane and disrupts membrane bound enzyme systems. How this is accomplished needs further elucidation.

Much of the information we possess concerning the regulation of chitin synthesis has been obtained through the use of organ culture systems. Our recent discovery that some insect cell lines synthesize chitin while others do not and that some are stimulated by 20–OH–E while others are not provides a new experimental system in which we can now study the process of chitin synthesis and the mechanisms which control it in ways that have hitherto not been possible.

6 REFERENCES

Becker, B. 1978. Effects of 20-hydroxyecdysone, juvenile hormone, Dimilin, and Captan on *in vitro* synthesis of peritrophic membranes in *Calliphora erythrocephala*. J. Insect Physiol. 24: 699–705.

Becker, B. 1980. Effects of polyoxin D on *in vitro* synthesis of peritrophic membranes in *Calliphora erythrocephala*. Insect Biochem. 10: 101–106.

Cohen, E. and Casida, J. 1980a. Properties of *Tribolium* gut chitin synthase. Pestic. Biochem. Physiol. 13: 121–128.

Cohen, E. and Casida, J. 1980b. Inhibition of *Tribolium* gut chitin synthase. Pestic. Biochem. Physiol. 13: 129–136.

DeLoach, J., Meola, S., Mayer, R. and Thompson, J. 1981. Inhibition of DNA synthesis by diflubenzuron in pupae of the stable fly *Stomoxys calcitrans* (L.). Pestic. Biochem. Physiol. 15: 172–180.

Ferkovich, S., Oberlander, H., Leech, C. and VanEssen, F. 1980. Hormonal control of chitin biosynthesis in imaginal disks. In: Invertebrate Systems In Vitro, Kurstak, E. Maramorosch, K. and Dubendorfer, A. (eds.). Elsevier/North-Holland, NY, pp. 209–216.

Fristrom, J. 1968. Hexosamine metabolism in imaginal disks of *Drosóphila melanogaster*. J. Insect Physiol. 14: 729–740.

Hajjar, N. and Casida, J. 1979. Structure-activity relationships of *benzoylphenyl* ureas as toxicants and chitin synthesis inhibitors in *Oncopeltus fasciatus*. Pestic. Biochem. Physiol. 11: 33–45.

Marks, E. 1972. Effects of ecdysterone on the deposition of cockroach cuticle *in vitro*. Biol. Bull. 142: 293–301.

Marks, E., Jang, E. and Stolee, R. 1984a (in press). Incorporation and Metabolism of N-acetylglucosamine by a lepidopteran cell line. In: Invertebrate Systems In Vitro, Kurstak, E. Maramorosch, K. and Oberlander, H. (eds.). Elsevier/North-Holland, NY.

Marks, E., Balke, J. and Klosterman H. 1984b. Evidence for chitin synthesis in an insect cell line. Arch. Insect Biochem. Physiol. 1: 225–230.

Marks, E. and Leopold, R. 1971. Deposition of cuticular substances in vitro by leg regenerates from the cockroach *Leucophaea madera* (F.). Biol. Bull. 140: 73–83.

Marks, E., Leighton, T., Leighton, F. 1982. Modes of action of chitin synthesis inhibitors. In: Insecticide Mode of Action. Coats, J. (ed.). Academic Press, NY, pp. 281–313.

Mayer, R., Chen, A. & DeLoach, J. 1980. Characterization of a chitin synthase from the stable fly *Stomoxys calcitrans* (L.) Insect Biochem. 10: 549–556.

Mitsui, T., Nobusawa, C. and Fukami, J. 1981. Inhibition of chitin synthesis by diflubenzuron in *Mamestra brassicae* (L.). J. Pesticide Sci. 6: 155–161.

Muzzarelli, R. 1977. Chitin. Pergamon Press, NY, pp. 155.

Oberlander, H. 1976. Hormonal control of growth and differentiation of insect tissues cultured *in vitro*. In Vitro 12, 225–235.

Oberlander, H. and Leach, C. 1974. Inhibition of chitin synthesis in *Plodia interpunctella*. Proc. First Int. Working Conf. on Stored-Product Entomology. Savannah, Georgia.

Richards, A. G. 1978. The chemistry of insect cuticle. In: Biochemistry of Insects, Rockstein, M. (ed.). Academic Press, pp. 205–232.

Richards, A. G. and Richards, P. 1977. The peritrophic membranes of insects. Ann. Rev. Entomol. 22: 219–240.

Reddiford, L., Kiguchi, K., Roseland, C., Chen, A. and Wolfgang, W. 1980. Cuticle formation and sclerotization *in vitro* by the epidermis of the tobacco hornworm *Manduca sexta*. In: Invertebrate Systems *in vitro*. Maramorosch, K., and Dubendorfor, A., (eds.). Elsevier/North Holland, NY, pp. 103–105.

Sowa, B. and Marks, E. 1975. An *in vitro* system for the quantitative measurement of chitin synthesis in the cockroach: inhibition by TH6040 and polyoxin D. Insect Biochem. 5: 855–859.

Surholt, B. 1975. Studies *in vivo* and *in vitro* on chitin synthesis during the larval-adult moulting cycle of the migratory locust *Locusta migratoria* L. J. Comp. Physiol. 102: 135–147.

Turnbull, I. and Howells, J. 1982. Effects of several larvicidal compounds on chitin biosynthesis by isolated larval integuments of the sheep blowfly *Lucilia coprina*. Aust. J. Biol. Sci. 35: 491–503.

Zacharuk, R. 1976. Structural changes of the cuticle associated with moulting. In: The Insect Integument. Hepburn, H. R. (ed.). Elsevier Scientific Pub. Co., Amsterdam, pp. 299–321.

Zaluska, H. 1958. Glycogen and chitin metabolism during development of the silkworm (*Bombyx mori* L.). Acta Biologiae Experimentalis 19: 340–351.

3. Interference with chitin biosynthesis in insects

E. Cohen

1 INTRODUCTION

Chitin is a naturally abundant highly organized amino-sugar bio-polymer. It is found in diverse taxonomic groups as a structural element largely for supportive purposes. Chitin is formed by certain diatom algae (Herth, 1979) and is prevalent in invertebrates, particularly in exo-skeletons and peritrophic membranes of arthropods (Neville, 1975) and along with the deacetylated form, chitosan, in cell walls of most fungal species (Muzzarelli, 1977). Chitin is absent in vertebrates and plants where the analogous structural polymers are collagen and cellulose, respectively.

Chitin is a major supportive component in insect cuticles, comprising up to 25–50% of their dry weight (Andersen, 1979). Together with various cuticular proteins it forms the bulk of the complex cuticular structures. These structures function as an exoskeleton to which skeletal muscles are anchored but also serve to protect insects against mechanical rupture, invasion of pathogens and most important against desiccation. Clearly chitin is of critical importance to insect survival and any interference with its periodic formation or degradation may result in defective cuticles and prove detrimental. The practical potential of such interference did not escape notice and chitin has been regarded as an excellent target for selective pesticide action (Misato et al., 1979). This chapter deals with chitin synthesis and deposition at the cellular and biochemical levels, and with the various sites which are or can be subjected to effective inter-ference. Emphasis will be placed on mode of action of various inhibitors which interact with the elaborate and complex events involved in chitin formation and regulation.

2 CHITIN

Chitin is a linear amino-sugar homopolyer composed of N-acetyl-D-glucosamine (GlcNAc) with β-1-4 linkages. This insoluble

Wright, J. E. and Retnakaran, A. (Eds), Chitin and Benzoylphenyl ureas. ISBN 978-94-010-8638-7.
© *1987, Dr W. Junk Publishers, Dordrecht.*

polysaccharide is formed by cells of ectodermal origin in arthropods and crystallizes into spatially organized, specifically oriented microfibrils which are integrated into integumental or peritrophic membrane structures. The crystallized forms of chitin were studies extensively using the X-ray diffraction technique (Rudall and Kenchington, 1973) and shown to consist of three types e.g., α, β and γ chitin. In the α and β forms the orientation of the chitin chains is opposite and parallel, respectively. In the γ type the arrangement of chains alternates. All three forms of crystallized chitin were detected in a wide range of insect species and they also can be found in different cuticular regions in the same organism.

2.1 *Synthesis*

A cascade of biochemical transformations converts glucose molecules into the chitin polymer (Figure 1). Phosphorylation, amination, acetylation and formation of UDP-N-acetyl-D-glucosamine (UDP-GlcNAc), are major steps catalyzed by cytoplasmic soluble enzymes. Up to the polymerization step the metabolites formed can be utilized also by other biochemical pathways. Chitin polymerization is catalyzed by a cell membrane embedded enzyme and the product by yet an unknown mechanism is transferred to the outside of the cells. Candy and Kilby (1962) detected in the desert locust, *Schistocerca gregaria* activity of enzymes involved in transformation of glucose to UDP-GlcNAc. In part due to technical difficulties demonstration of the polymerization step which is last in the chitin formation pathway, prove to be elusive (Surholt, 1975; Verdanis, 1976; Verloop and Ferrell, 1977; Hajjar and Casida, 1979). In contrast Glaser and Brown back in 1957 obtained a cell-free extract capable of chitin polymerization from the fungus *Neurospora crassa*. Since then similar systems have been extracted from a large number of yeast and filamentous fungi and their properties and kinetics have been thoroughly studied (Jan, 1974; Peberdy and Moore, 1975; Gooday and de Rousset-Hall, 1975; Muzzarelli, 1977; Ruiz-Herrera et al., 1977; Duran and Cabib, 1978; Braun and Calderone, 1979; Gooday, 1979; Selitrennikoff, 1979; Vermeulen et al. 1979). Only recently insect cell-free preparations capable of chitin polymerization have been obtained (Cohen and Casida, 1980a, 1982; Mayer et al., 1980a). This step is catalyzed by a particulate enzyme, chitin synthetase (CS) [UDP-2-acetamido-2-deoxy-D-glucose: Chitin 4-β-acetamidodeoxyglucosyltransferase, EC 2.4.1.16] and the reaction is:

$$(\text{GlcNAc})_n + \text{UDP-GlcNAc} \xrightarrow[\text{synthetase}]{\text{chitin}} (\text{GlcNAc})_{n+1} + \text{UDP}$$

The insect cell-free systems studied include gut enzymes from larvae of

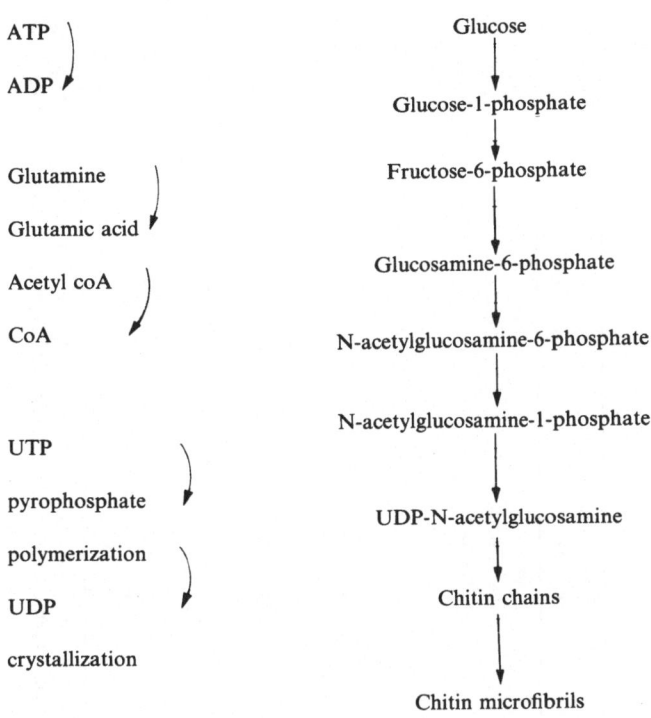

Figure 1. Pathway of Chitin Synthesis and Deposition

various *Tribolium* species (Cohen and Casida, 1980a); integumental enzymes extracted from larvae of the cabbage looper *Trichoplusia ni* (Cohen and Casida, 1982) and the sheep blowfly *Lucilia cuprina* (Turnbull and Howells, 1983), and from the cecropia moth *Hyalophora cecropia* pharate adults (Cohen and Casida, 1982) as well as an active CS enzyme from the stable fly *Stomoxys calcitrans* whole pupae extracts (Mayer et al., 1980a). Based largely on criteria of resistance to hot alkali treatment, degradation by chitinase and hydrolysis by concentrated hydrochloric acid to give glucosamine (GlcN), the product of the polymerization reaction was verified as chitin. Sources, properties and requirements of various arthropod CS enzymes are summarized in Table 1. The cell-free preparations are extracted from either the larval or the pupal stages where active chitin biosynthesis takes place. Being a particulate enzyme the CS activity was assayed using mitochondrial or microsomal fractions. A rather low temperature (22°C) was found optimal for the *Tribolium* gut CS and a decrease in enzymatic activity was apparent when incubation was carried out at elevated temperatures (Cohen and Casida, 1980a). The *S. calcitrans* and the brine shrimp *Artemia salina* enzymes were assayed at

Table 1. Systems of chitin synthetase in arthropods.[a]

Organism	Tribolium castaneum	Trichoplusia ni	Hyalophora cecropia	Stomoxys calcitrans	Artemia[b] salina
Stage	larva[c]	larva	pupa	pupa	larva
Tissue	whole[d]	integument	integument	whole[d]	whole[d]
Fraction	microsomal	mitochondrial	mitochondrial	mitochondrial	microsomal
CS assay					
Incubation time (min)	30	30	30	120	60–90
Temp. (°C)	22	22	22	30	37
pH	7.2	7.2	7.2	7.0	7.1
Mg^{2+} (mM)	17	10	10	NR	30
GlcNAc (mM)	17	18	18	Inhibition	NR
UDP inhibition, I_{50} (µM)	20	100	–	1000(38.5%)[f]	500(83.0%)
Trypsin stimulation (µg/ml)	13(40%)	–	–	40(28%)	100(70%)
Reference	Cohen & Casida (1980a, b)	Cohen & Casida (1982)	Cohen & Casida (1982)	Mayer et al. (1980a)	Horst (1981)

[a] Only systems which were significantly characterized are included. [b] A crustacean species. [c] Gut enzyme. [d] Enzyme extracted from the whole organism. [e] I_{50} — 50% inhibition. [f] In parentheses % inhibition or stimulation at the designated concentrations. NR — not required. GlcNAc- N-acetyl-D-glucosamine. UDP-uridine 5′-diphosphate.

30°C and 37°C, respectively. Buffer systems with pH close to the neutral range (7.0–7.2) were used. Like the fungal CS complex GlcNAc and magnesium ions stimulated the polymerizing activity of *Tribolium castaneum, T. ni* and *H. cecropia* enzymes. Yet GlcNAc was not required by *A. salina* CS and was even inhibitory in the case of *S. calcitrans* enzyme. The CS enzymes were inhibited by UDP which is released upon polymerization, but varies in their degree of sensitivity. Trypsin treatment increased the CS activity and such effect can be related to unmasking of hidden catalytic or allosteric sites of the enzyme, or alternatively to activation of possible zymogenic enzyme complexes existing along with active CS units. It is noteworthy that the *Tribolium* gut enzyme is stimulated up to 50% by the presence of 10% dimethyl sulfoxide in the incubation mixture (Cohen and Casida, 1983). The possible zymogenicity of the insect CS will be addressed later.

The process and mechanism of chitin polymerization have evolved independently in fungi and arthropods. Nevertheless, similarity in properties and requirements of their CS complexes may indicate genetic conservation during evolution.

2.2 *Fibrillogenesis*

In vitro formation of chitin microfibrils by cell-free preparations was investigated in fungi (Cabib and Farkas, 1971; Ruiz-Herrera et al., 1975; Bartnicki-Garcia et al., 1979) and in an insect (Cohen, 1982). A considerable amount of fibrous precipitate composed of chitin microfibrils and proteins was formed in *T. castaneum* CS incubation mixtures shortly after the onset of the polymerization reaction. The above implies rapid crystallization of newly formed chitin chains. When this precipitate was negatively stained and examined under the electron microscope, an extensive network of longitudinally oriented bundles of chitin microfibrils was observed. Small filaments coalesced forming thicker microfibrils which ranged from 10–80 nm in diameter. The microfibrils were attached to largely spheroidal particles (50–250 nm in diameter) which appeared as a core and shell. Thin filaments which were observed inside the core or attached to the outside of the particle might represent different stages of fibrillogenesis. These particles very much resemble the cytoplasmic vesicles termed chitosomes, which were extensively studied in the fungus *Mucor rouxii* by Bartnicki-Garcia and colleagues (Ruiz-Herrera et al., 1975; Bracker et al., 1976; Bartnicki-Garcia et al., 1979), and were observed in taxonomically diverse fungal groups as well (Bartnicki-Garcia et al., 1978). The chitosomes are believed to represent clusters of CS enzymes conveyed from the site of synthesis, apparently in the Golgi apparatus to appropriate locations in the plasma membrane. It is yet uncertain whether the insect chitosome-like structures (Cohen, 1982) are true cytoplasmic organelles or perhaps artificial vesicles formed by mechanical forces involved in preparing the microsomal fraction. However, electron microscopy study revealed the presence of chitosomes in the twospotted spider mite *Tetranychus urticae* (Mothes and Seitz, 1981).

The link between polymerization and fibrillogenesis is poorly defined and understood. One theory (Elorza et al., 1983) contends that the two processes are simultaneous events and that the nascent chitin polymers are self-assembled by hydrogen bond formation. On the other hand, Herth (1980) was able to demonstrate a gap between polymerization and microfibril formation. This gap allows dyes such as Calcofluor white and Congo red with high affinity to polysaccharides, to interact with the newly formed chitin polymers and disrupt their crystallization to microfibrils.

2.3 *Site of chitin synthesis*

The CS enzymes are not evenly distributed in the plasma membranes. They are restricted to the outer cell membranes which face areas of cuticle or peritrophic membrane formation. The study of Locke and Huie (1979) with the larger cana leafroller *Calpodes ethlius* indicates that the CS complexes are organized as clusters of multienzyme units in the cell membranes of epidermal cells. The units are presumably part of membrane plaques located on tips of microvilli. The proximity of the integumental plaques or similar areas in the gut epithelium together with the tightly packaged CS units are likely to ensure the formation of adjacent chitin polymers ready for crystallization. Such spatial arrangement of plaques and enzyme complexes appears to be critical for the integrity of the procuticle and the peritrophic membrane structures.

The cellular cytoskeleton emerges as a crucial factor in determining the synthesis and the arrangement of chitin polymers. The significant role of the microtubular system is evident from the study of Oberlander et al. (1983) with the Indian meal moth *Plodia interpunctella* cultured imaginal disks. Based on the disruptive effects of colcemid and vinblastine on microtubular proteins, the authors suggest a physical link between plasma plaques and microtubuli which probably affect chitin formation as well as the orientation of chitin chains and microfibrils. The cytoskeletal elements may be also involved in translocating packaged CS units to the exact site in the plasma membrane.

Figure 2. depicts major cellular elements associated with cuticle formation. Clusters of CS units perhaps in the form of chitosome-like particles are conveyed to the microvilli and integrate into the plasma membranes at the plaque area. It is conceivable that the CS enzyme spans the plasma membrane and its catalytic site faces the cytoplasm. Based on

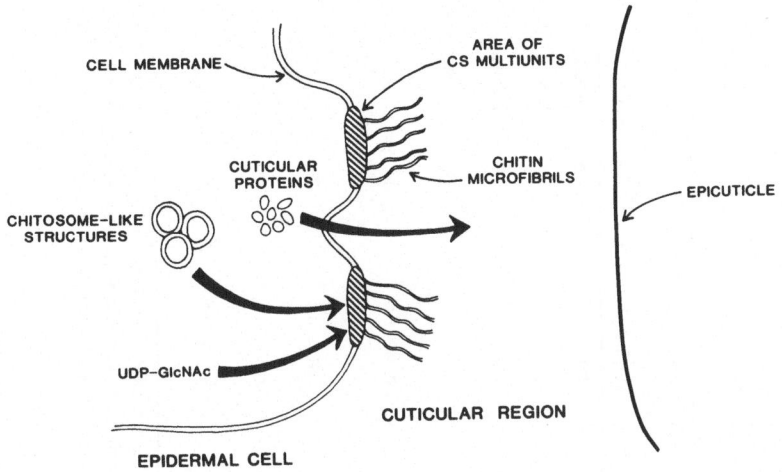

Figure 2. Schematic representation of major cellular components linked to cuticle formation. Area of CS multienzyme represents plasma membrane plaque described by Locke and Huie (1979).

treatment with glutaraldehyde on inner and outer surfaces of cell membranes and on the fact that these membranes are impermeable to the CS substrate, Cabib and Roberts (1982) demonstrated the above sidedness of the polymerizing enzyme in yeast cells. On the other hand, Mitsui et al. (1984) claim that the catalytic site of the CS is on the outer surface of cell membranes in mid-gut epithelial cells of the cabbage armyworm *Mamestra brassicae*, and that a transport system involves in transferring the UDP-GlcNAc across the cell membrane. However, the data they present are inadequate to support their suggestion. The CS units inserted in the cell membrane might be in a zymogenic state, and upon possible activation, and in the presence of sufficient levels of substrate and co-factors will start to synthesize chitin chains. These polymers by an unknown mechanism are extruded outside the cell membrane and crystallize to form microfibrils. Cuticular proteins secreted at the same time associate with the chitin microfibrils and form procuticular layers.

3 REGULATION

3.1 *Chitin formation*

Whole insects, mainly at the larval stage have been used to demonstrate rapid incorporation of injected radiolabeled precursors into chitin. Autoradiographic studies revealed a defined band of radioactivity localized inside the endocuticle, representing newly formed chitin polymers.

Table 2. Interference with chitin synthesis in insect organ cultures.

Organism	Tissue	Stage[b]	Labeled[c] precursor	Ecdysterone[d]	Inhibition I_{50}[a] (M)		Reference
					Polyoxin-D	Diflubenzuron	
Leucophaea[e] maderae	leg[f] regenerate	L	GlcN GlcNAc GlcN	R	7.5×10^{-7}	6.1×10^{-11}	Sowa & Marks (1975)
Oncopeltus fasciatus	isolated[f] abdomen	A	Glc GlcN	NR	1.5×10^{-5}	5.5×10^{-7}	Hajjar & Casida (1979)
Calliphora[g] erythrocephala	midgut	A	GlcNAc	R	1.9×10^{-5}	–	Becker (1978, 1980)
Stomoxys calcitrans	integument[h]	P	Glc GlcN GlcNAc	NR	1.3×10^{-5}	5.2×10^{-8}	Mayer et al. (1980)
Lucilia cuprina	integument	L	GlcN GlcNAc	NR	6.0×10^{-7}	7.0×10^{-7}	Turnbull & Howells (1982)
Manduca sexta	integument	L	Glc GlcN	NR	–	1.1×10^{-9}	Mitsui et al. (1980)
Chilo[i] suppressalis	integument	L	GlcN	R	1.0×10^{-5}	8.3×10^{-8}	Kitahara et al. (1983)

[a] I_{50} — 50% inhibition. [b] Insect stages, L — larva, P — pupa, A — adult. [c] precursor for chitin synthesis. Glc-glucose, GlcN — glucoasmine, GlcNAc — N-acetyl-D-glucosamine. [d] Ecdysterone requirement. R — required, NR — not required. [e] Over 95% inhibition by captan at 10^{-4} M. [f] Integumental tissue. [g] Formation of peritrophic membranes; I_{50} for captan 1.7×10^{-6} M. [h] Imaginal epidermis. [i] Parameter of inhibition was the thickness of cuticle; captan had no effect.

Insect organ cultures provide controlled systems for studying chitin synthesis. In such systems manipulations which involve nutrients, precursors, hormones and various metabolic inhibitors have facilitated and advanced important biochemical and physiological research. Cuticle formation and chitin biosynthesis have been investigated in different insect cultured systems using glucose (Glc), glucosamine (GlcN) or N-acetyl-D-glucosamine (GlcNAc) as precursors (Table 2). It appears that long-term organ cultures such as the Madeira cockroach, *Leucophaea maderae,* leg regenerates (Marks and Leopold, 1971; Sowa and Marks, 1975), larval integuments from the Asiatic rice borer *Chilo suppressalis* (Kitahara et al., 1983) or imaginal disks of *P. interpunctella* (Oberlander et al., 1978, 1980) require exogenous supply of the molting hormone for initiating and maintaining chitin synthesis and deposition, and cuticle formation. Synthesis of chitin is normally detected 24 hr following the hormone addition. Contrary to the above, short-term tissue cultures which include isolated abdomens of the large milkweed bug *Oncopeltus fasciatus* young adults (Hajjar and Casida, 1979) and integumental tissues from pupae of *S. calcitrans* (Mayer et al., 1980b), *L. cuprina* (Turnbull and Howells, 1982) or the tabacco hornworm *Manduca sexta* (Mitsui et al., 1980) did not need addition of exogenous ecdysterone for chitin formation. Apparently these cultured tissues were previously exposed to endogenous hormone which triggered synthesis of cuticular components including chitin. In the short-term tissue cultures, the radioactive precursors were rapidly incorporated into chitin and for practical purposes such systems are clearly advantageous.

The role played by the molting hormone in the overall cuticle formation and in chitin synthesis is well documented. Exogenously applied ecdysterone stimulated cuticle formation in cockroach leg regenerates (Marks and Leopold, 1970, 1971) and chitin synthesis and deposition in the silkworm *Bombyx mori* larval integument (Kimura, 1973). A dramatic 10-fold increase in chitin synthesis in response to a 24 hr pulse of exogenously applied ecdysterone was observed in *P. interpunctella* cultured imaginal disks (Oberlander et al., 1983). Furthermore, the same research group reported a dose-response relationship between levels of the hormone and the rate of chitin formation (Oberlander et al., 1978).

In a study using inhibitors of RNA and protein synthesis, Oberlander et al. (1980) showed that chitin synthesis in *Plodia* imaginal disks was dependent on protein synthesis during and after exposure to ecdysterone. Similar results were obtained in a crustacean species by Stevenson and Tung (1971).

Cells in the alimentary canal which involve in forming the peritrophic membrane are continuously engaged in chitin synthesis throughout the active stages of the insect life cycle. Therefore, mid-gut preparations were recommended for studying chitin synthesis (Surholt, 1975; Heinrich et al.,

1979). Mitsui et al. (1984) studied inhibition of chitin synthesis in cultured guts of *M. brassicae,* and recently a cell-free CS preparation was obtained from *Tribolium* gut (Cohen and Casida, 1980a). It was observed that the formation of the peritrophic membrane in adult dipteran species was also stimulated by ecdysterone (Peters, 1976; Becker, 1978). Although affecting formation of chitin, the molting hormone apparently does not act directly on the machinery of chitin synthesis and deposition. The insect growth hormone triggers a concerted biochemical events at the time of molting and metamorphosis, part of which includes chitin formation. However, interference with levels of insect hormones was suggested by Yu and Terriere (1975, 1977) as the mode of action of insecticidal benzoylphenyl ureas known to disrupt chitin formation.

3.2 *Chitin synthetase complex*

The control of CS at the cellular level in various fungal systems has been extensively studied (Muzzarelli, 1977; Bartnicki-Garcia et al., 1979; Gooday, 1979; Gooday and Trinci, 1980; Cabib and Roberts, 1982). A major leverage for control of chitin synthesis is related to the zymogenic state of CS units in chitosomes or to enzymes already embedded in cell membranes. *In vitro* activation of the fungal CS by trypsin or by natural proteolytic enzymes has been demonstrated in a number of studies (Ruiz-Herrera and Bartnicki-Garcia, 1976; Cabib and Farkas, 1971; Ruiz-Herrera et al., 1977; Braun and Calderone, 1978). Evidence for proteins acting as inhibitors of the proteolytic enzymes and must be removed prior to activation of the zymogenic enzymes (McMurrough and Bartnicki-Garcia, 1973; Ulane and Cabib, 1974; Braun and Calderone, 1979), and the direct inhibition of active fungal CS (Ulane and Cabib, 1974; Craig et al., 1981) further indicate the complexity of the CS regulatory mechanism. The existence of a zymogenic state of insect CS and its conversion to an active form by endogenous proteases has been recently proposed (Leighton et al., 1981). Although highly probable, no endogenous integumental proteases capable of CS activation have been demonstrated. Moreover, the evidence from available insect CS enzymes indicates that no prior proteolytic activation is required for considerable active *in vitro* chitin polymerization. Active CS were obtained when exogenous protease inhibitors were included throughout the extraction procedure (Cohen and Casida, 1980a). Also no proteolytic action was necessary for activation of chitin polymerization and microfibril formation by *T. castaneum* chitosome-like particles (Cohen, 1982). Unlike the fungal system where zymogenic CS is dramatically stimulated by trypsin, the insect enzyme is activated to a modest 28–70% (See Table 1). This relatively mild effect may be attributed to unmasking of hidden CS complexes (Cohen and

Casida, 1980a). In their electron microscopy study of *C. ethlius* epidermal cells Locke and Huie (1979) observed the disappearance of membrane plaques prior to pupation. These areas were pinocytosed and apparently digested by hydrolytic enzymes. The disappearance of membrane plaques coincided with a peak of the molting hormone. There was a transitional period at apolysis where no plaques could be detected. Subsequently new plaque areas were formed *de novo* ready to deposit a new cuticle. Since part of such plaques is believed to accommodate chitin polymerizing enzymes, degradation of such enzymes as well as their later synthesis and subsequent insertion into cell membranes are major cellular events involved in the machinery of chitin formation. It is most likely that the turnover of the macromolecular components responsible for chitin synthesis is under hormonal control. Thus, injection of ecdysterone into *H. cecropia* diapausing pupae which initiated CS activity (Cohen and Casida, 1982) might be related to formation of new enzyme units in the epithelial cells. However, in this case the possibility that the CS activity could be regulated *in situ* by the hormone should be also considered. Changes in physiological conditions such as starvation or entering the pupal stage were reported to be accompanied by reduced levels of CS activity (Cohen and Casida, 1982). Modulation of chitin synthesis rates might be associated with subtle regulatory effects at the molecular level.

Generally, regulation of the CS activity might function at three major levels.
1) An irreversible reaction which activates a presumed zymogenic CS and may occur before or after the incorporation of the synthetase unit into plasma membranes.
2) Reversible active-inactive forms of the polymerizing enzyme triggered by internal or external stimuli.
3) Modulation at the level of the active enzyme by mechanisms that may include accessibilities and concentrations of substrate and co-factors or removal of inhibitory components. The above is schematically illustrated as follows:

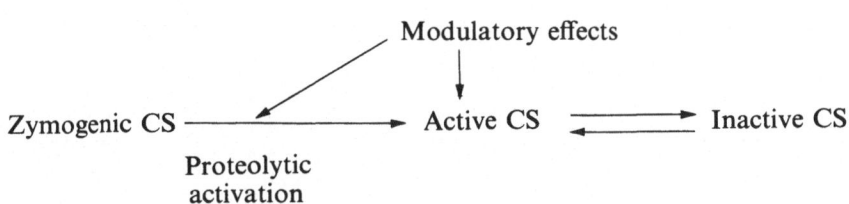

Theoretically, many points of interference to block or modify chitin polymerization exist. This subject will be discussed later in connection with the inhibitory effects of various compounds on chitin formation.

4 INHIBITION OF CHITIN SYNTHESIS

4.1 *Inhibitors*

The nucleoside peptide antibiotics and the benzoylphenyl ureas are two major groups of compounds acting primarily at the chitin synthesis target site. Several compounds of both groups have been developed and commercialized as selective pest control agents. The nucleoside peptide antibiotics are applied to control several fungal caused diseases. They have been thoroughly investigated and their precise biochemical lesion is known. The benzoylphenyl ureas, initially synthesized as weed control agents were found to be highly effective insecticides against a variety of insect pests. These compounds do not affect either fungal growth or fungal CS (Van Eck, 1979; Cohen and Casida, 1980b; Leighton et al., 1981; Cohen, unpublished results) and their exact mode of action in insects is still unresolved. One group of pyrimidine nucleoside peptides were isolated from *Streptomyces cacaoi* var. *asoensis* and named polyoxins (A–M) (Hori et al., 1971). The various polyoxins are minor modifications of a basic structure composed of pyrimidine, ribose and attached peptides (see Figure 3 for structure of polyoxin-D). The polyoxins are effective in suppress-

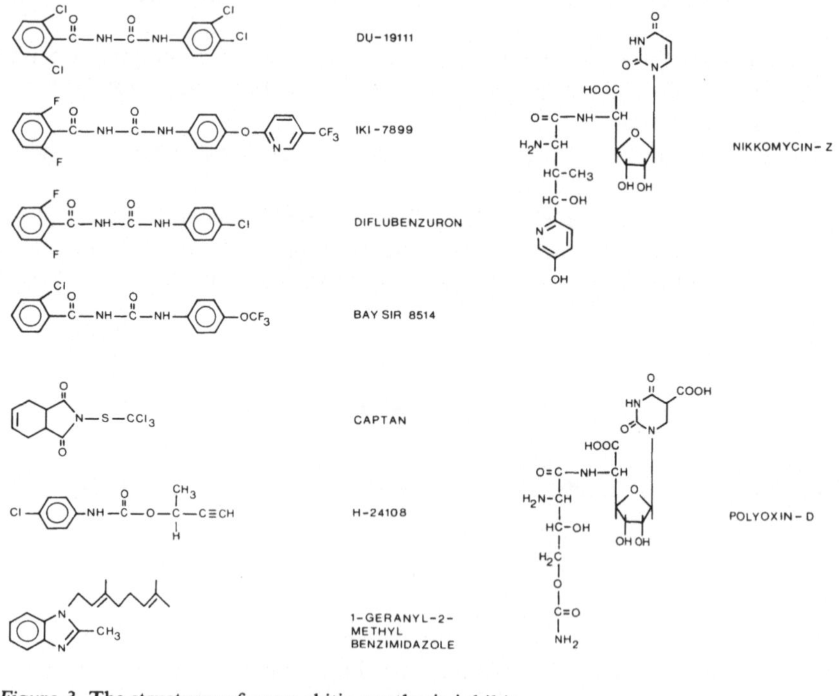

Figure 3. The structures of some chitin synthesis inhibitors.

ing the growth of various fungi and have been used as successful agricultural fungicides in Japan (Hori et al., 1974a). Polyoxin-B is especially effective in controlling *Alternaria kikuchiana* which causes the black spot disease in pears. Other polyoxins are potent fungicides against the rice sheath blight caused by *Pellicularia sasakii* and against additional fungal diseases in fruit trees and vegetables (Misato, 1982).

Nikkomycins are a new group of compounds recently isolated from culture broth of *Streptomyces tendae* (Dähn et al., 1976; Fiedler, 1981). Nikkomycin X and Z (Figure 3) are the major components in the complex mixture of closely related substances which are structurally similar to polyoxins (Müller et al., 1981). The main difference is that a pyridine ring is attached to the last amino acid. Due to their polar nature the antibiotics which disrupt chitin synthesis in fungal systems (Hori et al., 1974a; Bartnicki-Garcia and Lippman, 1972; Muzzarelli, 1977; Brillinger, 1979; Kobinata et al., 1980; Müller et al., 1981) are practically ineffective as insect control agents. Nevertheless, defective cuticles were observed in larvae of the Mexican bean beetle *Epilachna varivestis* fed bean leaves sprayed with 0.1% nikkomycin solution (Schlüter, 1982). The same antibiotic compound inhibited molting and disrupted the arrangement of chitin microfibrils in the procuticle of the mite *T. urticae* (Mothes and Seitz, 1982). Polyoxin A is insecticidal when injected into nymphs of the migratory grasshopper *Melanoplus sanguinipes* (Vardanis, 1978). Death occurred during the molting process due to interference with chitin formation. *In vitro* study by Vardanis (1976) using abdominal integuments confirmed the above observation. Cuticular growth was inhibited and abnormal cuticular structures were formed after injection of polyoxin D into larvae of the large cabbage white *Pieris brassicae* (Gijswijt et al., 1979). Like the peptidyl nucleoside antibiotics which are effective in blocking chitin synthesis is fungal system and have practical use in controlling various fungal caused diseases, the benzoylphenyl ureas assume a similar importance in insects. The considerable insecticidal properties of the benzoylphenyl urea compounds were discovered over a decade ago by scientists at the Philips Duphar Company, initially being interested in developing effective herbicides. Combining diuron and dichlobenil resulted in a benzoylphenyl urea compound DU-19111 [1-(2,6-dichlorobenzoyl)-3-(3,4-dichlorophenyl)urea] (Figure 3) which was highly insecticidal (van Daalen et al., 1972; Mulder and Gijswijt, 1973; Wellinga et al., 1973). Subsequent quantitative structure-activity relationship studies yielded diflubenzuron [1-(4-chlorophenyl)-3-(2,6-difluorobenzoyl)urea] (Figure 3) as the optimal compound of this series of benzoylphenyl ureas (Verloop and Ferrell, 1977). This compound chosen for further development has been commercialized under the trade name Dimilin. Other highly potent insecticidal benzoylphenyl ureas such as Bay Sir 8514 [1-(4-tri-fluoromethoxyphenyl)-3-(2-chlorobenzoyl)urea] and IKI-7899 [N-2,6-difluorobenzoyl- N'-4-(3-

chloro-5-trifluoromethylpyridine-2-yloxy)-3,5-dichlorophenyl urea] (Figure 3) have been since introduced (Zoebelein et al., 1980; Haga et al., 1982). The benzoylphenyl ureas are largely stomach poisons, control a wide variety of insect pests and have favorable toxicological properties with respect to acute toxicity to warm-blooded animals, phytotoxicity, accumulation in soil and biomagnification in the environment (Verloop and Ferrell, 1977). It has been suggested that those selective chemicals should be compatible with various IPM programs. However, although many non-target organisms are unaffected, the benzoylphenyl ureas are not entirely selective with regard to aquatic arthropods (Apperson et al., 1978; Farlow et al., 1978; Nimmo et al., 1980) or to certain beneficial insects (Wilkinson et al., 1978; Zungoli et al., 1983).

5 ACTION OF BENZOYLPHENYL UREAS

Insecticidal benzoylphenyl ureas are regarded as insect growth regulators because of their delayed effect and symptoms of mortality. Mortality occurs at the time of molting, normally 4–6 days following treatment (Verloop and Ferrell, 1977). Affected insects unable to ecdyse are moving within the exuviae, lose hemolymph, blacken and die apparently due to desiccation. Histological observations of treated insects indicated that the cuticular structure being weak and delicate could not withstand increased turgor during molting (Mulder and Gijswijt, 1973) and were defficient of their normal mechanical properties (Ker, 1977; Grosscurt and Andersen 1980). Deformation of the new cuticle was manifested in lacking or severe distortions of the endocuticular layers (Mulder and Gijswijt, 1973; Grosscurt, 1978; Gijswijt et al., 1979; Lim and Lee, 1982). Although cuticular proteins and chitin could be implicated in the effect of insecticidal benzoylphenyl ureas it soon became apparent that only chitin synthesis was drastically disrupted.

Post and Vincent (1973) reported that DU-19111 suppressed incorporation of labeled glucose into the cuticle of *P. brassicae* larvae and this effect was later confirmed by autoradiographic studies (Post et al., 1974), indicating chitin synthesis inhibition. Deul et al. (1978) showed that the inhibitory effect was fast and chitin synthesis in *P. brassicae* larvae was blocked by diflubenzuron in 15 min. Further research on the action of benzoylphenyl ureas on cuticle and peritrophic membrane formation as well as on inhibition of chitin biosynthesis has been extended to isolated organ cultures. Results from studies conducted with such organ cultures belonging to various taxonomic orders of insects are summarized in Table 2. Diflubenzuron as a representative of the insecticidal benzoylphenyl ureas is highly effective in blocking the incorporation of several labeled precursors into chitin. The I_{50} values for this pesticide range from 7.0×10^{-7}M in *L. cuprina* larval integument to 6.1×10^{-11}M in the case of the extremely sensitive system of *L. maderae* leg regenerates. Leighton et al. (1981) examined other effective benzoylphenyl ureas including the potent insecticide Bay Sir 8514 with the above system, and found their I_{50} values to range between 1.6×10^{-11}M and 9.4×10^{-11}M. It was suggested that the cockroach leg regenerates accumulated diflubenzuron from the medium and its actual local concentration was higher than the initial level (Sowa and Marks, 1975). Hajjar and Casida (1979) utilized isolated abdomens of *O. fasciatus* young adults for structure-activity relationship study of various benzoylphenyl urea compounds. They reported a good correlation between toxicity (LD_{50} values) and the degree of chitin inhibition, in the same insect. In addition to the integumental structures, diflubenzuron also suppressed synthesis of chitin during formation of peritrophic membranes in the migratory locust *Locusta migratoria* (Clarke et al., 1977), in the blowfly *Calliphora erythrocephala* (Becker, 1978) and in the cabbage armyworm *M. brassicae* (Mitsui et al., 1984).

6 ACTION OF PYRIMIDINE-NUCLEOSIDE PEPTIDES

Although the chitin synthesis inhibitory antibiotics are not considered as insecticides, histological and biochemical observations clearly point that their effect is very similar to that of the insecticidal benzoylphenyl ureas. Injection of polyoxin A into 5th instar nymphs of the grasshopper *M. sanguinipes* caused mortality at the time of molting (Vardanis, 1978). Histological examination of *P. brassicae* larvae injected with polyoxin-D revealed cuticular deformations similar to those caused by diflubenzuron (Gijswijt et al., 1979). Nikkomycin was found to damage the organized arrangement of chitin microfibrils in the procuticle of the mite *T. urticae* (Mothes and Seitz, 1982). Reduction in the weight of peritrophic membranes and disruption of their fine structure due to polyoxin-D was reported in *C. erythrocephala* (Becker, 1980). This compound inhibited incorporation of GlcN and GlcNAc into chitin formed by midgut epithelial cells of *M. brassicae* (Mitsui et al., 1984). Inhibition of chitin synthesis by polyoxins in various organ culture systems was reported (Table 2; Vardanis, 1976; Van Eck, 1979; Mitsui et al., 1980). Generally, compared with diflubenzuron, polyoxin-D is less effective in inhibiting chitin synthesis and its I_{50} values are around 10^{-5}M. This antibiotic is more inhibitory in the cockroach leg regenerate system ($I_{50} = 4 \times 10^{-9}$M. Leighton et al., 1981) and in the sheep blowfly integument ($I_{50} = 6 \times 10^{-7}$M, Turnbull and Howells, 1982).

7 INHIBITION OF CHITIN SYNTHETASE ACTIVITY

Fungicidal nucleoside peptide antibiotics such as polyoxins and nikkomycins are well known powerful competitive inhibitors of the fungal chitin polymerizing enzyme (Hori et al., 1974b; Bartnicki-Garcia and Lippman, 1972; Muzzarelli, 1977; Gooday, 1979; Kobinata et al., 1980). Since these antibiotics affect *in vivo* and *in vitro* insect systems producing symptoms of poisoning and lesions similar to that of diflubenzuron, it has been inferred that the precise mode of action of insecticidal benzoylphenyl ureas involves the chitin polymerization enzyme. The above assumption was supported by reports that UDP-GlcNAc accumulation accompanied inhibition of chitin synthesis by benzoylphenyl ureas (Verloop and Ferrell, 1977; Hajjar and Casida, 1979; Van Eck, 1979). Nevertheless, to clearly establish the mechanism of action of these compounds, insect cell-free systems capable of chitin polymerization were required. Such systems were for some time elusive allegedly due to methodological problems (Surholt, 1975; Vardanis, 1976; Verloop and Ferrell, 1977; Hajjar and Casida, 1979). Early reports on active insect CS preparations (Jaworski et al., 1963; Porter and Jaworski, 1965) could not be confirmed. As mentioned earlier in the review recent studies have made available a number of active cell-free CS preparations from several insect sources. These preparations include gut enzymes from various *Tribolium* species (Cohen and Casida, 1980a) and *M. brassicae* (Mitsui et al., 1984), integumental enzymes extracted from *T. ni, H. cecropia* (Cohen and Casida, 1982), *L. cuprina* (Turnbull and Howells, 1983) and *C. suppressalis* (Kitahara et al., 1983), and a CS preparation from *S. calcitrans* whole pupae Mayer et al., 1980. Also a CS from larvae of the crustacean *A. salina* was characterized (Horst, 1981). It appears that insecticidal benzoylphenyl ureas do not act directly on the polymerization step (Table 3). Diflubenzuron and Bay Sir

Table 3. Inhibition of chitin synthetase in arthropods.

Organism	Inhibition, I_{50a} (μM) Polyoxin-D	Diflubenzuron	Reference
Tribolium castaneum	4	NE	Cohen & Casida (1980b)
Trichoplusia ni	14	NE	Cohen & Casida (1982)
Hyalophora cecropia	300(14%)[b]	NE	Cohen & Casida (1982)
Chilo suppressalis	50	NE	Kitahara et al. (1983)
Stomoxys calcitrans	1000(40%)	NE	Mayer et al. (1980)
Artemia salina[c]	200(17%)	3(52–92%)[d]	Horst (1981)

[a] I_{50} — 50% inhibition. [b] In parentheses % inhibition at the designated concentrations. [c] A crustacean species. [d] % Inhibition depended on membrane preparation. NE — no effect.

8514, two potent *in vivo* chitin synthesis inhibitors do not affect enzymes extracted from *Tribolium, Trichoplusia, Stomoxys* and *Chilo*. Yet Turnbull and Howells (1983) reported that their integumental CS preparation was sensitive to diflubenzuron. However, their results are hard to interpret since inhibition did not exceed 50% and chitin was not the only product formed. The CS preparation of *A. salina* when preincubated with diflubenzuron displayed reduced chitin synthesis and appeared to be sensitive to benzoylphenyl ureas. It seems that although the polymerization step in insects is undoubtedly affected by those compounds, their precise biochemical lesion at this site remains unknown. Later in the review alternative possibilities for the mode of action of diflubenzuron will be considered and discussed.

7.1 *Interference by polyoxins and nikkomycins*

As has been shown in fungal systems, the nucleoside peptide antibiotics are also powerful chitin synthetase inhibitors in several insect cell-free systems (Table 3). Nikkomycin is extremely effective in blocking chitin polymerization in *T. castaneum* gut and *T. ni* integumental preparations with I_{50} values of $0.02\,\mu M$ and $0.06\,\mu M$, respectively (Table 4; Cohen and Casida, 1983). This antibiotic is 200 and 233 fold more active than polyoxin-D in the respective CS systems. The integumental CS preparation from *C. suppressalis* is also sensitive to polyoxin D. Yet for

Table 4. Inhibition of insect chitin synthetase systems.

Compound	Inhibition, I_{50a} (μM) *Tribolium castaneum*[b]	*Trichoplusia ni*[c]
Pyrimidine nucleoside peptides		
Polyoxin-D	4	14
Nikkomycin	0.02	0.06
Benzoylphenyl ureas		
Diflubenzuron	NE	NE
BAY SIR 8514	NE	NE
IKI-7899	NE	–
Benzimidazoles		
1-Geranyl-benzimidazole	50	–
1-Citronellyl benzimidazole	68	–
Captan[d]	6	30(99%)[e]
H-24108[f]	400(38%)	200(73%)

[a] I_{50} — 50% inhibition. [b] Gut enzyme. [c] Integumental enzyme. [d] Sulfenimide compound. [e] In parentheses % inhibition at the designated concentration. [f] Phenyl carbamate compound. NE — no effect.

unclear reasons the CS complexes from *S. calcitrans, H. cecropia* and from the crustacean *A. salina* are relatively insensitive to the antibiotic. It is noteworthy that these enzyme systems were also unaffected by the nucleotides UDP and UTP (Cohen and Casida, 1983). The nucleoside peptide antibiotics act as competitive inhibitors of the chitin polymerizing enzyme due to their close resemblance to the substrate UDP-GlcNAc. From kinetic studies Hori et al. (1974b) concluded that the nucleoside moiety in the substrate or inhibitors binds to a specific site in the enzyme and is essential for their biological activity. Conceivably, the peptidyl part in the antibiotics interacts with the catalytic site of the CS.

7.2 *Thiol blocking compounds*

Captan [1,2,3,6-tetrahydro-N-(trichloromethylthio)-phthalimide] (Figure 3) is an effective wide spectrum fungicide. This compound was found to suppress formation of the peritrophic membrane in blowflies with a low I_{50} value of 1.7×10^{-6}M (Becker, 1978). In the cockroach leg regenerate system, captan blocked chitin synthesis at a high concentration of 10^{-4}M (Marks and Sowa, 1976). On the other hand, cultured integuments of *C. suppressalis* were insensitive to captan (Kitahara et al., 1983). *Tribolium* gut and *Trichoplusia* integumental CS systems were strongly inhibited by this fungicide (Table 4) and by its close analogs captafol and folpet as well as by dichlofluanid (Cohen and Casida, 1980b, 1982). N-Ethylmaleimide which is a strong thiol blocking reagent is a weak inhibitor of *Tribolium* CS (Cohen and Casida, 1980b). Captan and its decomposition product thiophosgene (Peeples and Dalvi, 1978) are known to interact with thiol groups in fungal systems (Richmond and Somers, 1968; Siegel, 1970; Lukens, 1971). Dithiothreitol included in *Tribolium* CS incubation mixture protected the enzyme from the inhibitory effects of captan (Cohen and Casida, 1980b). This protection which involves the interaction of the fungicide with sulfhydryl groups of this compound must be removed when thiol-blocking compounds are assayed. It is interesting to note that fungal CS systems are apparently insensitive to the action of captan compared with their insect counterparts (Brillinger, 1979; Cohen and Casida, 1980b; Leighton et al., 1981). From studies with captan and related compounds it is inferred that the insect CS enzymes contain sensitive thiol groups which readily interact with thiol-blocking agents. Such interference leads to drastic inhibition of the chitin polymerization step.

7.3 *Benzimidazoles*

The available *Tribolium* gut CS radioassay (Cohen and Casida, 1980a) which is simple, reproducible and convenient can be useful in screening and evaluation of compounds acting as CS inhibitors. Structure-activity relationship study of inhibitory benzimidazole analogs was conducted using the above assay (Cohen et al., 1984). Certain benzimidazoles with a terpenoyl chain were found to be considerably toxic to houseflies and one of the compounds, 1-geranyl-2-methyl benzimidzole (Figure 3) blocked the molting of silkmoth larvae at a low dose of 0.5 μg per insect (Kuwano et al., 1982). Since the toxic symptoms resembled those of insecticidal benzoylphenyl ureas, this benzimidazole and a series of structurally-related compounds were assayed to evaluate their potency as CS inhibitors. It appears that a terpenoyl moiety such as geranyl, citronellyl or neryl are necessary for the inhibitory action (Table 4; Cohen et al., 1984). Among the compounds tested 1-geranyl- and 1-neryl-benzimidazole are the most potent inhibitors (I_{50}, 50 μM). Their activity, however, is 12-fold less than that of polyoxin-D. The mechanism of interference by the novel group of inhibitory benzimidzoles is not known. The steep slope of their inhibition curves compared with that of polyoxin-D (Figure 4) indicates a different mode of action.

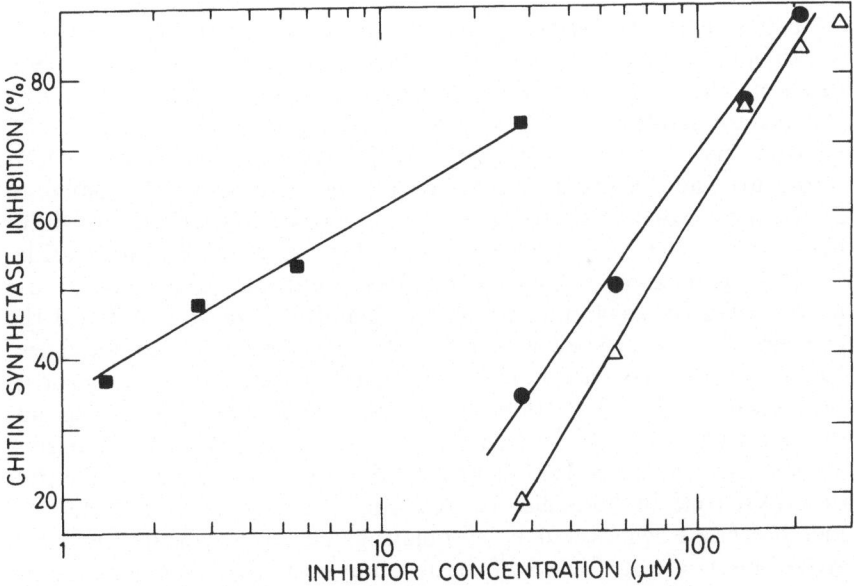

Figure 4. Inhibition curves of *Tribolium castaneum* gut chitin synthetase. ■ polyoxin D, ●-1-geranyl-benzimidazole; △-N-geranyl acetanilide.

7.4 *Other compounds*

Assorted compounds were assayed as potential CS inhibitors. Several phenyl carbamates like H-24108 [3-butyn-2-yl-N(4-chlorophenyl) carbamate] (Figure 3), an insect growth retardant (Schaefer et al., 1974), are weak inhibitors (40–70% inhibition at 300 μM) of *Tribolium* gut and *Trichoplusia* integumental CS enzymes (Cohen and Casida, 1980b, 1982). The herbicide barban, also a phenyl carbamate, displayed somewhat higher inhibitory effects (Cohen and Casida, 1980b). The mode of action of this group of compounds has not been defined. Plumbagin, a natural product from the African medicinal plant *Plumbago capensis*, which inhibits molting in various insects was assayed with the *T. ni* integumental CS system (Kubo et al., 1983). This compound is a weak insect CS inhibitor giving 30% inhibition at a relatively high level of 300 μM. The triazine CGA-19255 which inhibited GlcNAc incorporation into chitin in the cockroach leg regenerate system, and was regarded as a weak chitin synthesis inhibitor (Miller et al., 1979) was inactive in the *Tribolium* assay (Cohen and Casida, 1980b).

8 INTERFERENCE OF BENZOYLPHENYL UREAS WITH CHITIN SYNTHESIS

It has been established that the primary action of insecticidal benzoylphenyl urea compounds is at the site of chitin biosynthesis. Synthesis of cuticular or peritrophic membrane proteins as well as the tanning process are unaffected by these compounds (Hunter and Vincent, 1974; Ishaaya and Casida, 1974; Ker, 1977; Verloop and Ferrell, 1977; Grosscurt and Andersen, 1980). Several biochemical events related to cuticle formation and sclerotization were reported to be affected by diflubenzuron, yet they were regarded as secondary targets (Cohen and Casida, 1983). Diflubenzuron stimulated activities of phenoloxidase and chitinase in houseflies (Ishaaya and Casida, 1974) and at high doses reduced the catechol levels associated with sclerotization (Grosscurt and Andersen, 1980). Nevertheless, Ker (1977) demonstrated that the tanning process proceeded unaffected by this compound. Diflubenzuron was also implicated with targets that determine the molting hormone titer in insects. Yu and Terriere (1975, 1977) suggested that diflubenzuron alters the titer of ecdysterone in houseflies by reducing the activity of ecdysterone-metabolizing enzymes and by stimulating the mixed-function oxidases. Accordingly, elevated levels of the molting hormone might enhance the activities of chitinases and phenoloxidases, effects found in the study of Ishaaya and Casida (1974). Apparently this form of interference by diflubenzuron was not supported by studies which clearly showed that the inhibitor suppressed chitin synthesis in ecdysterone-independent organ cultures (Table 2). Addition of the hormone did not alter the rate of chitin synthesis in houseflies (Van Eck, 1979). Moreover, diflubenzuron did not alter the levels of ecdysterone in the milkweed bug nymphs (Hajjar and Casida, 1979) or in the stable fly pharate pupae (O'Neill et al., 1977; DeLoach et al., 1981). Also no changes in ecdysone or ecdysterone metabolism were observed (Hajjar and Casida, 1979).

Another primary target for benzoylphenyl ureas is associated with DNA synthesis. Diflubenzuron reduced the synthesis of DNA in the imaginal epidermal histoblasts of the stable fly *S. calcitrans* (Meola and Mayer, 1980; DeLoach et al., 1981). Since new adult epidermis was not formed, no incorporation of GlcNAc into chitin was detected. In this case the exact biochemical lesion was not defined and it is unknown whether the cytostatic or antimitotic effects of diflubenzuron were direct ones. The effect of diflubenzuron on DNA synthesis has been recently demonstrated in melanoma cells where it significantly reduced uptake of the four nucleosides (Mayer et al., 1984). The chemosterilizing effect of benzoylphenyl ureas (Wright and Harris, 1976; Oliver et al., 1977; Chang, 1979) might be related to inhibition of DNA synthesis (Mitlin et al., 1977). Although accepted that the primary target for insecticidal benzoylphenyl ureas is chitin synthesis, the precise biochemical lesion has not yet pinpointed. Diflubenzuron apparently acts at the polymerizing step since no effect on sugar metabolism or transport into the integument was detected (Hajjar and Casida, 1979). Furthermore, the accumulation of the substrate UDP-GlcNAc shown in a number of studies (Verloop and Ferrell, 1977; Hajjar and Casida, 1979; Van Eck, 1979) supports the assumption that chitin polymerization is directly affected by diflubenzuron. Since the available evidence clearly shows that the insect CS is insensitive to diflubenzuron and related benzoylphenyl ureas, several suggestions for their possible mode of action have been advanced (Cohen and Casida, 1983). One hypothesis favored by Leighten et al. (1981) postulates a proteolytic inhibition by diflubenzuron which prevents the conversion of a zymogenic CS into an active polymerase form. Inactive zymogenic CS complexes which require proteolytic activation has been described in a number of fungal systems, notably in the yeast *S. cerevisiae* (Cabib and Farkas, 1971) and in the yeast form of *M. rouxii* (Ruiz-Herrera et al., 1977). Addition of trypsin or endogenous fungal proteases were necessary to stimulate *in vitro* the zymogenic enzyme. The evidence for proteolytic activation of insect CS is circumstantial and involves inhibition of mammalian proteases by diflubenzuron (Leighton et al., 1981). No inhibition of relevant insect integumental proteases has been demonstrated. Moreover, diflubenauron did not inhibit integumental proteases in wing buds of the migratory locust nymphs (Cohen, unpublished results). It may well be that diflubenzuron inhibits specific proteases which act on a possible zymogenic insect CS. However, such a suggestion for the mode of action of diflubenzuron is questionable due to the irreversible nature of the proteolytic activation coupled with the fact that the inhibitor blocked rapidly (within 15 min) an ongoing chitin synthesis (Deul et al., 1978; Hajjar and Casida, 1979). It is hard, therefore, to implicate diflubenzuron in proteolytic activation since in cases where it was studied an already active enzyme functioned. Also, no prior proteolytic activation of the gut (Cohen and Casida, 1980a; Cohen, 1982) or integumental (Cohen and Casida, 1982) enzymes was required for chitin polymerization and fibrillogenesis. The inhibition of organophosphate sensitive proteases throughout the *Tribolium* CS preparation and assay which had no effect on the enzyme activity, indicated the absence of an endogenous protease activator (Cohen and Casida, 1980a). The stimulation of insect CS enzymes by trypsin (see Table 1) was rather small compared with similar effects on fungal zymogenic enzymes and it might be attributed to unmasking of catalytic or allosteric sites of the insect CS.

Another possibility for the action of benzoylphenyl ureas is related to interference with the accessibility of substrate and co-factors. Being an allosteric enzyme, the rate of chitin synthesis depends to a large extent on the level of the available substrate. Interference with the availability of UDP-GlcNAc and activators such as magnesium ions is an attractive possibility. Such a possibility was approached by Mitsui et al. (1984) using cabbage armyworm gut preparations. They proposed that the catalytic site of the CS enzyme is located on the outer surface of the membrane and to be available, the substrate must cross the plasma membrane barrier. Diflubenzuron was proposed to act on enzymatic systems responsible for translocation of UDP-GlcNAc across the plasma membranes. Possible transport systems such as Na^+-K^+ or $Ca^{2+}+Mg^{2+}$ ATPases were found to be insensitive to diflubenzuron. It should be added that it is more likely that the catalytic site of the CS enzyme is facing the cytoplasm and the location proposed by Mitsui et al. (1984) might represent a special case of the midgut epithelial cells. It is noteworthy that careful studies with yeasts using brief treatments with glutaraldehyde demonstrated the inner location of the CS complex (Cabib and Roberts, 1982).

Interference of diflubenzuron with chitin polymerization might be related to the inhibitory effect of UDP which is released during the polymerization of GlcNAc units. Uridine diphosphatase (UDPase) activity was detected in the fungus *Coprinus cinereus* (de Rousset-Hall and Gooday, 1975) and it has been proposed as a regulatory mechanism for chitin synthesis. A similar mechanism can be extended to the insect polymerizing enzyme. Insect CS is sensitive to UDP but not to UMP (Cohen and Casida, 1980b), and benzoylphenyl ureas might interfere with the presumed UDPase action resulting in a build up of inhibitory levels of UDP at the polymerization site. This possible mode of action has not been approached.

The "active metabolite" theory maintains that diflubenzuron is converted in the insect integument to an extremely active metabolite. This possibility appears unlikely due to the fact that the metabolism of diflubenzuron in insects is slow (Ivie and Wright, 1978; Cohen and Casida, 1980b). Also the fast response to diflubenzuron rules out the possibility of generating sufficient level of such a metabolite at the site of polymerization.

Inhibition of regulatory mechanisms linked to the polymerization step in chitin formation by benzoylphenyl ureas is an interesting possibility. Such regulatory mechanisms are vaguely defined and await further studies on the insect CS system. They may include a reversible shifting of active-inactive states of the CS enzyme, where diflubenzuron might stabilize the inactive state. It has been postulated by Cohen and Casida (1980b) that a delicate *in vivo* regulatory mechanism is disrupted upon CS extraction resulting in continuous synthesis of chitin in cell-free preparations. It is plausible that the inhibition of crustacean CS by diflubenzuron (Horst, 1981) is associated with such a regulatory mechanism which remained intact. Also an attractive possibility is the interference of benzoylphenyl ureas with a sensitive elaborate spatial organization of the CS or with attached protein complexes which are associated with either synthesis or extruding of nascent chitin polymers to the outer surface. Such interference may include allosteric changes in the CS or in other membrane elements which prevents normal chitin synthesis or translocation of newly formed chitin chains through the plasma membrane.

9 CONCLUSIONS

Chitin synthesis in insects is restricted to cuticle forming epidermal cells and to certain areas of epithelial cells in the gut which form the chitoprotein complex of the peritrophic membrane. Chitin synthesis by the gut epithelium appears to proceed continuously during the active phases of the insect life cycle. Synthesis of cuticular chitin by the epidermal cells is apparently subjected to complex regulatory mechanisms which determine location, time and quantity. The gross biochemical events involved in chitin biosynthesis are fairly known and recently some detailed studies utilizing cell-free chitin synthetase has shed light on the polymerization step. Yet many processes related to polymerization, microfibril formation or the molecular organization of the CS in the plasma membrane are poorly understood. Also fundamental knowledge with respect to control mechanisms at the molecular level, and effects of insect hormones and other physiological changes on chitin synthesis, is lacking.

The model in Figure 5 represents essential elements associated with chitin formation and deposition, and indicates possible sensitive sites for interference. This model is not complete and by no means incorporates all

66

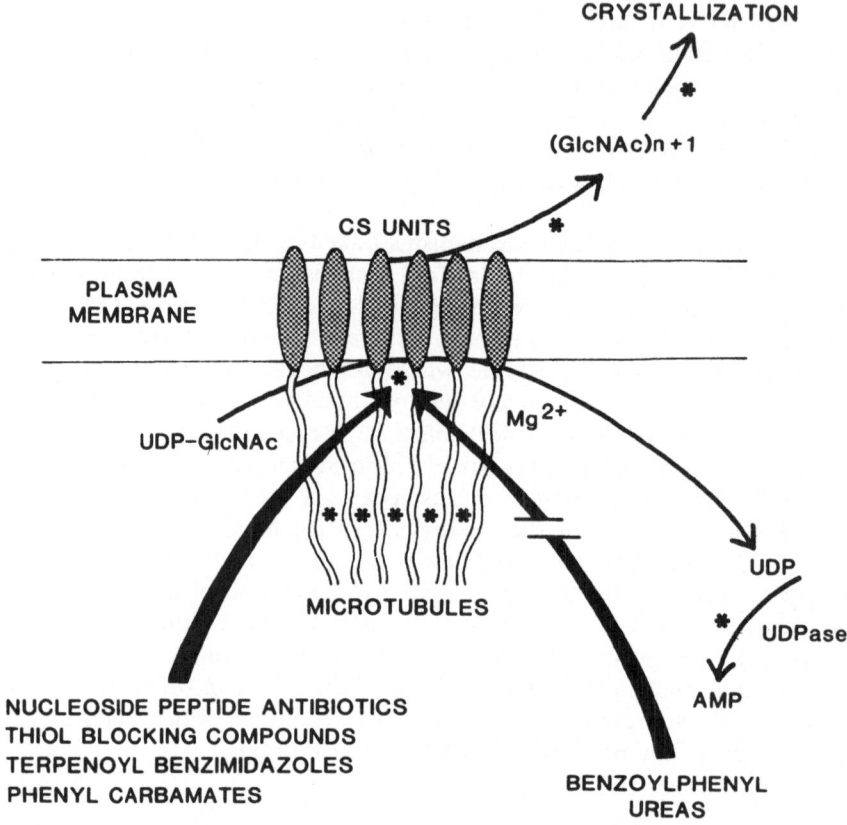

CRYSTALLIZATION

(GlcNAc)n + 1

CS UNITS

PLASMA
MEMBRANE

UDP–GlcNAc

Mg²⁺

MICROTUBULES

UDP

UDPase

AMP

NUCLEOSIDE PEPTIDE ANTIBIOTICS
THIOL BLOCKING COMPOUNDS
TERPENOYL BENZIMIDAZOLES
PHENYL CARBAMATES

BENZOYLPHENYL
UREAS

Figure 5. Diagram of the principal elements associated with chitin polymerization and fibrillogenesis. Bold arrows — Interference at the polymerization step. Asterisks — possible sites for interference with chitin formation.

factors which take part in the intricate phenomenon of chitin formation. Yet it can illustrate the complexity and possible roles of cellular components and biochemical events involved in the process of forming chitin microfibrils. The CS is a membrane enzyme located in the outer cell membranes. Many enzyme units are clustered in distinct areas, seemingly in membrane plaques, which are organized in close proximity. This arrangement ensures the crystallization of nascent chitin polymers and subsequently their assembly into microfibrils embedded in the matrix of cuticles or peritrophic membranes. It is conceivable that the CS enzyme spans the plasma membranes and its catalytic and binding sites are facing the cytoplasm. By completely unknown mechanism the nascent insoluble polymer formed is extruded to the outer surface. Crystallization of chitin chains and arrangement of the different chitin forms are also unresolved phenomena.

Chemical interference with chitin synthesis might occur at various levels notably the polymerization site. Compounds like the nucleoside peptide antibiotics compete with UDP-GlcNAc on the binding and catalytic sites of the CS enzyme. Terpenoyl benzimidazoles and phenyl carbamates act directly on the polymerase yet their precise biochemical lesion is unknown. The insect CS has sensitive thiol groups since sulfhydryl reactive compounds like captan, folpet or dichlofluanid are highly effective inhibitors. Being an allosteric enzyme, the quantitative aspect of chitin formation is closely dependent on levels of substrate at the site of synthesis. Interference with the availability of GlcNAc or co-factors like magnesium ions could modulate chitin polymerization. Regulation at this level presumably occurs *in vivo* when physiological conditions are altered. UDP which was shown to strongly inhibit the CS activity is apparently removed from the polymerization site. One possible way of its removal is a closely located UDPase enzyme. Inhibition of the presumed UDPase might be a possible site for interference. The turnover of the CS complexes occurs periodically and is tightly associated with high titers of the molting hormone. Whether the newly formed CS units are in a zymogenic state and undergo proteolytic activation remains unknown. The cytoskeletal elements in the epidermal cells might play a crucial role in chitin synthesis and microfibrillar orientation. Drugs which inhibit formation of microtubular proteins drastically disrupt chitin synthesis. The newly formed CS units, perhaps clustered in chitosome-like particles, might be attached to microtubuli which direct them into the proper areas in the cell membranes. It has been also postulated that the cytoskeletal proteins are attached to the membranal CS units and play an important role in the correct arrangement of chitin microfibrils.

The discovery of insecticidal benzoylphenyl ureas can be regarded as a milestone in insect chitin research. The fact that these compounds act on the chitin target has prompted a renewed interest in insect chitin synthesis, a research area which has been relatively dormant. The benzoylphenyl ureas do not interfere directly with the polymerizing enzyme as has been assumed and their precise biochemical lesion is still ellusive. Several possibilities which might be implicated in the mode of action of these chitin synthesis inhibitors were previously discussed and their likelihood evaluated. Obviously basic research at the biochemical and ultrastructural levels must be expanded and advanced for a better understanding of the organizational pattern, function and regulation of the insect CS complex. Such research is essential if the elucidation of the exact mode of action of diflubenzuron and possibly other unrelated compounds which interfere with chitin synthesis and fibrillogenesis is expected. The applicative potential which might derive from the above mentioned research is considerable so far as devising and developing new agrochemicals acting on a selective target such as chitin. At the same time such compounds could assist in the

68

chemical dissection of the chitin synthesis phenomenon. It may present an excellent example where basic and applied research can be interwoven for mutual benefit.

10 ACKNOWLEDGMENT

The author thanks Mrs. Carolyn Tibbetts from the Department of Entomological Sciences, the University of California, Berkeley for drawing the illustrations of the chitin synthesis complex.

11 REFERENCES

Andersen, S. O. 1979. Biochemistry of insect cuticle. Ann. Rev. Entomol. 29: 24–61.

Apperson, C. S., Schaefer, C. N., Colwell, A. E., Werner, G. H., Anderson, N. L. Dupras, Jr. E. F. and Longanecker, D. R. 1978. Effects of diflubenzuron on *Chaoborus astictopus* and nontarget organisms and persistence of diflubenzuron in lentic habitats. J. Econ. Entomol. 71: 521–527.

Bartnicki-Garcia, S., Bracker, C. E., Reyes, E. and Ruiz-Herrera, J. 1978. Isolation of chitosomes from taxonomically diverse fungi and synthesis of chitin microfibrils *in vitro*. Exp. Mycol.2: 173–192.

Bartnicki-Garcia, S. and Lippman, E. 1972. Inhibition of *Mucor rouxii* by polyoxin-D: Effect on chitin synthetase and morphological development. J. Gen. Microbiol. 71: 301–309.

Bartnicki-Garcia, S., Ruiz-Herrera, J. and Bracker, C. E. 1979. Chitosomes and chitin synthesis. In Fungal Walls and Hyphal Growth. Burnett, J. H. and Trinci, A. P. J. Eds. Cambridge University Press. Cambridge, pp. 149–168.

Becker, B. 1978. Effect of 20-hydroxy-ecdysone, juvenile hormone, Dimilin, and captan on *in vitro* synthesis of peritrophic membranes in *Calliphora erythrocephala*. J. Insect Physiol. 24: 699– 705.

Becker, B. 1980. Effect of polyoxin-D on *in vitro* synthesis of peritrophic membranes in *Calliphora erythrocephala*. Insect Biochem. 10: 101–106.

Bracker, C. F., Ruiz-Herrera, J. and Bartnicki-Garcia, S. 1976. Structure and transformation of chitin synthetase particles (chitosomes) during microfibril synthesis *in vitro*. Proc. Natl. Acad. Sci. U.S.A. 73: 4570–4574.

Braun, P. C. and Calderone, R. A. 1978. Chitin synthesis in *Candida albicans*: Comparison of yeast and hyphal forms. J. Bacteriol. 133:1472–1477.

Braun, P. C. and Calderone, R. A. 1979. Regulation and solubilization of *Candida albicans* chitin synthetase. J. Bacteriol. 140: 666–670.

Brillinger, G. Y. 1979. Metabolic products of microorganisms 181. Chitin synthase from fungi, a test model for substances with insecticidal properties. Arch. Microbiol. 121: 71–74.

Cabib, E. and Farkas, V. 1971. The control of morphogenesis: An enzymatic mechanism for the initiation of septum formation in yeast. Proc. Natl. Acad. Sci. U.S.A. 68: 2052–2056.

Cabib, E. and Roberts, R. 1982. Synthesis of the yeast cell wall and its regulation. Ann. Rev. Biochem. 51: 763–793.

Candy, D. J. and Kilby, B. A. 1962. Studies on chitin synthesis in the desert locust. J. Exp. Biol. 39: 129–140.

Chang, S. C. 1979. Laboratory evaluation of diflubenzuron, penfluron, and Bay Sir 8514 as female sterilants against the house fly. J. Econ. Entomol. 72: 479–481.

Clarke, L., Temple, G. H. R. and Vincent, J. F. V. 1977. The effects of a chitin inhibitor-Dimilin- on the production of peritrophic membrane in the locust, *Locusta migratoria*. J. Insect Physiol. 23: 241–246.

Cohen, E. 1982. In vitro chitin synthesis in an insect: formation and structure of microfibrils. Europ. J. Cell Biol. 16: 284–294.

Cohen, E. 1985. Chitin synthetase activity and inhibition in different insect microsomal preparations. Experientia. 41: 470–472.

Cohen, E. and Casida, J. E. 1980a. Properties of Tribolium gut chitin synthetase. Pestic. Biochem. Physiol. 13: 121–128.

Cohen, E. and Casida, J. E. 1980b. Inhibition of Tribolium gut chitin synthetase. Pestic. Biochem. Physiol. 13: 129–136.

Cohen, E. and Casida, J. E. 1982. Properties and inhibition of insect integumental chitin synthetase. Pestic. Biochem. Physiol. 17: 301–306.

Cohen, E. and Casida, J. E. 1983. Insect chitin synthetase as a biochemical probe for insecticidal compounds. In Pesticide Chemistry. Human Welfare and the Environment. Miyamoto, J. and Kearney, P. C. eds. Pergamon Press. Oxford. pp. 25–32.

Cohen, E., Kuwano, E. and Eto, M. 1984. The use of Tribolium chitin synthetase assay in studying effects of benzimidazoles with a terpene moiety and related compounds. Agric. Biol. Chem. 48: 1617–1620.

Craig, G. D., Campbell, J. McA. and Peberdy, J. F. 1981. Endogenous chitin synthase inhibitor in Aspergillus nidulans. Trans. Br. Mycol. Soc. 77: 579–585.

Dähn, U., Hagenmaier, H., Höhne, H., König, W. A., Wolf, G. A. and Zähner, H. 1976. Stoffwechselprodukte von Mikroorganismen. 154. Mitteilung. Nikkomycin, ein neuer Hemmstoff der Chitinsynthese bei Pilzen. Arch. Microbiol. 107: 143–160.

DeLoach, J. R. Meola, S. M., Mayer, R. T. and Thompson, J. M. 1981. Inhibition of DNA synthesis by diflubenzuron in pupae of the stable fly Stomoxys calcitrans (L). Pestic. Biochem. Physiol. 15: 172–180.

de Rousset-Hall, A. and Gooday, G. W. 1975. A kinetic study of a solubilized chitin synthetase preparation from Coprinus cinereus. J. Gen. Microbiol. 89: 146–154.

Deul, D. H., DeJong, B. J. and Kortenbach, J. A. M. 1978. Inhibition of chitin synthesis by two 1-(2,6-disubstituted benzoyl)-3-phenylurea insecticides. II Pestic. Biochem. Physiol. 8: 98–105.

Duran, A. and Cabib, E. 1978. Solubilization and partial purification of yeast chitin synthetase. Confirmation of the zymogenic nature of the enzyme. J. Biol. Chem. 253: 4419–4425.

Elorza, M. V., Rico, H. and Sentandreu, R. 1983. Calcofluor white alters the assembly of chitin fibrils in Saccharomyces cerevisiae and Candida albicans cells. J. Gen. Microbiol. 129: 1577–1582.

Farlow, J. E., Breaud, T. P., Steelman, C. D. and Schilling, P. E. 1978. Effects of the insect growth regulator diflubenzuron on non-target aquatic populations in a Louisiana intermediate marsh. Environ. Entomol. 7: 199–204.

Fiedler, H.-P. 1981. Quantitation of nikkomycins in biological fluids by ion-pair reversed-phase high-performance liquid chromatography. J. Chromatog. 204: 313–318.

Gijswijt, M. J., Deul, D. H. and DeJong, B. J. 1979. Inhibition of chitin synthesis by benzoyl-phenylurea insecticides, III. Similarity in action in Pieris brassicae (L.) with polyoxin-D. Pestic. Biochem. Physiol. 12: 87–94.

Glaser, L. and Brown, D. H. 1957. The synthesis of chitin in cell free extracts of Neurospora crassa. J. Biol. Chem. 228: 729–742.

Gooday, G. W. 1979. Chitin synthesis and differentiation in Coprinus cinereus. In Fungal Walls and Hyphal Growth. Burnett, J. H. & Trinci, A. P. J., eds. Cambridge University Press. Cambridge. pp. 203–223.

Gooday, G. W. and de Rousset-Hall, A. 1975. Properties of chitin synthetase from Coprinus cinereus. J. Gen. Microbiol. 89: 137–145.

Gooday, G. W. and Trinci, A. P. J. 1980. Wall structure and biosynthesis in fungi. In: Eukaryotic Microbial Cell. Soc. Gen. Microbiol. Sym. 30: 207–252.

Grosscurt, A. C. 1978. Diflubenzuron: Some aspects of its ovicidal and larvicidal mode of action and an evaluation of its practical possibilities. Pestic. Sci. 9: 373–386.

Grosscurt, A. C. and Andersen, S. O. 1980. Effects of diflubenzuron on some chemical and mechanical properties of the elytra of Leptinotarsa decemlineata. Proceedings of the Koninklijke Nederlandse Akademie van Wetenschappen. Ser. C. Vol. 83: 143–150.

Haga, T. Toki, T. Koyanagi, T. and Nishiyama, R. 1982. Structure-activity relationships of a series of benzoyl-pyridyloxyphenyl urea derivatives. Paper II d-7 of the Fifth International Congress of Pesticide Chemistry. Abstracts. Kyoto, Japan. August 29-September 4.

70

Hajjar, N. P. and Casida, J. E. 1979. Structure-activity relationships of benzoylphenyl ureas as toxicants and chitin synthesis inhibitors in *Oncopeltus fasciatus*. Pestic. Biochem. Physiol. 11: 33–45.

Heinrich, P., Surholt, B. and Zebe, E. 1979. Autoradiographische Untersuchungen zum Einbau von ^{14}C-Glukose and ^{3}H-N-Acetyl-Glukosamin bei der Wanderheuschrecke *Locusta migratoria*. Cytobiologie. 9: 45–58.

Herth, W. 1980. Calcofluor white and Congo red inhibit chitin microfibril assembly of Poteri oochromonas: Evidence for a gap between polymerization and microfibril formation. J. Cell Biol. 87: 442–450.

Herth, W. and Zugenmaier, P. 1977. Ultrastructure of the chitin fibrils on the centric diatom *Cyclotella cryptica*. J. Ultrastr. Res. 61: 230–239.

Hori, M., Eguchi, J., Kakiki, K. and Misato, T. 1974a. Studies on the mode of action of polyoxins. VI Effect of polyoxin-B on chitin synthesis in polyoxin-sensitive and resistant strains of *Alternaria kikuchiana*. J. Antibiot. 27: 260–266.

Hori, M., Kakiki, K. and Misato, T. 1974b. Further study on the relation of polyoxin structure to chitin synthetase inhibition. Agric. Biol. Chem. 38: 691–698.

Hori, M., Kakiki K., Suzuki, S., and Misato, T. 1971. Studies on the mode of action of polyoxins. Part III. Relation of polyoxin structure to chitin synthetase inhibition. Agric. Biol. Chem. 35: 1280–1291.

Horst, M. N. 1981. The biosynthesis of crustacean chitin by a microsomal enzyme from larval brine shrimp. J. Biol. Chem. 256: 1412–1419.

Hunter, E. and Vincent, J. F. V. 1974. The effects of a novel insecticide on insect cuticle. Experientia. 30: 1432–1433.

Ishaaya, I. and Casida, J. E. 1974. Dietary TH 6040 alters composition and enzyme activity of housefly larval cuticle. Pestic. Biochem. Physiol. 4: 484–490.

Ivie, G. W. and Wright, J. E. 1978. Fate of diflubenzuron in the stable fly and house fly. J. Agric. Food Chem. 26: 90–94.

Jan, Y. N. 1974. Properties and cellular localization of chitin synthetase in *Phycomyces blakesleeanus*. J. Biol. Chem. 249: 1973–1979.

Jaworski, E., Wang, L. and Marco, G. 1963. Synthesis of chitin in cell-free extracts of *Prodenia eridania*. Nature (London) 198: 790.

Ker, R. F. 1977. Investigation of locust cuticle using the insecticide diflubenzuron. J. Insect Physiol. 23: 39– 48.

Kimura, S. 1973. The control of chitin deposition by ecdysterone in larvae of *Bombyx mori*. J. Insect Physiol. 19: 2177–2181.

Kitahara, K., Nakagawa, Y., Nishioka, T. and Fujita, T. 1983. Cultured integument of *Chilo suppressalis* as a bioassay system of insect growth regulators. Agric. Biol. Chem. 47: 1583–1589.

Kobinata, K., Uramoto, M., Nishii, M., Kusakabe, H., Nakamura, G. and Isono, K., 1980. Neopolyoxins A, B, and C, new chitin synthetase inhibitors. Agric. Biol. Chem. 44: 1709–1711.

Kubo, I., Uchida, M. and Klocke, J. A. 1983. An insect ecdysis inhibitor from the African medicinal plant. *Plumbago capensis* (Plumbaginaceae); a naturally occurring chitin synthetase inhibitor. Agric. Biol. Chem. 47: 911–913.

Kuwano, E., Sato, N. and Eto, M. 1982. Insecticidal benzimidazoles with a terpenoid moiety. Agric. Biol. Chem. 46: 1715–1716.

Leighton, T., Marks, E. and Leighton, F. 1981. Pesticides: Insecticides and fungicides as chitin synthesis inhibitors. Science. 213: 905–907.

Lim, S. J. and Lee, S. S. 1982. The toxicity of diflubenzuron to *Oxya japonica* (Willemse) and its effect on moulting. Pestic. Sci. 13: 537–544.

Locke, M. and Huie, P. 1979. Apolysis and the turnover of plasma membrane plaques during cuticle formation in an insect. Tissue and Cell 11: 277–291.

Lukens, R. J. 1971. Chemistry of Fungicide Action. Springer-Verlag, New York. pp. 59, 60 and 94.

Marks, E. P. and Leopold, R. A. 1970. Cockroach leg regeneration: Effects of ecdysterone in vitro. Science. 167: 61–62.

Marks, E. P. and Leopold, R. A. 1971. Deposition of cuticular substances *in vitro* by leg regenerates from the cockroach, *Leucophaea maderae* (F.). Biol. Bull. 140: 73–83.

Marks, E. P. and Sowa, B. A. 1976. Cuticle formation in vitro. In the Insect Integument. Hepburn, H. R., ed., Elsevier. Amsterdam. pp. 339–357.

Mayer, R. T., Chen, A. C. and DeLoach, J. R. 1980a. Characterization of a chitin synthase from the stable fly, *Stomoxys calcitrans* (L.) Insect Biochem. 10: 549–556.

Mayer, R. T., Meola, S. M., Coppage, D. L. and DeLoach, J. R. 1980b. Utilization of imaginal tissues from pupae of the stable fly for the study of chitin synthesis and screening of chitin synthesis inhibitors. J. Econ. Entomol. 73: 76–80.

Mayer, R. T., Netter, K. J., Leising, H. B. and Schachtschabel, D. O. 1984. Inhibition of the uptake of nucleosides in cultured Harding-passey melanoma cells by diflubenzuron. Toxicology. 30: 1–6.

McMurrough, I. and Bartnicki-Garcia, S. 1973. Inhibition and activation of chitin synthesis by *Mucor rouxii* cell extracts. Arch. Biochem. Biophys. 158: 812–816.

Meola, S. M. and Mayer, R. T. 1980. Inhibition of cellular proliferation of imaginal epidermal cells by diflubenzuron in pupae of the stable fly. Science. 207: 985–987.

Miller, R. W., Corley, C., Cohen, C. F., Robbins, W. E. and Marks, E. P. 1979. Efficacy and mode of action of CGA-19255, Paper 50 of Pesticide Chemistry Division. Amer. Chem. Soc. National Meeting, Honolulu, Hawaii, 2 to 6 April.

Misato, T. 1982. Present status and future prospects of agricultural antibiotics. J. Pestic. Sci. 7:301–305.

Misato, T., Kakiki, K. and Hori, M. 1979. Chitin as a target for pesticide action: Progress and prospect. In Advances in Pesticide Science. Geissbuhler, H., Brooks, G. T. and Kearney, P. C., eds. Pergamon Press, Oxford. pp. 458–464.

Mitlin, N., Wiygul, G. and Haynes, J. W. 1977. Inhibition of DNA synthesis in boll weevils (*Anthonomus grandis* Boheman) sterilized by Dimilin. Pestic. Biochem. Physiol. 7: 559–563.

Mitsui, T., Nobusawa, C. and Fukami, J.-I. 1984. Mode of inhibition of chitin synthesis by diflubenzuron in the cabbage armyworm, *Mamestra brassicae* L. J. Pestic. Sci. 9: 19–26.

Mitsui, T., Nobusawa, C., Fukami, J.-I., Colins, J. and Riddiford, L. M. 1980. Inhibition of chitin synthesis by diflubenzuron in *Manduca* larvae. J. Pestic. Sci. 5: 335–341.

Mothes, U. and Seitz, K. A. 1981. A possible pathway of chitin synthesis as revealed by electron microscopy in *Tetranychus urticae* (Acari, Tetranychidae). Cell Tissue Res. 214: 443–448.

Mothes, U. and Seitz, K.-A. 1982. Action of the microbial metabolite and chitin synthesis inhibitor nikkomycin on the mite *Tetranychus urticae*; an electron microscope study. Pestic. Sci. 13: 426–441.

Mulder, R. and Gijswijt, M. J. 1973. The laboratory evaluation of two promising new insecticides which interfere with cuticle deposition. Pestic Sci. 4: 737–745.

Müller, H., Furter, R., Zähner, H. and Rast, D. M. 1981. Metabolic products of microorganisms 203. Inhibition of chitosomal chitin synthetase and growth of *Mucor rouxii* by nikkomycin Z, nikkomycin X, and polyoxin A: A comparison. Arch. Microbiol. 130: 195–197.

Muzzarelli, R. A. A. 1977. Chitin. Pergamon Press, Oxford.

Nimmo, D. R., Hamaker, T. L., Moore, J. C. and Wood, R. A. 1980. Acute and chronic effects of Dimilin on survival and reproduction of *Mysidopsis bahia*. Aquatic Toxicology, ASTM STP 707. Eaton, J. G., Parrish, P. R. and Hendricks, A. C., Eds., American Society for Testing Materials. pp. 366–376.

Neville, A. C. 1975. Biology of Arthropod Cuticle. Springer-Verlag, New York.

Oberlander, H., Ferkovich, S., Leach, E. and Van Essen, F. 1980. Inhibition of chitin biosynthesis in cultured imaginal discs: Effect, of alpha-amanitin, actinomycin-D, cycloheximide, and puromycin. Wilhelm Roux's Arch. Dev. Biol. 188: 81–86.

Oberlander, H., Ferkovich, S. M., Van Essen, F. and Leach, C. E. 1978. Chitin biosynthesis in imaginal discs cultured in vitro. Wilhelm Roux's Arch. Dev. Biol. 185: 95–98.

Oberlander, H., Lynn, D. E. and Leach, C. E. 1983. Inhibition of cuticle production in imaginal discs of *Plodia interpunctella* (cultured *in vitro*): Effects of colcemid and vinblastine. J. Insect Physiol. 29: 47–53.

Oliver, J. E., DeMilo, A. B., Brown, R. T. and McHaffey, D. G. 1977. AI3-63223: A highly effective boll weevil sterilant. J. Econ. Entomol. 70: 286–288.

O'Neill, M. P., Holman, G. M. and Wright, J. E. 1977. β-Ecdysone levels in pharate pupae of the stable fly, *Stomoxys calcitrans* and interaction with the chitin inhibitor difluben-zuron. J. Insect Physiol. 23: 1243–1244.

Peeples, A. and Dalvi, R. R. 1978. Toxicologic studies of N- trichloromethylthio-4-cyclohexene-1,2-dicarboximide (Captan): Its metabolism by rat liver drug-metabolizing enzyme system. Toxicology. 9: 341–351.

Peperdy, J. F. and Moore, P. M. 1975. Chitin synthase in *Mortierella vinacea*: Properties, cellular location and synthesis in growing cultures. J. Gen. Microbiol. 90: 228–236.

Peters, W. 1976. Investigation on the peritrophic membranes of Diptera. In The Insect Integument. Hepburn, H. R., ed. Elsevier. Amsterdam, pp. 515–543.

Porter, C. A. and Jaworski, E. G., 1965. Biosynthesis of chitin during various stages in the metamorphosis of *Prodenia eridania*. J. Insect Physiol. 11: 1151–1160.

Post, L. C., DeJong, B. J. and Vincent, W. R. 1974. 1-(2,6-disubstituted benzoyl)-3-phenyl urea insecticides: Inhibitors of chitin synthesis. Pestic. Biochem. Physiol. 4: 473–483.

Post, L. C. and Vincent, W. R. 1973. A new insecticide inhibits chitin synthesis. Naturwissenschaften. 60: 431–432.

Richmond, D. V. and Somers, E. 1968. Studies on the fungitoxicity of captan. VI. Decomposition of ^{35}S-labelled captan by *Neurospora crassa* conidia. Ann. Appl. Biol. 62: 35–43.

Rudall, K. M. and Kenchington, W. 1973. The chitin system. Biol. Rev. 48: 597–636.

Ruiz-Herrera, J. and Bartnicki-Garcia, S. 1976. Proteolytic activation and inactivation of chitin synthetase from *Mucor rouxii*. J. Gen. Microbiol. 97: 241–249.

Ruiz-Herrera, J., Lopez-Romero, E. and Bartnicki-Garcia, S. 1977. Properties of chitin synthetase in isolated chitosomes from yeast cells of *Mucor rouxii*. J. Biol. Chem. 252: 3338–3343.

Ruiz-Herrera, J., Sing, V. O., Van der Woude, W. J. and Bartnicki-Garcia, S. 1975. Microfibril assembly of granules of chitin synthetase. Proc. Natl. Acad. Sci. USA. 72: 2706–2710.

Schaefer, C. H., Wilder, W. H., Mulligan, III, F. S. and Dupras, Jr., E. F. 1974. Insect development inhibitors: Effects of Altosid®, TH6040 and H 24108 against mosquitoes (*Diptera: Culicidae*). Calif. Mosquito Control Assoc. 42: 137–139.

Schlüter, U. 1982. Ultrastructural evidence for inhibition of chitin synthesis by nikkomycin. Wilhelm Roux's Arch. Dev. Biol. 191: 205–207.

Selitrennikoff, C. P. 1979. Chitin synthase activity from the *slime* variant of *Neurospora crassa*. Biochim. Biophys. Acta 571: 224–232.

Siegel, M. R. 1970. Reactions of certain trichloromethyl sulfenyl fungicides with low molecular weight thiols. *In vitro* studies with glutathione. J. Agric. Food Chem. 18: 819–822.

Sowa, B. A. and Marks, E. P. 1975. An *in vitro* system for the quantitative measurement of chitin synthesis in the cockroach: Inhibition by TH 6040 and polyoxin-D. Insect Biochem. 5: 855–859.

Stevenson, J. R. and Tung, D. A., 1971. Inhibition by actinomycin D of the initiation of chitin biosynthesis in the crayfish. Comp. Biochem. Physiol. 39B: 559–567.

Surholt, B. 1975. Studies *in vivo* and *in vitro* on chitin synthesis during the larval-adult moulting cycle of the migratory locust, *Locusta migratoria* L. J. Comp. Physiol. 102: 135–147.

Turnbull, I. F. and Howells, A. J. 1982. Effects of several larvicidal compounds on chitin biosynthesis by isolated larval integuments of the sheep blowfly *Lucilia cuprina*. Aust. J. Biol. Sci. 35: 491–503.

Turnbull, I. F. and Howells, A. J. 1983. Integumental chitin synthase activity in cell-free extracts of larvae of the Australian sheep blowfly, *Lucilia cuprina*, and two other species of Diptera. Aust. J. Biol. Sci. 36: 251–262.

Ulane, R. E. and Cabib, E. 1974. The activating system of chitin synthetase from *Saccharomyces cerevisiae*. Purification and properties of an inhibitor of the activating factor. J. Biol. Chem. 249: 3418–3422.

Van Daalen, J. J., Meltzer, J., Mulder, R. and Wellinga, K. 1972. A selective insecticide with a novel mode of action. Naturwissenschaften 59: 312–313.

Van Eck, W. H. 1979. Mode of action of two benzoylphenyl ureas as inhibitors of chitin synthesis in insects. Insect Biochem. 9: 295–300.

Vardanis, A. 1976. An *in vitro* assay system for chitin synthesis in insect tissue. Life Sci. 19: 1949–1956.

Vardanis, A. 1978. Polyoxin fungicides: Demonstration of insecticidal activity due to inhibition of chitin synthesis. Experientia. 34: 228–229.

Verloop, A. and Ferrell, C. D. 1977. Benzoylphenyl ureas — A new group of larvicides interfering with chitin deposition. In Pesticide Chemistry in the 20th Century. Plimmer, J. R., ed. ACS Sym. Ser., No. 37. Amer. Chem. Soc., Washington, D. C., pp. 237–270.

Vermeulen, C. A., Raeven, M. B. J. M. and Wessels, J. G. H. 1979. Localization of chitin synthase activity in subcellular fractions of *Schizophyllum commune* protoplasts. J. Gen. Microbiol. 114: 87–97.

Wellinga, K., Mulder, R. and Van Daalen, J. J. 1973. Synthesis and laboratory evaluation of 1-(2,6-disubstituted benzoyl)-3-phenyl ureas, a new class of insecticides. II. Influence of the acyl moiety on insecticidal activity. J. Agr. Food Chem. 21: 993–998.

Wilkinson, J. D., Biever, K. D., Ignoffo, C. M., Pons, W. J., Morrison, R. K. and Seay, R. S. 1978. Evaluation of diflubenzuron formulations on selected insect parasitoids and predators. J. Ga. Entomol. Soc. 13: 227–236.

Wright, J. E. and Harris, R. L. 1976. Ovicidal activity of Thompson-Hayward TH 6040 in the stable fly and horn fly after surface contact by adults. J. Econ. Entomol. 69: 728–730.

Yu, S. J. and Terriere, L. C. 1975. Activities of hormone metabolizing enzymes in house files treated with some substituted urea growth regulators. Life Sci. 17: 619–626.

Yu, S. J. and Terriere, L. C. 1977. Ecdysone metabolism by soluble enzymes from three species of Diptera and its inhibition by the insect growth regulator TH-6040. Pestic. Biochem. Physiol. 7: 48–55.

Zoebelein, G., Hamman, I. and Sirrenberg, W. 1980. BAY SIR 8514, a new chitin synthesis inhibitor. Z. Ang. Ent. 89: 289–297.

Zungoli, P. A., Steinhauer, A. L. and Linduska, J. J. 1983. Evaluation of diflubenzuron for Mexican bean beetle (*Coleoptera: Coccinellidae*) control and impact on *Pediobius foveolatus* (*Hymenoptera: Eulophidae*). J. Econ. Entomol. 76: 188–191.

4. Mode of action and insecticidal properties of diflubenzuron

A.C. Grosscurt and B. Jongsma

ABBREVIATIONS

a.i. = active ingredient; dfb = diflubenzuron; CS = chitin synthase; Dol-P = dolichol-phosphate; GlcNAc = N-acetyl-D-glucosamine; IDI = insect development inhibitor; IGR = insect growth regulator; LC_{50} = concentration required to kill 50% of the test organisms; UDP = uridine-diphosphate; UDPGlcNAc = uridine-diphosphate-N-acetyl-glucosamine; UMP = uridine-monophosphate; WP = wettable powder.

1 INTRODUCTION

The insect integument consists of an epidermal cell layer with the overlying cuticle which it has secreted. The two major components of the insect cuticle are protein and chitin. Chitin, a polymer of acetylglucosamine, always appears to occur in association with protein. The ratio of protein to chitin differs in cuticles of different species. The conformation of the cuticle is maintained by (among other things) interactions both between protein molecules (tanning) and between protein and chitin. The cuticle can be divided into a thin outermost layer, the epicuticle, and the thicker procuticle of which the outermost layers are secreted first. In the procuticle a heavily tanned exocuticle, a lightly tanned endocuticle and occasionally an untanned mesocuticle can be distinguished.

Growth of juvenile stages is limited by the cuticle, since it is only capable of limited stretching. For marked increase in size the cuticle must be shed and replaced. Before shedding of the old cuticle (ecdysis) the endocuticle detaches from the epidermis (apolysis). The endocuticle and, if present, the mesocuticle are dissolved and the degradation products are subsequently resorbed by the epidermis for recycling in the next instar. Apolysis is initiated by the hormone ecdyson. In adult insects cuticle

Wright, J. E. and Retnakaran, A. (Eds), Chitin and Benzoylphenyl ureas. ISBN 978-94-010-8638-7.
© *1987, Dr W. Junk Publishers, Dordrecht.*

deposition and tanning occur during a relatively short post-moulting period.

Occurrence of chitin is mainly restricted to arthropods, fungi and nematodes. It does not occur in vertebrates. Chemical compounds which specifically interfere with chitin deposition therefore offer possibilities for development as pesticides with a high qualitative selectivity towards vertebrates.

The first insecticides with an activity primarily based on interference with chitin deposition were the benzoyl ureas. Their insecticidal activity was first described by Philips-Duphar research workers (Van Daalen et al. 1972). Of this group of compounds, diflubenzuron (dfb) was the first commercially available insecticide. Diflubenzuron is the common name for 1-(4-chlorophenyl)-3-(2,6-di-fluoro-benzoyl)urea. Code names include: PH 60-40, TH 6040, DU 112307, ENT-29054, OMS 1804. The registered trade mark of Duphar B.V. of diflubenzuron formulations is Dimilin[®].

In practice dfb is mainly used for control of species belonging to the orders of the Diptera, Lepidoptera and Coleoptera. Limited possibilities are found in the orders of the Hemiptera, Hymenoptera, Orthoptera and Acarina. In the latter order effects are only found in the Eriophyidae (e.g. *Phyllocoptruta oleivora*, the citrus rust mite (McCoy 1978) and *Aculus schlechtendali*, the apple rust mite (Duphar B.V. unpublished)).

Insecticides interfering with physiological processes which when disturbed lead to lethal effects only with subsequent growth and development as a prerequisite, are generally classified as IGR's (insect growth regulators) or IDI's (insect development inhibitors). The benzoyl ureas can best be listed as IDI's since, in contrast to IGR's like juvenoids, they do not regulate growth but rather inhibit a vital process, viz. cuticle deposition.

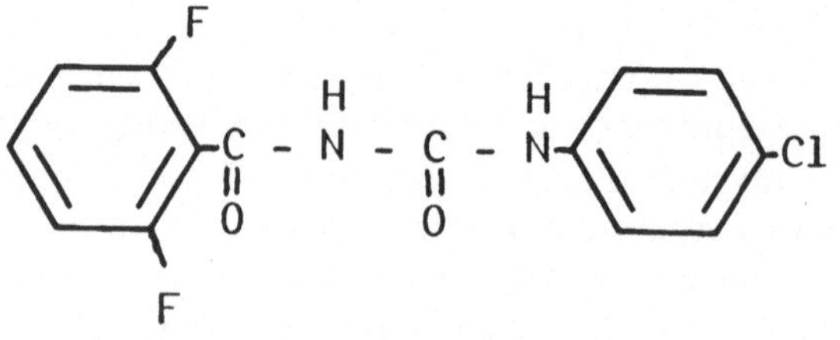

Figure 1. Chemical structure of diflubenzuron.

2 MODE OF ACTION OF THE LARVICIDAL AND OVICIDAL ACTIVITIES OF DIFLUBENZURON

Effects of dfb generally become obvious as an incapacity to moult or hatch eggs. Macroscopic effects are also seen in some cases in the period between treatment with dfb and the next moult. For example, when dfb treatment of larvae of *Leptinotarsa decemlineata* (Colorado potato beetle) starts immediately after moulting, the normal wrinkles in the cuticle gradually disappear and they develop a swollen appearance. Normal locomotion is also seriously interfered with.

Microscopic observation of dfb treated larvae has shown a disturbance of formation of the lamellate deposition of the procuticle. After fixation and staining, globular material can be found instead. The degree of distortion is different in the various regions of the cuticle when immature insects are treated with dfb. This phenomenon has been illustrated with locust cuticles by Ker (1977). Tissues of treated larvae reaching the next moult show that apolysis and resorption of the malformed cuticle normally occurs. However, after deposition of the epicuticle, formation of the procuticle is disturbed again. As a consequence, the new instar is either not able to ecdyse due to lack of rigidity of its exoskeleton or it dies shortly after ecdysis.

Histochemical studies reveal that the globular material which is formed after disturbance of the proper cuticle formation disappears after treatment with pronase but that chitinase has no effect (Smit 1977). These results suggest that the globules contain proteins but no chitin.

In oviparous insects phenomena attending the ovicidal activity caused either by contact activity on eggs or by treatment of adult female insects are similar. The larva in the egg develops fully but is unable to leave the egg. At marginal dosages, when eclosion is only partly inhibited, high mortality of the first larval instar may occur (Crystal 1978; Grosscurt 1978). Electron microscopic observations of embryos of *Leptinotarsa decemlineata*, where the females were fed dfb treated food, showed distortions in the cuticle which were comparable to those previously described for larvae (Grosscurt 1978a).

In the larviparous tsetse fly (*Glossina morsitans morsitans*) phenomena after intoxication with dfb are different from those in oviparous insects. After topical application to adult female tsetse flies, egg hatch and larval development, which take place in the uterus, proceed normally. However, after larviposition the apparently normal 3rd instar larvae fail to pupate. From these observations it was speculated that no chitin synthesis occurs in the early developmental stages of the tsetse fly, but only at puparium formation shortly after larviposition (Jordan and Trewern 1978; Jordan et al. 1979).

A reduction in fertility was also reported in cases where untreated females were confined with dfb contaminated males (Moore and Taft 1975; McGregor and Kramer 1976; Wright and Spates 1976; McLaughlin 1978). This phenomenon was more closely studied by Moore et al. (1978) with *Anthonomus grandis* (cotton boll weevil). Their data show that dfb is apparently transferred between the sexes by physical contact. However, transfer by copulation is highly improbable. These observations and an earlier study by Grosscurt (1976) with *Musca domestica* (housefly) also make it unlikely that failure of the eggs to hatch is caused by sterility of the male.

A number of hypotheses has arisen from the large number of biochemical studies concerning the mode of action of dfb at cellular level. The most plausible hypothesis appears to be that the primary effect of dfb is an inhibition of synthesis of deposition of chitin, because complete inhibition of deposition can be measured shortly after application of dfb. For example, after injection of *Pieris brassicae* larvae this effect was observed after 15 minutes (Deul et al. 1978). Furthermore, a build up of UDPGlcNAc, a precursor of chitin, was observed at the same time as the inhibition of chitin deposition (Hajjar and Casida 1979; Van Eck 1979).

After injection into insect larvae a close similarity in symptoms was observed between dfb and Polyoxin D (Vardanis 1978; Gijswijt et al. 1979). Poloxin D is an antibiotic compound used as a foliar fungicide, which is a competitive inhibitor of chitin synthase (CS) in fungi and causes accumulation of UDPGlcNAc (Endo and Misato 1969).

The reaction inhibited in fungi by Polyoxin D can be represented as:

$$\text{UDP-GlcNAc} + (\text{GlcNAc})_n \rightarrow (\text{GlcNAc})_{n+1} + \text{UDP}$$

Polyoxin D shows its insecticidal activity only after injection. Brillinger (1979) compared dfb, Polyoxin D and Nikkomycin (a compound similar to Polyoxin D) and found that only dfb did not inhibit the CS isolated from the fungus *Coprinus cinereus*. This observation means either that the CS in insects and fungi are different or that dfb blocks another step in synthesis and incorporation of chitin in the cuticle. The chemical structure of UDPGLcNAc, Polyoxin D and Nikkomycin are given in Figure 2.

Because of the lack of an active cell-free preparation of the CS complex isolated from insects, dfb could until 1980 only be tested in vivo, in tissue culture or in incubated insect parts such as abdominal integuments. All these tests, which have in common the presence of whole cells, show dfb interfering with cuticle synthesis and deposition or the incorporation of labelled precursors. A review of these tests was given by Hammock and Quistad (1981).

In 1980 both Cohen and Casida and Mayer et al., described for the first time a homogenate of insect tissues in which synthesis of chitin was shown. They used guts of *Tribolium castaneum* and homogenates of *Stomoxys*

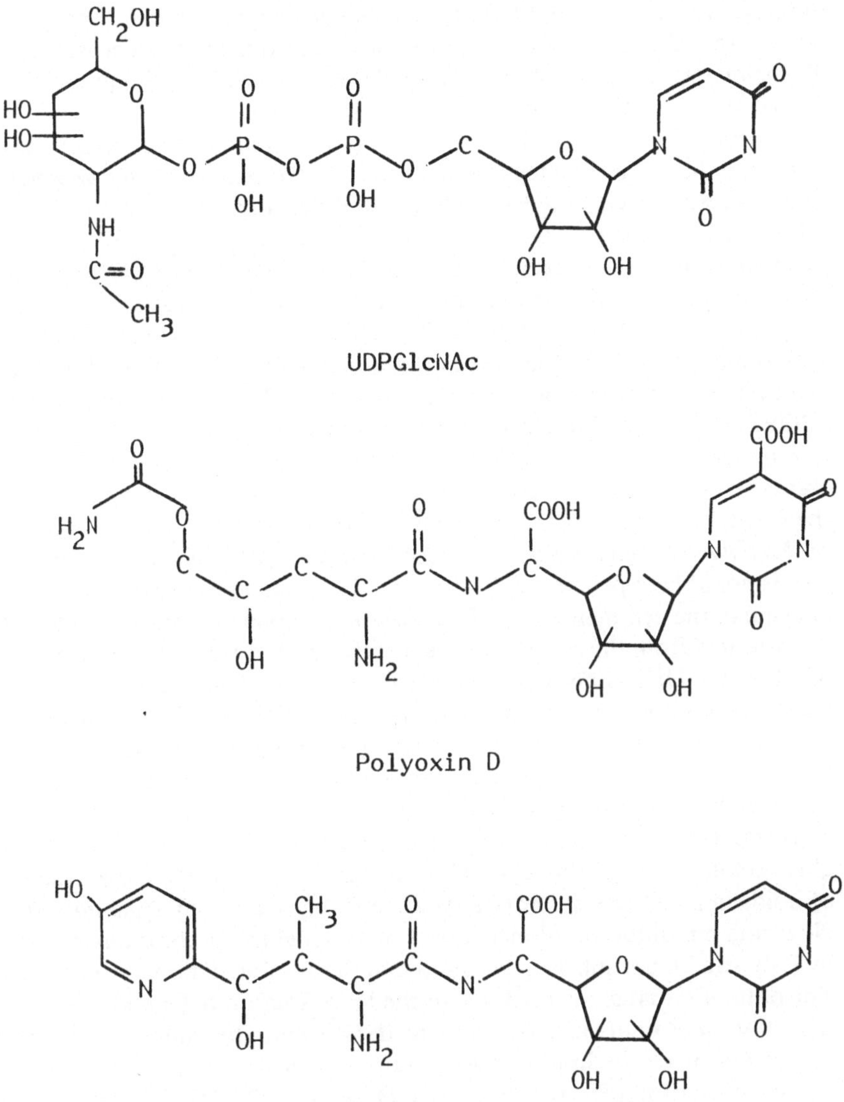

Figure 2. Chemical structures of UDPGlcNAc, Polyoxin D and Nikkomycin Z.

calcitrans pupae, respectively. Later on CS was also isolated from the integument of *Trichoplusia ni* larvae and wing tissue of *Hyalophora cecropia* pupae (Cohen and Casida 1982). Synthesis of chitin in these preparations was inhibited by Polyoxin D and Nikkomycin, but not by dfb. (Nikkomycin was not tested on *S. calcitrans*). Likewise, Cohen (1982) demonstrated that the formation of chitin microfibrils by microsomes of *T. castaneum* was not inhibited by 0.3 mM dfb while in the presence of 10 μM Polyoxin D no fibrous material was formed.

Recently Turnbull and Howells (1983) presented their results with homogenates of larval integuments from *Lucilia cuprina, Musca domestica* and *Calliphora erythrocephala*. They found that dfb and Polyoxin D both inhibited the incorporation of radioactive precursors by 50–70% in a product which was characterized as chitin. This result might be explained by the observation that the chitin synthesis was mainly associated with a fraction which sedimented at 1000 g. This means that it was associated with relatively large fragments of cells, which still might contain the chitin synthesizing complex in an organized structure. The fact that only partial inhibition was found might be due to enzymatic activity of CS which was not integrated in the membrane structure any more. The findings of Vardanis (1979) are of interest in relation to this point. He concluded that the CS in his insect system, consisting of intact cells in abdominal halves of *Melanoplus sanguinipes*, was accessible to UDPGlcNAc, a substrate that is not able to penetrate into the cell. This indicates the presence of the enzyme in the cell membrane. The enzyme complex, however, could only be saturated from the cell side of the membrane and the incorporation rate of label with UDPGlcNAc as a substrate was low when compared to those with glucosamine, which can penetrate into the cell. From the above results it can be concluded that there is only an inhibitory effect of dfb as long as whole cells or at least integrated parts of cells are present in the insect chitin synthesizing system.

Horst (1981) measured the incorporation of labelled N-acetyl-D-glucosamine from UDPGlcNAc into a chitinous product after extraction of lipids in a cell-free system derived from the crustacean *Artemia salina*. He found no inhibition of incorporation of label in the chitinous product by Polyoxin D, but suprisingly found that dfb inhibited this incorporation. The data of Horst show the label in the lipid fraction to be about 80% of the total label built in. However, no data about the inhibition of label incorporation in the lipid fraction were published.

The effects of dfb and Polyoxin D on CS of different origins are summarized in Table 1.

In a few experiments with the system described by Horst (1981) we estimated the total rate of label incorporation from UDPGlcNAc, including the label incorporated in the lipid fraction. We found strong inhibition by 10 μM Polyoxin D but only marginal inhibition by large doses of dfb up to 500 μM. The data are presented in Table 2.

The inhibition at the highest concentration of dfb (actually about 0.3 mM due to maximum solubility) is significant at a level of $P = 0.995$ but might be due to nearly complete inhibition of incorporation of label in some chitinous material.

Table 1. In vivo activity and inhibition of CS in vitro of dfb and Polyoxin D

Test species belonging to	Biological activity in vivo[1]		Inhibition of CS in vitro	
	dfb	Polyoxin D	dfb	Polyoxin D
Fungi	−	+	−	+
Crustaceans	+	unknown	+	−
Insects	+	+/−[2]	+/−[3]	+

[1] − = not active; + = active
[2] active by injection, not active by oral application
[3] active in whole cells and large membrane fragments, not active in cell free systems.

Table 2. Total incorporation of label (dpm, n = 3; including label in the lipid fraction) after incubation at 37°C of a homogenate from Artemia salina with 100 nCi ^{14}C-UDPGlcNAc with or without Polyoxin D or dfb

Incubation time (hours)	10 μM Polyoxin D		10 μM dfb	
	−	+	−	+
0	75	139	97	55
0.75	1433	217	1099	1043
1.5	2325	229	1823	1931

	dfb; quantity to give a maximum concentration of		
	0	50 μM	500 μM
0	97	113	109
0.75	488	440	352
1.5	658	617	513

The combined results suggest that, in contrast with the results concerning the chitinous product, the coupling of N-acetyl-D-glycosamine to lipoid material is inhibited by Polyoxin D and not by dfb. A further conclusion from these data might be that in this preparation from A. salina the site of action of dfb is intact and, moreover, that this site is different from the point of attack of Polyoxin D, because Polyoxin D does not inhibit the formation of chitinous material. Another reason for the observation that Polyoxin D does not inhibit the CS might be that the active centre or the accessibility for Polyoxin D to it differs from that in insect species. This would imply that in A. salina the CS is different from the corresponding enzyme in insects, but that this does not hold true for other sugar handling enzymes.

The observation that dfb does not inhibit the incorporation of sugars in lipid material is in accordance with already published observations that no other carbohydrate handling systems than the chitin synthesizing complex in Arthropods are influenced by dfb. Neither effects were found on biosynthesis and deposition of hyaluronic acid and chondroitin sulfate in mice, rats, and chicken (Crookshank et al. 1978; Bentley et al. 1979) nor on N-acetyl-hexosaminyl-transferases in higher animals (Stoolmiller

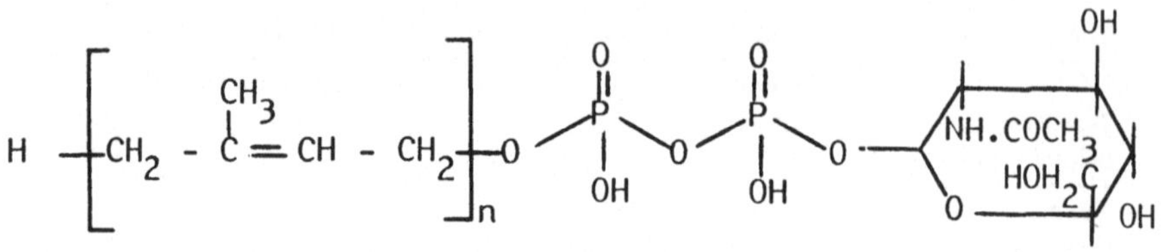

Figure 3. N-acetyl-glucosamine linked to a Dolichol carrier by two phosphate groups; for animals n = 17 to 21 units.

1978). Nor was it found that dfb produced any effect on the synthesis of viral glycoproteins in which the following reactions were studied:

in vitro: 1. Dol-P + UDPGlcNAc → Dol-P-P-(GlcNAc)$_2$ + UDP + UMP

2. Glycoprotein + UDPGlcNAc → Glycoprotein-GlcNAc + UDP

The chemical structure of Dol-P-P-GlcNAc is presented in Figure 3.

in vivo: 3. incorporation of ^{3}H-glucosamine in lipid bound oligo saccharides and in proteins.

(R. Datema, personal communication).

Dolichol-phosphates are recognized as lipid intermediates in the transport of sugars, in insects as well as in mammals (Quesada Allué et al. 1975, 1976). GlcNAc is handled as Dolichol-P-P-GlcNAc. Tunicamycin, like Polyoxin D, inhibited the incorporation of GlcNAc in chitin in a tissue culture of abdominal parts of *Triatoma infestans* and *Galleria mellonella* larval epidermal tissue (Quesada Allué 1982). It has no effect on cell-free insect chitin synthesizing systems (Cohen and Casida 1980; Mayer 1980). The chemical structure of Tunicamycin is presented in Figure 4. The structure shows that Tunicamycin does not only contain an uridine entity but also a GlcNAc group and an aliphatic chain.

The compound has its specific site of action in the reaction:

Dol-P + UDP-GlcNAc → Dol-P-P-GlcNAc + UMP

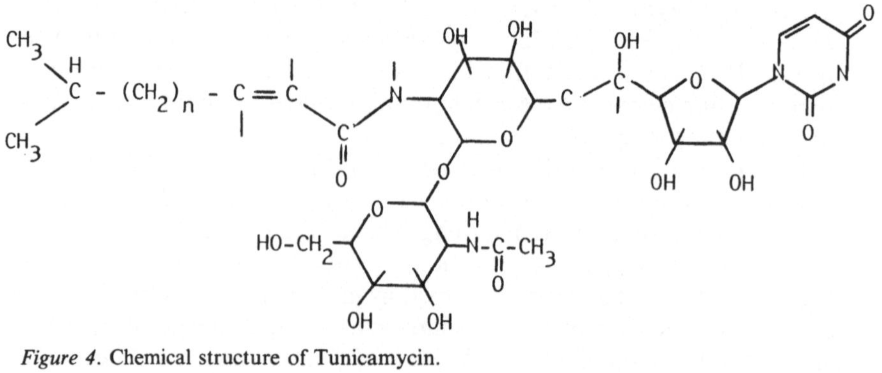

Figure 4. Chemical structure of Tunicamycin.

(Parodi and Leloir 1979). The enzymes concerned in the formation of the dolichol-linked saccharides are located in the membranous fraction of the cell (Elbein 1981; Horst 1983).

It is assumed, however, that in fungi dolichol derivatives do not play a role in chitin synthesis, because addition of Dolichol-phosphate to solubilized yeast enzyme failed to enhance the activity. Moreover, Tunicamycin was almost without effect (Cabib 1981). Tunicamycin also competitively inhibits the CS in *Neurospora crassa* (Selitrennikoff 1979) but in this special case it might be acting as a substrate analogue of UDPGlcNAc and not as an inhibitor in the formation of Dolichol intermediates. This might be a point of difference between chitin synthesis in fungi and in insects. From the above mentioned results of Brillinger (1979) it was deduced that either the enzymes in fungi and in insects are different, or that dfb attacks at another site in the chitin synthesis pathway. We suggest that chitin synthesis in fungi is inhibited by Polyoxin D and Nikkomycin by inhibition of the GlcNAc polymerizing CS, but that chitin synthesis in insects is inhibited by Polyoxin D, Nikkomycin and Tunicamycin at the level of the lipid intermediates. In both systems the inhibition could be the result of competition with the substrate, viz. GlcNAc. It might be possible that in *A. salina* as in fungi lipid intermediates, the synthesis of which is inhibited in cell-free preparations by Polyoxin D but only slightly by dfb, do not play a part in chitin synthesis. Horst (1983) showed the formation of lipid-linked oligosaccharides, both in vivo and in microsomal preparations of *A. salina*. However, it was not shown that these oligosaccharides are involved in chitin synthesis. If they are involved, it is not clear why Polyoxin D does not inhibit chitin synthesis in this case.

Summarizing, we may conclude that chitin synthesis is a complex system and for the moment the most likely point of attack of dfb seems to be the last reaction viz. the polymerization of GlcNAc to chitin. This reaction possibly takes place in membranes. Little is known about the actual processes that take place in the membrane. Several authors have proposed a scheme for chitin synthesis (Candy and Kilby 1962; Deul et al. 1978; Turnbull and Howells 1982). In our opinion such a scheme should include, at least for chitin synthesis in insects, the handling of GlcNAc by Dolichol derivatives.

However, as stated above, it is unlikely that dfb is active at the level of the sugar handling Dolichol intermediates. In Figure 5, the scheme for chitin synthesis in fungi, insects and crustaceans and the sites of inhibition of the several compounds discussed is presented.

In addition to Dolichol phosphates the microtubular system may also have some relation to chitin synthesis. Oberlander et al. (1983) studied chitin synthesis in imaginal wing disks of *Plodia interpunctella*. Chitin synthesis was inhibited by Colcemid and Vinblastine. These substances disturb the proper functioning of the microtubular system, but do not

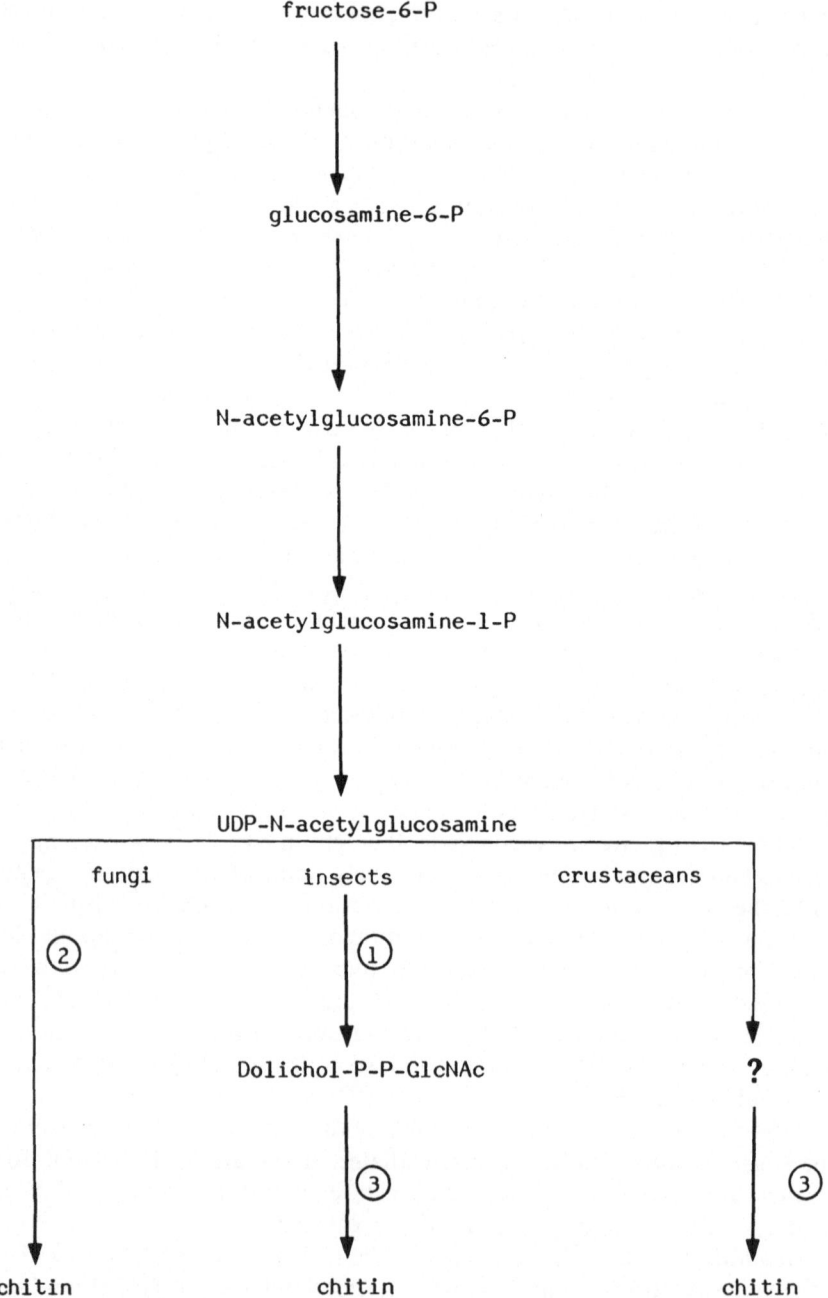

Figure 5. Proposed scheme for chitin synthesis in fungi, insects and crustaceans. 1: inhibition by Polyoxin D, Nikkomycin and Tunicamycin 2: inhibition by Polyoxin D and Nikkomycin (sometimes also by Tunicamycin) 3: inhibition by dfb; in cell-free preparations from insects only as long as the enzyme is part of an organized structure in a large cell fragment.

inhibit protein synthesis. It is suggested that microtubules play a role in cell functions such as maintaining cell shape, transport within the cell, feeding, movement of chilia and flagella and compartmentalization of the cell (Stephens and Edds 1976; Berlin et al. 1979; Hyams and Stebbings 1979; Margolis and Wilson 1981). It was also observed that microtubules mirrored in orientation the recently deposited cellulose microfibrils in plant cell walls. Evidence has been presented that microtubules in plant cells influence the direction in which the cellulose synthase complex moves during microfibril generation. The microtubules do not propel the complexes through the plasma, but act rather as guides. An intact plasma membrane seems to be required for active cellulose synthesis (Ledbetter 1981).

Besides hypotheses based on the inhibition of polymerization of monomers a number of alternative hypotheses has been put forward to explain the action of dfb. Yu and Terrière (1975, 1977) suggested that it impedes metabolism of ecdysone, which increases levels of chitinase and phenol oxidase. Redfern et al. (1982) found that dfb had no effect on the production of ecdysones in last instar nymphs of *Oncopeltus fasciatus*, but that the decline of the titer prior to adult acdysis was slightly retarded. These results therefore only partly agree with the above mentioned suggestion. However, O'Neill et al. (1977) found that the titer of β-ecdysone in pharate *Stomoxys calcitrans* pupae was not altered by dfb within the first 10 h after pupation. Furthermore, Hajjar and Casida (1979) provided evidence that dfb also did not alter in vivo metabolism of either α- or β-ecdysone. Measurements of the levels of chitinase and phenol oxidase are also conflicting. Increased levels of these enzymes upon dfb treatment have been shown in *Musca domestica* (Ishaaya and Casida 1974). However, this effect was not found in *Galleria mellonella* (Hegazy 1984).

Observations by Deul et al. (1978) also make it less probable that ecdyson has a primary effect. They showed that in larvae of *P. brassicae* dfb did not affect chitinase activity, and furthermore, chitin deposition was already completely blocked in early 5th instar larvae in which chitinase activity was practically absent.

Leighton et al. (1981) have suggested that dfb prevents chitin synthesis by interfering with the proteolytic activation of a CS zymogen. Turnbull and Howells (1983) have also maintained that this explanation seems best to fit their results. A chymotrypsin-like protease was the most likely candidate to activate the zymogen (Marks et al. 1982). We examined the activity of chymotrypsin with or without dfb after incubation times of up to four hours. The tests were performed with two substrates, N-acetyltyrosine ethyl ester and the much bigger casein. The results are summarized in Tables 3 and 4.

From Tables 3 and 4 it is obvious that we find no clear inhibition by incubating chymotrypsin up to four hours with maximum concentrations of dfb of 250 times the solubility.

Table 3. Activity of chymotrypsin estimated with N-acetyl-tyrosine ethyl ester following two methods of incubation of enzyme with dfb, in % of activity without dfb at time zero (= start of incubation). Decreasing absorption at 237 nm is taken as a measure for activity; 0.2 ml of enzyme in 1 mM HCl incubated or not with dfb added to 3.0 ml of a substrate solution containing 23.7 mg in 50 ml 1/15 M phosphate buffer pH = 7.0

A. No stirring during incubation with dfb; average results of 3 replicates at t = 0 and 2 replicates after incubation

Enzyme incubated with dfb, μg ml^{-1}	% Activity after incubation time (hours)	
	2	4
0	86	87
9.3	94	86
38.6	92	83

B. Continuous stirring during incubation with dfb; average results of 2 replicates.

Enzyme incubated with dfb, μg ml^{-1}	% Activity after incubation time (hours)	
	0.5	1
0	81	64
1.0	85	86
3.1	85	82
10.1	85	83
32.5	91	83

Table 4. Activity of chymotrypsin estimated with casein according to Rick (1974); in % activity of incubation without dfb (n = 3)

Enzyme incubated with dfb, μg ml^{-1}	% Activity after incubation time (hours)	
	1.5	3
0	100	100
5.3	98	100
57.2	96	102

The possibility of action via a potent metabolite which can be formed in vivo is not very likely as metabolism of dfb is very low, and in a highly susceptible species like *P. brassicae* it is not even detectable (Duphar internal information).

The overall results strongly suggest that dfb acts at some transport process needed for polymerization of GlcNAc to chitin or deposition of the formed chitin chains. This process most likely takes place in the cell membrane.

3 SUSCEPTIBLE STAGES AND ROUTES OF INTOXICATION

Depending on the insect species, in practice dfb acts as an insecticide by showing one, or a combination, of the following activities:

Larvicidal activity by ingestion. This is the main activity of the compound.

Larvicidal activity by contact. Generally, dfb toxicity by cuticular application is low. Exceptions to this rule are *Spodoptera littoralis* (Ascher and Nemny 1976) and *Spodoptera exigua* (Granett et al. 1983).

Ovicidal activity by contact. This activity is mainly responsible for practical control of some leaf blotch miners (Gracillariidae) such as *Leucoptera scitella, Lithocolletis blancardella* and *Stigmella malella* (apple pygmy moth), and also for control of *Laspeyresia pomonella* (codling moth) (Hoying and Riedl 1980; Elliott and Anderson 1982).

Ovicidal activity by treatment of adult female insects. This "through the female" effect is used in practice for control of Curcilionidae (weevils) e.g. *Anthonomus grandis* (cotton boll weevil) (Wright and Roberson, 1981). Though dfb treatment of adult insects is not lethal, effects can be found when treatment starts shortly after pupal eclosion. These effects are discussed in chapter 7.

4 SOME FACTORS AFFECTING THE LARVICIDAL ACTIVITY

4.1 *Intrinsic factors*

Under "mode of action" it has already been discussed that after reaching the target area inside the insect body, dfb acts very rapidly on inhibition of chitin deposition. However, lethal effects of a dfb treatment are not instantaneous. At normal rates the mortality of larvae is generally delayed until the next moult. When marginal rates are used it can even be extended to one of the next moults. This phenomenon stresses the requirement for a longer observation period than usual with quick-acting insecticides.

Generally dfb is active on all larval instars of susceptible species. If the efficacy of the compound on distinct instars differs, the susceptibility decreases with increasing age. This implies that the general advice for insect control in practice is to apply dfb at an earlier stage in the life cycle than conventional insecticides. An exception to the general rule on instar susceptibility is *Choristoneura fumiferana* (eastern spruce budworm). In this species older instars are more susceptible to dfb than younger ones (Retnakaran and Smith 1974; Granett and Retnakaran 1977).

For larvae, the length of the exposure period is a very important factor. Histological observations showed that when dfb was administered to larvae for a limited period, chitin synthesis was able to resume. The effect of dfb is visible as a narrow band of distortions. It is succeeded by normal layers of endocuticular tissue (Grosscurt 1978a).

This observation can explain the lower level of final mortality after brief exposure to the compound during a certain instar, as shown for example with the aquatic larvae of *Simulium* spp. living in fast flowing water (Lacey and Mulla 1977; McKague et al. 1978). For foliage feeding insects, a category which is one of the major practical outlets for dfb, this factor does not have a cramping effect on its efficacy due to the long persistance of dfb on leaves.

A first example of large difference in susceptibility to dfb in closely related species was found in the family of the Tortricidae (leafrollers). Or this group *Adoxophyes orana* (summer fruit tortrix), *Archips rosana* (rose tortrix moth) and *Pandemis heparana* showed a low susceptibility in the field, while *Archips podana, Spilonota ocellana* (eye-spotted bud moth), *Hedya nubiferana* and *Pammene rhediella* (fruitlet mining tortrix) could be adequately controlled in orchards. In subsequent laboratory experiments with 4th instar larvae, *Archips podana* proved to be highly susceptible to dfb (LC$_{50}$ of 0.85 mg a.i. kg^{-1} diet) whereas *A. rosana* and *A. orana* were only moderately susceptible (LC$_{50}$'s of 200 and 120 mg a.i. kg^{-1} diet, respectively) (Van der Molen 1975). These results indicate that the large differences in control of these species in the field are at least partly due to considerable intrinsic differences in susceptibility. In forest lepidoptera some striking differences in susceptibility to dfb also occur. Retnakaran et al. (1980) compared *Choristoneura fumiferana* (eastern spruce budworm), which is relatively insensitive to dfb, with the highly susceptible species *Malacosoma disstria* (forest tent caterpillar). Differences in the amount of body chitin and rates of chitin synthesis did not account for the difference in susceptibility to dfb. However, using radioactive labelled dfb, *M. disstria* retained a relatively higher amount of the ingested radioactivity than did *C. fumiferana*. The same phenomenon was found by comparing *Choristoneura occidentalis* (Western spruce budworm) and *Orgyia pseudotsugata* (Douglas-fir tussock moth). In this case *O. pseudotsugata*, which is about 100 times more susceptible in laboratory experiments, retained about 16 times more dfb than *C. occidentalis* (Granett et al. 1980). It was also found that the percentage dfb retention was negatively correlated with the percentage relative metabolism in these two species. These results suggest that the cause of the species difference in percentage retained dfb may not only be due to differences in absorption, but also to relative metabolism rates.

4.2 *External factors*

Information about the effect of temperature on the larvicidal activity of dfb is limited to Simuliidae. In this family the relationship seems to be species-dependent.

With *S. verecundum*, which was collected in Canada, no significant difference was found in activity at 22 and 11°C (McKague et al. 1978) whereas with *S. vittatum*, collected in California, USA, the activity gradually increased at temperatures from 15 to 24°C (Lacey and Mulla 1978). However, the results might be influenced by the different concentrations and exposure periods (1 mg litre^{-1}/15 min. and 0.02 mg litre^{-1}/60 min. respectively).

Upon ingestion by larvae, the particle size of dfb is strongly negatively correlated with the biological activity (Mulder and Gijswijt 1973). This effect might be caused by the low solubility of dfb in water (between 0.1–0.2 mg/litre at 24°C), combined with a low dissolution rate. The latter can be illustrated by data from the commercial WP formulation (particle size 2–5 μm). When this formulation is suspended in water in a concentration of 0.5 mg/litre, it takes about 1 h to arrive at a dissolved state of 0.1 mg/litre and about 4 h to reach saturation.

5 SOME FACTORS AFFECTING THE OVICIDAL CONTACT ACTIVITY

5.1 *Intrinsic factors*

The ovicidal activity of dfb by direct contact with insect eggs is species-dependent. It is likely that after oviposition the properties of the chorion largely influence the capacity of the insecticide to penetrate into the egg. Therefore, the spectrum of activity by topical application and by contamination of females is not always identical. E.g. no contact activity was found with eggs of *M. domestica*, not even after dipping for 90 minutes in a suspension in water with 1000 mg a.i. litre^{-1}. However, egg eclosion was fully prevented after the females had ingested food containing 100 mg a.i. kg^{-1} (Grosscurt 1976). This lack of ovicidal activity of dfb by topical application appears to be a common feature in all Muscidae. However, ovicidal effects by topical application of dfb have been demonstrated in other families belonging to the Diptera, viz. Culicidae (Busvine et al. 1976; Miura et al. 1976). Psilidae (Overbeck 1979) and Simuliidae (Lacey and Mulla 1977). Furthermore, this type of ovicidal effect was found in the following families of Lepidoptera: Gracillariidae (Grosscurt 1980; Injac 1981), Lymantriidae (Forgash et al. 1978), Noctuidae (Ascher and Nemny 1974; Büchi 1978), Olethreutidae (Hoying and Riedl 1980; Elliott and Anderson 1982), Pieridae (Grosscurt 1977) and Tortricidae (Ascher et al. 1978). In Coleoptera it was found with Chrysomelidae (Grosscurt 1976; Weiss 1977), Curculionidae (Ottens and Todd 1979) and Coccinellidae (Holst 1974).

The ovicidal contact activity is negatively correlated with the age of the egg at the time of treatment (Ascher and Nemny 1974; Miura et al. 1976; Grosscurt 1977; Lacey and Mulla 1977; Elliott and Anderson 1982).

5.2 *External factors*

The ovicidal contact activity is positively correlated with the relative humidity of the environment (Grosscurt 1977). Ascher et al. (1978), using eggs of *Lobesia botrana* (vine moth) also reported positive correlation with temperature.

The formulation of the active ingredient plays an important role. The activity of a formulation increases when the particle size of the active ingredient decreases. Depending on the insect species and the crop, the activity can also be increased by addition of surfactants (Grosscurt 1977; Elliott and Anderson 1982). In experiments with *Laspeyresia pomonella* (codling moth) the increase of ovicidal activity of dfb after addition of the surfactant Tween 20, was more pronounced in 3 day-old eggs than in 0 to 2.5 day-old eggs. Addition of Tween 20 to the dfb suspension also improved the residual activity of dfb on foliage and immature apples but not on waxy mature apples (Elliott and Anderson 1982).

6 SOME FACTORS AFFECTING THE OVICIDAL ACTIVITY BY TREATMENT OF ADULT FEMALE INSECTS

6.1 *Intrinsic factors*

An ovicidal activity through the female can be obtained by oral uptake or by contact activity of dfb. In laboratory experiments with 25 benzoyl ureas, with variations in lipophilic properties, both larvicidal and ovicidal activities either by adult feeding or by injection of female adults of *Musca domestica* were correlated (Grosscurt and Tipker 1980). The ovicidal activity after injection proved to be the basic activity in both ovicidal activity by adult feeding as well as in larvicidal activity by feeding. However, in either case this activity was coupled to a lipophilic parameter (expressed as Π in the octanol/water system). A remarkable finding also made in this study was that as lipophilicity of the compounds increased, only the gut wall became an important barrier but that transport into the egg was apparently much less influenced.

In principle the spectrum of ovicidal activity by treatment of females is broader than by topical application of eggs (see chapter 5.1). However, the limiting factor for "through the female" effects is the capacity for uptake of dfb, both oral and by contact with the female. In practice, this type of application is therefore mainly limited to Coleoptera and Diptera.

In two Diptera species belonging to the Anthomyiidae, viz. *Hylemya antiqua* (onion maggot) and *Hylemya brassicae* (cabbage maggot) a remarkable difference in susceptibility was found. After injection of dfb into females of *H. antiqua* even dosages of 10 µg per fly resulted in a slight ovicidal effect only. With *H. brassicae* a dosage of 0.3 µg per fly resulted in complete prevention of egg hatch. Oral application of dfb showed similar difference in susceptibility (Grosscurt et al. 1980).

When oral treatment is discontinued, in species with a synchronous egg maturation, egg eclosion in newly laid egg batches gradually increases. The rate of this reversibility is negatively correlated with the dosage in the previous treatment (Grosscurt 1976), but positively with the age of the adults (Holst 1974).

Cochliomyia hominivorax (screw-worm) is a species with synchronous maturation of the eggs in the ovaries. In this species the ovicidal effect of a 1 day oral feeding period increased with the maturity of the eggs at the moment of treatment (Crystal 1978). This observation can be explained by considering two factors, viz. the fact that dfb cannot exert an effect before the onset of chitin synthesis during embryonic development and the rapid excretion of dfb after discontinuation of the treatment.

With regard to resistance development and cross-resistance to dfb in *Musca domestica*, the ovicidal activity through the female shows a pattern that completely differs from the one observed in larvae. Generally, in larvae only low levels of both types of resistance occur. However, in *M. domestica* the ovicidal cross resistance can be much more severe. Furthermore, selection in the larval breeding medium results in high levels of ovicidal resistance without any simultaneous significant increase in larval resistance. This indicates that resistance mechanisms in larvae and adults are not mutually linked and that they mainly find expression in adults (Grosscurt 1980).

6.2 *External factors*

Most studies to assess a relationship between the formulation of dfb and the ovicidal activity through the female were done with *Anthonomus grandis* (cotton boll weevil). For practical control of this insect dfb is normally applied together with oil. Laboratory and greenhouse experiments, using either foliar application or topical application of the female weevils, revealed a significant increase of the ovicidal effect through the female by using cottonseed oil as a carrier instead of water (Lloyd et al. 1977; McLaughlin 1977). These small scale studies have not been clearly confirmed in subsequent field studies by other investigators. Most of these studies have not been published. An exception is a study by Burris et al. (1982). They found that the addition of 0.5 gal of cottonseed oil to dfb in a dosage of 1 oz a.i./acre resulted in a (not significant) increase in yield of cottonseed of 11%. The question of the necessity of an oil addition can also be answered from the other side, i.e. by lowering the quantity of the oil. In studies by Hopkins et al. (1982) it proved to be possible to lower the quantity of Sunoil 7N® or Dimoil® from 9.4 to 2.3 liter per hectare without any significant reduction in efficacy. However, in these studies a dfb formulation without addition of an oil was marked by absent. As far as we know, no attention has ever been paid to the effects of other external factors on the ovicidal activity through the female.

7 NON LETHAL EFFECTS ON ADULTS

7.1 *Effects on the exoskeleton*

In adult insects, cuticle deposition and a simultaneous increase in rigidity and decrease in mechanical penetrability of the cuticle take place during a limited postmoult period. Interference of dfb with these processes can seriously affect the structure and function of the adult exoskeleton. We can list these effects as follows:

7.1.1 *Structural changes in the adult cuticle*

According to Hunter and Vincent (1974) and Ker (1977) neither protein deposition nor tanning in the exoskeleton of adult locusts seems to be affected by dfb. However, these authors did not present quantitative data as far as the effect on tanning is concerned. After feeding dfb to newly-hatched adults, it was observed that the elytra of the Colorado potato beetle (*Leptinotarsa decemlineata*) remained weak. Mechanical properties of the elytra were measured by assessing their penetrability. It was found that blocking by dfb of the process by which the penetrability decreases in the postmoult period showed the same kinetics as inhibition of chitin synthesis. Furthermore, histological observations showed that distinct mesocuticular layers are still deposited after dfb treatment (Grosscurt 1978b). As mentioned earlier, no distinct growth takes place in soft larval cuticles after dfb treatment. Using the same fixation methods apparently only globular masses can be observed. These observations might indicate that stabilization of protein is different in soft and hard cuticles.

Subsequently it was found that in elytra of *L. decemlineata* the amount of protein was unaffected by a dfb treatment. The yield of ketocatechols, a parameter which is related to the degree of tanning of the structural proteins, was only partially inhibited. As the sensitivity to dfb of the latter effect was much less than that of the inhibition of chitin deposition, it was also concluded that interference of dfb with tanning was probably of a secondary nature (Grosscurt and Andersen 1980).

Until recently the general concept of hardness of insect cuticles was that hardening is due to impregnation and tanning, but not to chitin. However, from the above mentioned studies with dfb as a tool for the exclusive inhibition of chitin deposition, the role of chitin in the achievement of rigidity in the cuticle proved to be more crucial than has been hitherto assumed.

7.1.2 *Functional deficiencies in the adult cuticle*

The cuticle plays an important role in supporting the insect. In addition, the presence of a hard cuticle (with jointed appendages) is necessary for accurate walking and flying, and also provides protection against predators and parasites.

A number of functional deficiencies, following early post-moult feeding of dfb, have in fact been reported in the literature.

7.1.2.1 *Reduced walking ability*. — Effects on walking after post-moult feeding with dfb range from poorly coordinated movements as observed with the cotton boll weevil, *Anthonomus grandis* (Earle et al. 1979) to breaking of the femoro-tibial joints in the locust *Schistocerca gregaria* (Ker 1977), or even complete detachment of the hind legs of the grasshopper *Oxya japonica* (Lim and Lee 1982). The poorly coordinated movements of *A. grandis* also resulted in impaired mating ability.

7.1.2.2 *Reduced flight ability*. — Feeding of dfb to newly-emerged adults of the cotton boll weevil (*Anthonomus grandis*) reduced flight ability (Earle and Simmons 1979).

7.1.2.3 *Deficiency in the function of the cuticle as a protective shield*. — In section 7.1.1 we discussed a blocking of the increase in mechanical resistance of the elytra of *L. decemlineata* upon dfb treatment. Mechanical resistance will presumably be related to the breaking strength and may be an indicator of the lower ability of the cuticle to act as a protective shield against predators and parasites (Grosscurt and Andersen 1980). After an early post-emergence dfb treatment, adult insects probably also remain more vulnerable to penetration of entomopathogenic fungi than untreated adults. Effects can be expected to be similar to an observation in larvae of the tobacco hornworm (*Manduca sexta*) where entry of the pathogenic fungus *Metarhizium anisopliae* was facilitated by a combined treatment with dfb (Hassan and Charnley 1983).

7.1.2.4 *Effects on fecundity*. — Effects of dfb on fecundity have been studied in several ways, viz. by treatment of last larval instars or pupae and observation of effects in the subsequent adult survivors, and by treatment of adult females either from eclosion onwards or starting after a period of maturation.

Using the first method, a reduction in fecundity was found in the olive fruit fly, *Dacus oleae* (Fytizas 1976) and the large milkweed bug, *Oncopeltus fasciatus* (Hajjar and Casida 1979). No reduction was found with the stable fly, *Stomoxys calcitrans* and the house fly, *Musca domestica* (Wright and Spates 1976). When adults were treated from eclosion onwards no effects on fecundity were found with the horn fly, *Haematobia irritans* (Kunz and Bay 1977), *Musca domestica* (Grosscurt 1976), and the pink bollworm, *Pectinophora gossypiella* (Flint and Smith 1977).

However, apparent reduction in fecundity was found in the Mediterranean fruit fly, *Ceratitis capitata*, by Arambourg et al. (1977). This effect was more closely studied by Sarasua and Santiago-Alvarez (1983). After oral treatment of females, a reduction in fecundity could only be achieved when applied during the first day after adult eclosion. When females showing reduced fecundity were dissected, the numbers of fully developed eggs in the ovaries were not significantly decreased when compared with untreated females. However, in many cases eggs were shown to be blocking the common oviduct. The latter effect was also combined with a strongly reduced ability to evaginate the ovipositor. Histological observations in dfb treated females showed the common distortions in the structure of the integument. The most severely affected abdominal segment was the 8th, which is responsible for evagination of the ovipositor.

Based on these observations, the assumption that the known mode of action of dfb, viz. inhibition of chitin deposition, is also responsible for the observed reduction in fecundity of some species seems valid. The reduced fecundity in female adults of certain species after preceding treatment of immature stages can be explained in the same way by assuming persistence of an adequate quantity of the compound during maturation.

This assumption is also supported by the absence of a reduction in fecundity in all other cases where dfb treatment started after a period of maturation.

7.2 *Effects of peritrophic membranes*

Effects of diflubenzuron were also found on the composition of peritrophic membranes. A peritrophic membrane is a cylindrical sheath that surrounds the food mass in the midgut and sometimes also extends into the hindgut. It is composed of chitin fibres in close association with proteins. Clarke et al. (1977) examined peritrophic membranes of adult locusts (*Locusta migratoria*). After feeding these locusts with diflubenzuron-treated food, an increase in the fibrous appearance and a reduction in weight of the membrane were observed. This probably also means a weakening of the membrane. These phenomena can be explained by a reduction in chitin contents, also resulting in a reduced capacity to stabilize protein. Consequently, both the amount of chitin and the amount of protein deposited are reduced. Becker (1978) found a reduction of the dry weight and the length of peritrophic membranes of adult *Calliphora erythrocephala* (blowfly) after treatment with diflubenzuron. It is most commonly assumed that the function of the peritrophic membrane is to protect the midgut epithelium from abrasion by food particles. However, this assumption is not always valid, since a peritrophic membrane seems to be lacking in some species that ingest solid food whereas it is present in many juice feeders (Richards and Richards 1977). For this reason the physiological significance of damage to the peritrophic membrane is difficult to interpret.

8 REFERENCES

Arambourg, Y., Pralavorio, R. and Dolbeau, C. 1976. Premieres observations sur l'action du diflubenzuron (PH 6040) sur la fécondité, la longévité et la viabilité des oeufs de *Ceratitis capitata* Wied. (Dipt. Trypetidae). Revue de Zoologie Agricole 76: 118–126.

Ascher, K. R. S., Gurevitz, E. and Eliyahu M. 1978. The effect of diflubenzuron on eggs of the vine moth *Lobesia* (*Polychrosis*) *botrana* Den. & Schiff. (Lepidoptera: Tortricidae). Phytoparasitica 6: 25–27.

Ascher, K. R. S. and Nemny, N. E. 1974. The ovicidal effect of PH 60-40 (1-(4-chlorophenyl)-3-(2,6-difluorobenzoyl)-urea) in *Spodoptera littoralis* Boisd. Phytoparasitica 2: 131–133.

Ascher, K. R. S. and Nemny, N. E. 1976. Contact activity of diflubenzuron against *Spodoptera littoralis* larvae. Pestic. Sci. 7: 447–452.

Becker, B. 1978. Effects of 20-hydroxy-ecdysone, juvenile hormone, Dimilin and Captan on in vitro synthesis of peritrophic membranes in *Calliphora erythrocephala*. J. Insect Physiol. 24: 699–705.

Bentley, J. P. Weber, G. H. and Gould, D. 1979. The effect of diflubenzuron feeding on glycosaminoglycan and sulfhemoglobin biosynthesis in mice. Pestic. Biochem. Physiol. 10: 162–167.

Berlin, R. D., Caron, J. M. and Oliver, J. M. 1979. Microtubules and the structure and function of cell surfaces. In: Microtubules; Roberts, K. & Hyams, J. S. (eds.), Academic Press, New York, London, pp. 444–485.

Brillinger, G. U. 1979. Metabolic products of microorganisms 181. Chitin synthase from fungi, a test model for substances with insecticidal properties. Arch. Microbiol. 121: 71–74.

Büchi, R. 1978. Ovizide und larvizide Wirkung von Dimilin auf den Maiszünsler, *Ostrinia nubilalis* (Hbn.). Z. ang. Ent. 86: 67–71.

Burris, G., Clower, D. F., Pavloff, A. M. and Rogers, R. L. 1982. A three year summary of cotton entomology research with Dimilin and related materials. In: 1982 Proc. Beltwide Cotton Prod. Res. Conf., Las Vegas, Nevada. National Cotton Council of America, Memphis, pp. 202–204.

Busvine, J. R., Rongsriyam, Y. and Bruno, D. 1976. Effects of some insect development inhibitors on mosquito larvae. Pestic. Sci. 7: 153–160.

Cabib, E. 1981. The enzymatic synthesis of chitin and its regulation. In: Plant Carbohydrates Vol. II. Extracellular Carbohydrates; Tanner, W. A. & Loewus, F. A. (eds.), Springer-Verlag Berlin, Heidelberg, New York, pp. 395–415.

Candy, D. J. and Kilby, B. A. 1962. Studies on chitin synthesis in the desert locust. J. Exp. Biol. 39: 129–140.

Clarke, L., Temple, G. H. R. and Vincent, J. F. V. 1977. The effects of a chitin inhibitor – Dimilin – on the production of peritrophic membrane in the locust, *Locusta migratoria*. J. Insect Physiol. 23: 241–246.

Cohen, E. 1982. In vitro chitin synthesis in an insect: formation and structure of microfibrils. Eur. J. Cell Biol. 16: 289–294.

Cohen, E. and Casida, J. E. 1980. Inhibition of *Tribolium* gut chitin synthetase. Pestic. Biochem. Physiol. 13: 129–136.

Cohen, E. and Casida, J. E. 1982. Properties and inhibition of insect integumental chitin synthetase. Pestic. Biochem. Physiol. 17: 301–306.

96

Crookshank, H. R., Sowa, B. A., Kubena, L., Holman, G. M., Smalley, H. E. and Morison, R. 1978. Effect of diflubenzuron (Dimilin; TH-6040) on the hyaluronic acid concentration in chicken combs. Poultry Sci. 57: 804–806.

Crystal, M. M. 1978. Diflubenzuron-induced decrease of egg hatch of screwworms (Diptera: Calliphoridae). J. Med. Entomol. 15: 52–56.

Deul, D. H., De Jong, B. J. and Kortenbach, J. A. M. 1978. Inhibition of chitin synthesis by two 1-(2,6-disubstituted benzoyl)-3-phenyl-urea insecticides. Pestic. Biochem. Physiol. 8: 98–105.

Earle, N. W., Nilakhe, S. S. and Simmons, L. A. 1979. Mating ability of irradiated male boll weevils treated with diflubenzuron or penfluron. J. Econ. Entomol. 72: 334–336.

Earle, N. W. and Simmons, L. A. 1979. Boll weevil: ability to fly affected by acetone, irradiation, and diflubenzuron. J. Econ. Entomol. 72: 573–575.

Elbein, A. D. 1981. The role of lipid-linked saccharides in the biosynthesis of complex carbohydrates. In: Plant Carbohydrates Vol. II. Extracellular Carbohydrates; Tanner, W. A. & Loewus, F. A. (eds.), Springer-Verlag Berlin, Heidelberg, New York, pp. 166–193.

Elliott, R. H. and Anderson, D. W. 1982. Factors influencing the activity of diflubenzuron against the codling moth, *Laspeyresia pomonella* (Lepidoptera: Olethreutidae). Can. Ent. 114: 259–268.

Endo, A. and Misato, T. 1969. Polyoxin D, a competitive inhibitor of UDP-N-Acetylglucosamine: chitin N-acetylglucosaminyl transferase in *Neurospora crassa*. Biochem. Biophys. Res. Commun. 37: 718–722.

Flint, H. M. and Smith R. L. 1977. Laboratory evaluation of TH 6040 against the pink bollworm. J. Econ. Entomol. 70: 51–53.

Forgash, A. J., Respicio, N. C. and Khoo, B. K. 1978. Contact action of diflubenzuron on eggs and larvae of gypsy moth, *Lymantria dispar* L. (Lepidoptera: Lymantriidae). Journ. New York Ent. Soc. 86: 287.

Fytizas, E. 1976. L'action du TH 6040 sur la métamorphose de *Dacus oleae* Gmel. (Diptera: Trypetidae). Z. ang. Entomol. 81: 440–444.

Granett, J., Bisabri-Ershadi, B. and Hejazi, M. J. 1983. Some parameters of benzoylphenyl urea toxicity to beet armyworms (Lepidoptera: Noctuidae). J. Econ. Entomol. 76: 399–402.

Granett, J. and Retnakaran, A. 1977. Stadial susceptibility of eastern spruce budworm, *Choristoneura fumiferana* (Lepidoptera: Tortricidae), to the insect growth regulator Dimilin®. Can. Ent. 109: 893–894.

Granett, J., Robertson, J. and Retnakaran, A. 1980. Metabolic basis of differential susceptibility of two forest Lepidopterans to diflubenzuron. Ent. exp. & appl. 28: 295–300.

Grosscurt, A. C. 1976. Ovicidal effects of diflubenzuron on the housefly (*Musca domestica*). Med. Fac. Landbouww. Rijksuniv. Gent 41: 949–963.

Grosscurt, A. C. 1977. Mode of action of diflubenzuron as an ovicide and some factors influencing its potency. Proc. 9th Br. Insectic. Fungic. Conf. 1: 141–147.

Grosscurt, A. C. 1978a. Diflubenzuron: some aspects of its ovicidal and larvicidal mode of action and an evaluation of its practical possibilities. Pestic. Sci. 9: 373–386.

Grosscurt, A. C. 1978b. Effects of diflubenzuron on mechanical penetrability, chitin formation, and structure of the elytra of *Leptinotarsa decemlineata*. J. Insect Physiol. 24: 827–831.

Grosscurt, A. C. 1980. Larvicidal and ovicidal resistance to diflubenzuron in the housefly (*Musca domestica*). Proc. Kon. Ned. Akad. v. Wetensch., Amsterdam 83C, 127–141.

Grosscurt, A. C., Abels, R. and Deul, D. H. 1980. Unpublished report Duphar B.V.

Grosscurt, A. C. and Andersen, S. O. 1980. Effects of diflubenzuron on some chemical and mechanical properties of the elytra of *Leptinotarsa decemlineata*. Proc. Kon. Ned. Akad. v. Wetensch., Amsterdam 83C, 143–150.

Grosscurt, A. C. and Tipker, J. 1980. Ovicidal and larvicidal structure-activity relationships of benzoyl ureas on the housefly (*Musca domestica*). Pestic. Biochem. Physiol. 13: 249–254.

Gijswijt, M. J., Deul, D. H. and De Jong, B. J. 1979. Inhibition of chitin synthesis by benzoyl-phenylurea insecticides, III. Similarity in action in *Pieris brassicae* (L.) with Polyoxin D. Pestic. Biochem. Physiol. 12: 87–94.

Hajjar, N. P. and Casida, J. E. 1979. Structure-activity relationships of benzoylphenyl ureas as toxicants and chitin synthesis inhibitors in *Oncopeltus fasciatus*. Pestic. Biochem. Physiol. 11: 33–45.

Hammock, B. D. and Quistad, G. B. 1981. Benzoylphenyl ureas – mode of action. In: Progress in Pesticide Biochemistry Vol. I; Hutson, D. H. & Roberts, T. R. (eds.), John Wiley & Sons, Chichester, New York, Brisbane, Toronto, pp. 52–62.

Hassan, A. E. M. and Charnley, A. K. 1983. Combined effects of diflubenzuron and the entomopathogenic fungus *Metarhizum anisopliae* on the tobacco hornworm *Manduca sexta*. Proc. 10th Int. Congr. Plant Prot. 2: 790.

Hegazy, G. 1984. Ultrastructure of the integument of the sixth larval instar of *Spodoptera littoralis* Boisd. and *Galleria mellonella* L.: Changes associated with moulting and diflubenzuron treatment. Thesis State University of Gent.

Holst, H. 1974. Die fertilitätsbeeinflussende Wirkung des neuen Insektizids PDD 60-40 bei *Epilachna varivestis* Muls. (Col. Coccinellidae) und *Leptinotarsa decemlineata* Say. (Col. Chrysomelidae). Z. Pfl. Krankh. Pfl. Schutz 81: 1–7.

Hopkins, A. R., Moore, R. F. and James, W. 1982. Efficacy of diflubenzuron diluted in three volumes of oils on boll weevil progeny. J. Econ. Entomol. 75: 385–396.

Horst, M. N. 1981. The biosynthesis of crustacean chitin by a microsomal enzyme from larval brine shrimp. J. Biol. Chem. 256: 1412–1419.

Horst, M. N. 1983. The biosynthesis of crustacean chitin. Isolation and characterization of polyprenol-linked intermediates from brine shrimp microsomes. Arch. Biochem. Biophys. 223: 254–263.

Hoying, S. A. and Riedl, H. 1980. Susceptibility of the codling moth to diflubenzuron. J. Econ. Entomol. 73: 556–560.

Hunter, E. and Vincent, J. F. 1974. The effects of a novel insecticide on insect cuticle. Experientia 30: 1432–1433.

Hyams, J. S. and Stebbings, H. 1979. Microtubule associated cytoplasmic transport. In: Microtubules; Roberts, K. & Hyams, J. S. (eds.), Academic Press, New York, London, pp. 448–510.

Injac, M. 1981. Results of laboratory investigations of the ovicidal effect of diflubenzuron on eggs of different age of leaf miners of apple *Leucoptera scitella* Zell. and *Lithocolletis blancardella* F. Zastita bilja 32: 241–249.

Ishaaya, J. and Casida, J. E. 1974. Dietary TH-6040 alters composition and enzyme activity of housefly larval cuticle. Pestic. Biochem. Physiol. 4: 484–490.

Jordan, A. M. and Trewern, M. A. 1978. Larvicidal activity of diflubenzuron in the tsetse fly. Nature 272: 719–720.

Jordan, A. M., Trewern, M. A., Borkovec, A. B. and Demilo, A. B. 1979. Laboratory studies on the potential of three insect growth regulators for control of tsetse fly, *Glossina morsitans morsitans* Westwood (Diptera: Glossinidae). Bull. ent. Res. 69: 55–64.

Ker, R. F. 1977. Investigation of Locust cuticle using the insecticide diflubenzuron. J. Insect Physiol. 23: 39–48.

Kunz, S. E. and Bay, D. E. 1977. Diflubenzuron: effects on the fecundity, production, and longevity of the horn fly. Southwest. Entomol. 2: 27–31.

Lacey, L. A. and Mulla, M. S. 1977. Larvicidal and ovicidal activity of Dimilin® against *Simulium vittatum*. J. Econ. Entomol. 70: 369–373.

Lacey, L. A. and Mulla. M. S. 1978. Factors affecting the activity of diflubenzuron against Simulium larvae (Diptera: Simuliidae). Mosquito News 38: 264–268.

Ledbetter, M. C. 1981. The role of microtubules in plant cell wall growth. In: Recent advances in Phytochemistry Vol. 16: Creasy, L. L. & Hrazdina, G. (eds.), Plenum Press, New York, pp. 125–150.

Leighton, T., Marks, E. and Leighton, F. 1981. Pesticides: Insecticides and fungicides are chitin synthesis inhibitors. Science 213: 905–907.

Lim, S. J. and Lee, S. S. 1982. The toxicity of diflubenzuron to *Oxya japonica* (Willemse) and its effects on moulting. Pestic. Sci. 13: 537–544.

Lloyd, E. P., Wood, R. H. and Mitchell, E. B. 1977. Boll weevil: suppression with TH-6040 applied in cottonseed oil as a foliar spray. J. Econ. Entomol. 70: 442–444.

Margolis, R. L. and Wilson, L. 1981. Microtubules treadmills – possible molecular machinery. Nature 293: 705–711.

98

Marks, E. P., Leighton, T. and Leighton, F. 1982. Mode of action of chitin synthesis inhibitors. In: Insecticide mode of action; Coats, J. R. (ed.); Academic Press, New York, London, pp. 281–313.

Mayer, R. T., Chen, A. C. and Deloach, J. R. 1980. Characterization of a chitin synthase from the stable fly, *Stomoxys calcitrans* (L.). Insect Biochem. 10: 549–556.

McCoy, C. W. 1978. Activity of Dimilin on the developmental stages of *Phyllocoptruta oleivora* and its performance in the field. J. Econ. Entomol. 71: 122–124.

McGregor, H. E. and Kramer, K. J. 1976. Activity of Dimilin (TH 6040) against Coleoptera in stored wheat and corn. J. Econ. Entomol. 69: 479–480.

McKague, A. B., Pridmore, R. B. and Wood, P. M. 1978. Effects of Altosid and Dimilin on black flies (Diptera: Simuliidae): Laboratory and field tests. Can. Ent. 110: 1103–1110.

McLaughlin, R. E. 1977. Dose-response of the boll weevil to topical formulations of TH-6040. J. Georgia Entomol. Soc. 12: 369–373.

McLaughlin, R. E. 1978. Contact transfer of diflubenzuron (Dimilin®) by boll weevils and the relation of site of application and effect on egg hatch. Ent. Exp. & Appl. 23: 171–176.

Miura, T., Schaefer, C. H., Takahashi, R. M. and Mulligan, F. S. 1976. Effects of the insect growth inhibitor, Dimilin®, on hatching of mosquito eggs. J. Econ. Entomol. 69: 655–658.

Moore, R. F., Leopold, R. A. and Taft, H. M. 1978. Boll weevils: mechanism of transfer of diflubenzuron from male to female. J. Econ. Entomol. 71: 587–590.

Moore, R. F. and Taft, H. M. 1975. Boll weevils: chemosterilization of both sexes with busulfan plus Thompson-Hayward TH 6040. J. Econ. Entomol. 68: 96–98.

Mulder, R. and Gijswijt, M. J. 1973. The laboratory evaluation of two promising new insecticides which interfere with cuticle deposition. Pestic. Sci. 4: 737–745.

Oberlander, H., Lynn, D. E. and Leach, C. E. 1983. Inhibition of cuticle production in imaginal discs of *Plodia interpunctella* (cultured in vitro): effects of Colcemid and Vinblastine. J. Insect Physiol. 29: 47–53.

O'Neill, M. P., Holman, G. M. and Wright, J. E. 1977. β-Ecdysone levels in pharate pupae of the stable fly, *Stomoxys calcitrans*, and interaction with the chitin inhibitor diflubenzuron. J. Insect Physiol. 23: 1243–1244.

Ottens, R. J. and Todd, J. W. 1979. Effects of diflubenzuron on reproduction and development of *Graphognathus peregrinus* and *G. leucoloma*. J. Econ. Entomol. 72: 743–746.

Overbeck, H. 1979. Zur Wirkung von Dimilin auf das Eistadium der Möhrenfliege, *Psila rosae* F. (Diptera: Psilidae). Nachrichtenbl. Deut. Pflanzenschutzd. 31: 99–102.

Parodi, A. J. and Leloir, L. F. 1979. The role of lipid intermediates in the glycosylation of proteins in the eucaryotic cell. Biochim. Biophys. Acta 559: 1–37.

Quesada Allué, L. A. 1982. The inhibition of insect chitin synthesis by Tunicamycin. Biochem. Biophys. Res. Commun. 105: 312–319.

Quesada Allué, L. A., Belocopitow, E. and Marechal, L. R. 1975. Glycosyl transfer to an acceptor lipid from insects. Biochem. Biophys. Res. Commun. 66: 1201–1208.

Quesada Allué, L. A., Marechal, L. R. and Belocopitow, E. 1976. Chitin synthesis in *Triatoma infestans* and other insects. Acta Physiol. Latinoam. 26: 349–363.

Redfern, R. E., Kelly, T. J., Borkovec, A. B. and Hayes, D. K. 1982. Ecdysteroid titers and molting aberrations in last-stage *Oncopeltus* nymphs treated with insect growth regulators. Pestic. Biochem. Physiol. 18: 351–356.

Retnakaran, A., Granett, J. and Robertson, J. 1980. Possible physiological mechanisms for the differential susceptibility of two forest Lepidoptera to diflubenzuron. J. Insect Physiol. 26: 385–390.

Retnakaran, A. and Smith, L. 1975. Morphogenetic effects of an inhibitor of cuticle development on the spruce budworm, *Choristoneura fumiferana* (Lepidoptera: Tortricidae). Can. Ent. 107: 883–886.

Richards, A. C. and Richards, P. A. 1977. The peritrophic membranes of insects. Ann. Rev. Entomol. 22: 219–240.

Rick, W. 1974. Chymotrypsin. In: Methods of Enzymatic Analysis Vol. 2; Bergmeyer, H. E. (ed.); Verlag Chemie Weinheim, Academic Press, New York, London, pp. 1006–1009.

Sarasua, M. J. and Santiago-Alvarez, C. 1983. Effect of diflubenzuron on the fecundity of *Ceratitis capitata*. Ent. Exp. & Appl. 33: 223–225.

Selitrennikoff, C. P. 1979. Competitive inhibition of *Neurospora crassa* chitin synthetase activity by Tunicamycin. Arch. Biochem. Biophys. 195: 243–244.

Smit, W. 1977. Unpublished report. Zoological Laboratory, University of Amsterdam, The Netherlands.

Stephens, R. E. and Edds, K. T. 1976. Microtubules: Structure, chemistry and function. Physiol. Review 56: 709–777.

Stoolmiller, A. C. 1978. Toxicological study on the effect of diflubenzuron (1-(4-chlorophenyl)-3-(2,6-difluorobenzoyl)-urea) on rat C-6 astrocytoma cells in vitro. Gen. Pharmacol. 9: 11–16.

Turnbull, I. F. and Howells, A. J. 1982. Effects of several larvicidal compounds on chitin biosynthesis by isolated larval integuments of the sheep blowfly *Lucilia cuprina*. Austr. J. Biol. Sci. 35: 491–503.

Turnbull, I. F. and Howells, A. J. 1983. Integumental chitin synthase activity in cell-free extracts of larvae of the Australian sheep blowfly, *Lucilia cuprina*, and two other species of Diptera. Austr. J. Biol. Sci. 36: 251–262.

Van Daalen, J. J., Meltzer, J., Mulder, R. and Wellinga, K. 1972. A selective insecticide with a novel mode of action. Naturwissenschaften 59: 312–313.

Van der Molen, J. P. 1975. Unpublished report IPO, Wageningen, The Netherlands.

Van Eck, W. H. 1979. Mode of action of two benzoylphenyl ureas as inhibitors of chitin synthesis of insects. Insect Biochem. 9: 295–300.

Vardanis, A. 1978. Polyoxin fungicides: Demonstration of insecticidal activity due to inhibition of chitin synthesis. Experientia 34: 228–229.

Vardanis, A. 1979. Characteristics of the chitin synthesizing system of insect tissue. Biochem. Biophys. Acta 588: 142–147.

Weiss, M. 1977. Zur Wirkung von Dimilin auf die Imagines and Eier des Erlenblattkäfers, *Agelastica alni* L. (Coleoptera: Chrysomelidae). Anz. Schädlingskde, Pflanzenschutz, Umweltschutz 50: 161–164.

Wright, J. E. and Roberson, J. 1981. Laboratory evaluation of a method of sterilizing the boll weevil. J. Econ. Entomol. 74: 696–697.

Wright, J. E. and Spates, G. E. 1976. Reproductive inhibition activity of the insect growth regulator TH 6040 against the stable fly and the housefly: effects on hatchability. J. Econ. Entomol. 69: 365–368.

Yu, S. J. and Terrière, L. C. 1975. Activities of hormone metabolizing enzymes in house flies treated with some substituted urea growth regulators. Life Science 17: 619–624.

Yu, S. J. and Terrière, L. C. 1977. Ecdyson metabolism by soluble enzymes from three species of Diptera and its inhibition by the insect growth regulator TH-6040. Pestic. Biochem. Physiol. 7: 48–55.

5. Chitin biosynthesis after treatment with benzoylphenyl ureas

B. Mauchamp and O. Perrineau

1 INTRODUCTION

Benzoylphenyl ureas such as diflubenzuron and its analogs, are very effective in controlling a broad spectrum of insect pests. This class of chemicals are primarily stomach toxicants of larvae (Wellinga et al. 1973a, b) but are also known to affect adults (Soltani et al. 1984) and eggs (Wright and Spates 1976, Wright and Harris 1976, Fraragalla et al. 1980, Grosscurt 1980). These compounds either disrupt larval development or prevent egg-hatch, presumably by interfering with cuticle deposition (Mulder and Gijswijt 1973, Ishaaya and Casida 1974) and chitin synthesis (Post and Vincent 1973, Post et al. 1974, Deul et al. 1978, Gijswijt et al. 1979). These chitin synthesis inhibiting growth regulators appear to be insect or arthropod specific and have relatively low acute toxic effects on non-target species such as birds, fishes and mammals. Although they are currently being used for crop protection against many pests of economic importance, further studies on the mode of action are necessary to explain the secondary effects that are unrelated to chitin synthesis such as nucleic acid metabolism (Meola and Mayer 1980, DeLoach et al. 1981) and molting hormone metabolism (Yu and Terriere 1975, Schefle and Kuchenmeister 1981, Soltani et al. 1984). In order to understand the mode of action of the benzoylphenyl ureas at the cellular level, we focussed our studies on the efficiency of inhibition of chitin synthesis during cuticle development by diflubenzuron.

2 EFFICIENCY OF TREATMENT WITH BENZOYLPHENYL UREAS

We observed that the results obtained after treatment were greatly dependent on the quantity of diflubenzuron reaching the target tissues, which in turn was directly related to the mode and timing of application.

Wright, J. E. and Retnakaran, A. (Eds), Chitin and Benzoylphenyl ureas. ISBN 978-94-010-8638-7.
© *1987, Dr W. Junk Publishers, Dordrecht.*

Benzoylphenyl ureas are reported as larvicides that are effective upon ingestion. Also, they are mainly active against phytophagous insects in the orders of Lepidoptera and Orthoptera. Since these compounds are not systemic they cannot be used to control sucking insects or cryptic feeders. We performed our experiments on Lepidoptera (*Pieris brassicae, Mamestra brassicae, Spodoptera exigua* and *Manduca sexta*) and Orthoptera (*Locusta migratoria, Schistocerca gregaria* and *Gryllus domesticus*). In many studies reported on the dietary effects of diflubenzuron, the compound was incorporated in the diet at several concentrations (Baumler and Salama 1976, Ishaaya and Casida 1974, Redfern et al. 1977, Oliver et al. 1976, Retnakaran 1980 and Mitsui et al. 1980). The larvae were allowed to feed on the treated diet until the effects became apparent and at this stage it was difficult to determine the exact quantity of compound ingested. Field trials were usually carried out by spraying suspensions of the material as small particles and allowing the insects to feed continuously on the sprayed plants. It has been demonstrated that the smaller the particle size the greater the efficiency of the treatment (Mulder and Gijswijt 1973). In order to know the precise amount of the compound ingested and its timing we applied the compound as an acetonic solution on a small piece of leaf or diet. The larvae were first allowed to completely consume the pieces of treated leaves or diet after which they were allowed to feed on untreated food.

To test penetration through the integument, larvae were topically treated with various levels of a 1% DMSO-acetonic solution. The topical application allowed us to determine the effectiveness of the material on pupae as well as sucking insects such as the bugs *(Pyrrhocoris apterus, Dysdercus fasciatus)* and aphids *(Aphis fabae* and *Myzus persicae)*. An alternative method of topical application that has been reported is the dipping of insects in acetonic solutions for a few seconds (Faragalla et al. 1980, Soltani et al. 1983). Topical applications in general were found to be about 10 times less effective than the ingestion method. Injection of larvae and pupae with suspensions of the material as small particles was also carried out for some of the assays.

These studies allowed us to determine the quantity required to obtain 100% mortality at the following molt. We carried out our studies on chitin synthesis under these conditions.

3 MORPHOLOGICAL EFFECTS OF BENZOYLPHENYL UREAS ON INSECTS

The severity of the morphological effects depend on the mode of application and timing of treatment. Susceptibility of the insects appears to vary with age. It was also noted that in the codling moth, *Laspeyresia*

pomonella (Hoying and Riedl 1980), only the eggs appeared to be susceptible; in *Pieris brassicae* and in *Mamestra brassicae* the third larval instar
was less susceptible than the last larval instar. In our studies we considered
only the treatment during the last larval instar. It has been reported that
treated larvae in general, seemed unaffected during the feeding period and
that apolysis occurred at the usual time. Our results showed that the larvae
blackened gradually and died and that the behavior of treated insects was
unchanged, as we could observe the wandering stage followed by the pupal
spinning in *Pieris brassicae* or the digging behavior in *Spodoptera, Mamestra* and *Manduca* larvae. If *Manduca* larvae were fed from the beginning
of the fifth instar, larval cuticle and bleeding to death. When treatment
occurred less than one day before the wandering stage, the larvae were
incapable of ecdysis and remained within the intact cuticle. Observations
reported by several authors (Mulder and Gijswijt 1973, Ker 1977,
Mauchamp 1980, Mitsui et al. 1980) indicated that the effects on the
integument depends on the mode of treatment. With reduced concentrations of the compound in spite of massive inhibition of larval cuticle
deposition, the pupae ended up with a tan, pharate, pupal cuticle. In
treated pupae, pharate adult development takes place and the adult could
be seen through the pupal cuticle, especially the growth and pigmentation
of the scales. The pharate adult however, was unable to split the cuticle for
successful eclosion. Ker (1977) reported the effects of diflubenzuron on
locust cuticle where he indicated that glucosamine was present in hydroly-
sates of treated cuticle and raised the question of interaction between
chitin and cuticle matrix. Further investigations are necessary to explain
the nature of the material deposited after treatment.

4 MICROSCOPIC DETECTION OF CHITIN IN TREATED ANIMALS

Evidence of the interference of diflubenzuron on cuticle deposition has
been presented by Gijswijt et al. (1979) showing the similarity of effects
with Polyoxin D. Light microscopy has clearly shown the presence of a
large space between the cuticle deposited before the treatment and the
epidermal cells containing dense globules with tiny fibrillar material.
Mallory's or Masson's triple stains are not specific enough to give precise
data on the nature of the deposited material. We developed a method to
localize chitin by fluorescence (Mauchamp and Schrevel 1979) using binding with fluorescent wheat germ agglutinin (WGA). This lectin has been
demonstrated to be able to recognize specifically N-acetylglucosamine
(Nagata and Burger 1974) and chitin (Mauchamp and Schrevel 1979). It
has been used to detect chitin in insects (Figure 2) and also in fungi (Figure
1) and yeasts (Horrisberger and Vonlanthen 1977, Molano et al. 1980).

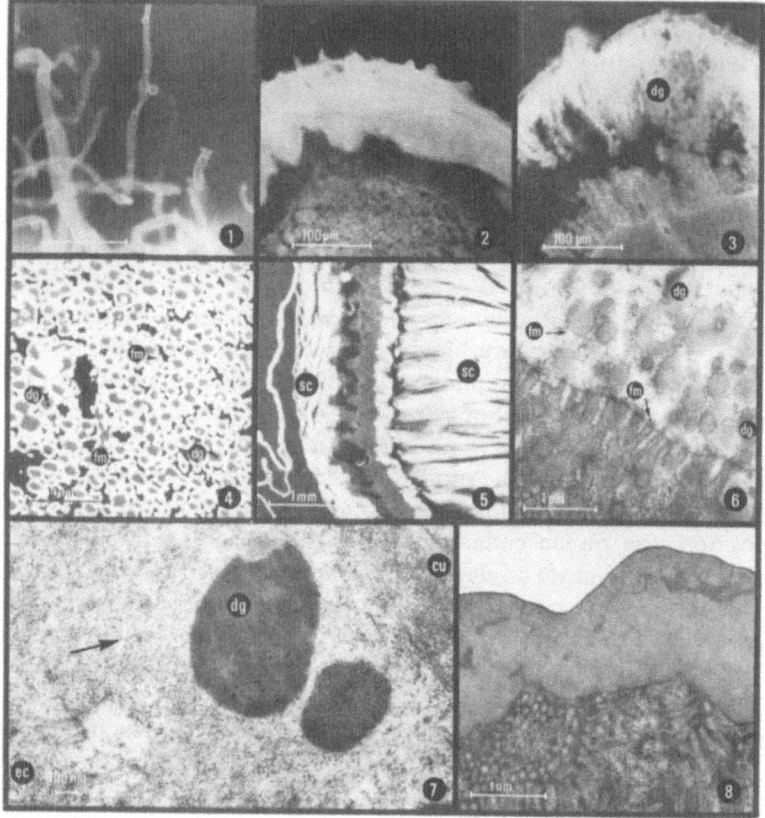

Figure 1. Fluorescent detection of fungal chitin by labeling with WGA-FITC. Bar = 1 mm.

Figure 2. Fluorescent detection of chitin of insect cuticle by labeling with WGA-FITC. Bar = 100 μm.

Figure 3. Fluorescent detection of material deposited after diflubenzuron treatment. (dg) = dense globules surrounded by fluorescent material. Bar = 100 μm.

Figure 4. Computer analysis of fluorescence detection of dense globules (dg) and strongly fluorescent surrounding material (fm). Bar = 100 μm.

Figure 5. Fluorescent detection of pharate adult cuticle and scales (sc) obtained after diflubenzuron treatment at the beginning of the pupal instar. Bar = 1 mm.

Figure 6. Electron microscopic appearance of material produced by epidermal cells after treatment with diflubenzuron. Dense globules (dg) come from exocytosis vesicles that open at the basis of microvillis. Fibrillar material (fm) is located on membranes and around dense globules. Bar = 1 μm.

Figure 7. Space between epidermal cell apex (ec) and larval cuticle (cu) deposited before diflubenzuron treatment; translucent microfibrils (− →) appear between dense globules (dg). Bar = 100 nm.

Figure 8. Dense pharate pupal tan cuticle produced by epidermal cells. Bar = 10 μm.

We indicated earlier that treated *Manduca* or *Pieris* larvae deposited a tan coloured pharate-pupal-cuticle. This cuticle was removed and incubated for 24 hr in 50% KOH at 100°C. The insoluble material was first washed with distilled water until the ph was neutral, then washed with 0.1 M TrisHCl buffer PH 7.4, following which a 1/10000 dilution of Evans blue was added. After rinsing the excess stain. the cuticle was observed under a Leitz Orthoplan fluorescence microscope. Small pieces of insoluble material displayed very high fluorescence indicating the presence of chitin. Incubation in the presence of N-acetylglucosamine prevented the fluorescence.

This fluorescence method can also be used on histological sections of larvae. Treated larvae were fixed with Carnoy's fixative, embedded in paraffin, sectioned at $7 \mu m$ thickness and collected on glass slides. The paraffin was removed and sections were hydrated and rinsed with 0.1 M TrisHCl buffer pH 7.4. They were then incubated for 30 min. in Fluorescent WGA, washed several times and observed under a fluorescence microscope. Fluorescence was observed as in the control on the cuticle deposited before the treatment but, positive labeling was also detected in the space inbetween the epidermal cells and the cuticle laid prior to treatment, containing globules (Figure 3). With thinner sections (less than $5 \mu m$) we noticed fluorescence at the periphery and between the globules. Computer analyses of these pictures confirmed that the highest fluorescence was located around the dense globules (Figure 4). No change in fluorescence intensity was detected in sections treated by protease; however if they were treated with chitinase (1 mg/ml for 24 h.) a large part of the fluorescence disappeared. Pharate-pupal-tan-cuticle deposited in spite of diflubenzuron treatment also showed intensive fluorescence. Similar observations were also reported by Soltani et al. (1984) in *Tenebrio molitor*.

When diflubenzuron was injected into pupae, growth of the scales was not prevented and adult-pharate-cuticle was deposited. The scales and cuticle were highly fluorescent (Figure 5).

These observations indicate that chitin was produced by epidermal cells even after treatment with diflubenzuron.

By electron microscopic observation it is possible to discriminate two kinds of materials produced by epidermal cells when the insect is treated during the last larval instar. If treatment occurred at the beginning of the fifth instar, affected cuticle revealed dense globules secreted by exocytosis with vesicles appearing at the base of microvilli and, by contrast, microfibrils bound to the apical membrane released between the globules (Figure 6). The exocytosis vesicles were produced by coalescence of Golgi vesicles. These microfibrillar structures did not disappear after treatment with alkali. At higher magnification the presence of microfibrils with an internal structure of electron transparent rods surrounded by a dark matrix was revealed (Figure 7). Filshie (1982) and Giraud-Guille (1984) reported that electron dense matrix was protein and the electron trans-luscent microfibrils were chitin. Diameter of these transluscent rods (3 nm) was in agreement with that indicated by these authors.

When the larvae were treated just before apolysis which corresponds to the end of the deposition of larval cuticle, pharate pupal tan cuticle is secreted. This newly secreted cuticle looks like a dense layer without laminae. This structure was similar to that of the inner epicuticle (Figure 8), but thicker. We also found a similar structure in scales formed by trichogen cells in both control and treated pupae.

5 PHYSICO-CHEMICAL DETECTION OF CHITIN IN TREATED ANIMALS

The histochemical test used for chitin detection can be compared to an immunochemical test in which the antibody would be the wheat germ agglutin (WGA). The WGA is therefore a sensitive probe for the detection and localization of chitin in several tissues.

In spite of the high specificity of this test we performed physio-chemical tests to confirm the presence of chitin. Fluorescent material was treated by hot dilute aqueous alkali and the residue was dissolved in cold 12N hydrochloric acid (1 h at 4°C). The viscous suspension of partly degraded chitin was precipitated by the addition of 50% ethanol. Chitin and chitosan were further analyzed by x-ray diffraction and infrared spectro-scopy. These non-destructive methods gave analogous results as obtained with the controls.

Insoluble material was treated with hot 6N hydrochloric acid. Chitin was hydrolyzed and converted into deacetylated saccharides. Also in the hydrolysate it was possible to detect glucosamine and acetic acid. Acetic acid was detected by gas chromatography on polar columns with flame ionization detector. However in many cases we obtained a partial hydrolysis and it was possible to detect oligosaccharides of glucosamine and acetylglucosamine. Microanalytical methods such as HPLC or HPLC coupled with mass spectrometer gave a specific and absolute detection of products obtained after hydrolysis (Mauchamp, unpublished data). The high sensitivity of these methods allowed identification of chitin in very small samples (Hackman and Goldberg 1981). HPLC was also used to analyze benzoylphenyl ureas in the epidermal cells.

These analyses made it possible to confirm that the material detected by fluorescence was chitin.

6 CONCLUSION

Chitin is widely distributed, and is one of the main components of arthropod cuticles and the fungal cell walls. Since benzoylphenyl ureas did not prevent budding process of yeasts or fungal growth (Van Eck 1979, Cohen and Casida 1980, Mauchamp and Leroux, unpublished data) we investigated the reasons for these compounds inhibiting chitin synthesis in insects. Furthermore, using cell free chitin synthetase systems, it was demonstrated that the benzoylphenyl ureas did not prevent chitin fibrillogenesis (Mayer et al. 1980, Cohen and Casida 1984). Cohen and Casida (1984) offered five hypotheses to explain the mode of action of benzoylphenyl ureas.

Our experiments have led us to consider two of these hypotheses, either a "disruption of accessibility of substrate and activators" or "disruption of a regulatory mechanism associated with the polymerization step in chitin formation". We observed clearly that proteins and chitin microfibrils were not associated after treatment with benzoylphenyl ureas. Association of proteins and precursors of microfibrillar forms of the chitin could occur in membranes of Golgi vesicles, this complex would then be transported to the apex of the cells where vesicle membranes fused with the plasma membranes before chitin fibrillogenesis. The Golgi vesicles have to be compared with "chitosomes" (Bartnicki-Garcia et al. 1978, 1979, Cohen 1982) or chitosomal system observed in the oocytes of *Tetranychus urticae* (Mothes and Seitz 1981). We demonstrated that membrane proteins acting as a lectin with high affinity to acetylglucosamine, could have a role in the transport of this saccharide through the cells (Mauchamp 1984). Association between precursors of chitin and specific protein can also be, prevented by treatment with tunicamycin which is an inhibitor of glycoprotein synthesis. With tunicamycin, cuticle formation can be prevented (Quesada-allue 1982, Mauchamp unpublished data), but no such effect could be shown in cell-free chitin synthetase system (Cohen and Casida 1984). In conclusion, we suggest that benzoylphenyl ureas in general and diflubenzuron in particular, act at the level of protein association during the first step of the chitin formation, a process that could be specific in arthropods.

7 REFERENCES

Baumler, W. and Salama, H. S. 1976. Some biochemical changes induced by Dimilin in the Gypsy moth *Porthretia dispar* L., Z. Ang. Ent. 81: 304–310.
Bartnicki-Garcia, S., Bracker, C. E., Reyes, E. and Ruiz-Herrera, J. 1978. Isolation of chitosomes from taxonomically diverse fungi and synthesis of microfibrils in vitro. Exp. Mycol. 2: 173–192.

108

Bartnicki-Garcia, S., Lippman, E. and Heick, J. 1979. Evidence for a poly peptide acceptor in fungal chitin biosynthesis. In Abstracts of the Annual Meeting of the American Society for Microbiology. p. 106. Los Angeles.

Cohen, E. 1982. In vitro chitin synthesis in an insect: formation and structure of microfibrils. Eur. J. Cell Biol. 16: 289–294.

Cohen, E. and Casida, J. 1980. Inhibition of Tribolium gut chitin synthetase. Pest. Biochem. Physiol. 13: 129–136.

Cohen, E. and Casida, J. 1983. Insect chitin synthetase as a biochemical probe for insecticidal compounds. Int. Congr. Pesticide Chemistry, 5th Proceedings, Ed. J. Miyamoto. Vol. 3: 25–32.

Deloach, J. R., Meola, S. M., Mayer, R. T. and Thompson, J. M. 1981. Inhibition of DNA synthesis by Diflubenzuron in pupae of the stable fly Stomoxys calcitrans L., Pest. Biochem. Physiol. 15: 172–180.

Deul, D. H., DeJong, B. J. and Kortenbach J. A. M. 1978. Inhibition of chitin synthesis by two 1-(2,6-Disubstituted Benzoyl)-3-phenylurea insecticides II. Perst. Biochem. Physiol. 8: 98–105.

Filshie, B. K. 1982. Fine structure of the cuticle of insects and other Arthropods. In: Insect Ultrastructure (eds King and Akai) Plenum Publishing Corporation. 1: 281–312.

Fraragalla, A. A., Berry E. C. and Guthrie, W. D. 1980. Ovicidal activity of Diflubenzuron on European corn borer egg masses. J. Econ. Entomol. 73: 573–574.

Gijswijt, M. J., Deul, D. H. and DeJong, B. J. 1979. Inhibition of chitin synthesis by Benzoylphenylurea insecticides: III Similarity in action in Pieris brassicae (L) with Polylxin D. Pest. Biochem. Physiol. 12: 87–94.

Giraude-Guille, M. M. 1984. Fine structure of the chitin-protein system in the crab cuticle. Tissue and Cell. 16: 75–92.

Grosscurt, A. C. 1980. Some physiological aspects of the insecticidal action of Difluben-zuron, an inhibitor of chitin synthesis. Ph.D. Thesis, Netherlands.

Hackman, R. H. and Goldberg M. 1981. A method for determinations of microgram amounts of chitin in Arthropod cuticles. Anal. Biochem, 110: 277–280.

Horrisberger, M. and Vonlanthen, M. 1977. Localization of mannan and chitin on thin sections of budding yeasts with gold markers. Arch. Microbiol. 115: 1–7.

Hoying, S. A. and Riedl, H. 1980. Susceptibility of the codling moth to Diflubenzuron. J. Econ. Entomol. 73: 556–560.

Ishaaya, I. and Casida, JU. 1974. Dietary TH 60–40 alters composition and enzyme activity of the house fly larval cuticle. Pestic. Biochem. Physiol. 4: 484–490.

Ker, R. F. 1977. Investigation of locust cuticle using the insecticide Diflubenzuron. J. Insect Physiol. 23: 39–48.

Mauchamp, B. 1980. Aspects ultrastructuraux, biochimiques et endocrines de la differen-ciation des formations epidermiques chez Pieris brassicae L. Thesis, Paris. Publ. Lab. Zool. ENS, 16: 250 p., 38 pl.h.t.

Mauchamp, B. 1982. Purification of an N-acetyl-D-glucosamine specific lectin (P. B. A.) from epidermal cell membranes of Pieris brassicae L. Biochimie. 64: 1001–1008.

Mauchamp, B. and Schrevel, J. 1977. Observation en microscopie a fluorescence de la cuticule des insectes: une methode faisant appel aux proprietes specifiques de la WGA vis-a-vis des glycoconjugues de la chitin. C. R. Acad. Sci. Paris 285: 1107–1110.

Mauchamp, B. and Hubert, M. 1984. Internalization of plasma membrane glucoconjugates and plasma membrane lectin into epidermal cells during pharate adult wing development of Pieris brassicae L.: correlation with resorption of molting fluid components. Biol. Cell. 50: 285–294.

Mayer, R. T., Chen, A. C. and DeLoach J. R. 1980. Characterization of a chitin synthase from the stable fly, Stomoxys calcitrans (L.) Insect Biochem, 10: 549–556.

Meola, S. M. and Mayer, R. T. 1980. Inhibition of cell proliferation of the imaginal epi-dermal cells by Diflubenzuron in pupae of the stable fly Stomoxys calcitrans L.) Science 207: 985–989.

Mitsui, T., Nobusawa, C., Kukami, T. I., Collins, J. and Riddiford, L. M. 1980. Inhibition of chitin synthesis by Diflubenzuron in Manduca larvae. J. Pestic. Science. 5: 335–341.

Molano, J., Bowers, B. and Cabib, E. 1980. Distribution of Chitin in the yeast cell wall. An ultrastructural and chemical study. J. Cell Biol. 85: 199–212.

Mothes, U. and Seitz K. A. 1981. A possible pathway of chitin synthesis as revealed by electron microscopy in *Tetranychus urticae* (Acari, Tetranychidae) Cell Tissue Res. 214: 443–448.

Mulder, R. and Gijswijt, M. J. 1973. The laboratory evaluation of two promising new insecticides which interfere with cuticle deposition. Pest. Sci. 4:737–745.

Nagata, Y. and Burger M. M. 1974. Wheat germ agglutinin and molecular characterisitics and specificity of sugar binding. J. Biol. Chem. 249: 3116–3122.

Oliver, J. E., DeMilo, A. B., Cohen, C. F., Shortino, T. J. and Robbins, W. E. 1976. Insect growth regulators. Analogues of TH-6038 and TH-6040. J. Agric. Food Chem. 24: 1065–1068.

Post, L. C. and Vincent, W. R. 1973. A new insecticide inhibits chitin synthesis. Naturwissen. 9: 431–432.

Post, L. C., DeJong and Vincent, W. R. 1974. 1-(2,6-Disubstituted benzoyl)-3-phenylurea insecticides: inhibitors of chitin synthesis. Pest. Biochem. Physiol. 4: 473-483.

Quesada-Allue L. A. 1982. The inhibition of insect chitin synthesis by Tunicamysin. Biochem. Biophys. Res. Comm. 105: 312-319.

Redfern, R. E., DeMilo, A. B. and Oliver J. E. 1977. Analogues of TH-6038 and TH-6040. Growth regulating effects on the fall armyworm. Botyu-Kagaku. 42: 89–91.

Retnakaran, A. 1980. Effect of 3 new moult-inhibiting insect growth regulators on the spruce budworm. J. Econ. Entomol. 73: 520–524.

Scheffel, H. and Kuchenmeister, J. 1981. Influence of Diflubenzuron on moult initiation and chitin biosynthesis in the centipede *Lithobius forficatus* L. In: Sehnal, F., Zabza, A., Menn, J. J. and Cymborowski, B. (eds), Regulation of insect development and behavior. Technical University of Wroclaw. p. 1138.

Soltani, N. 1982. Effects d'un insecticide, le Diflubenzuron sur *Tenebrio molitor* (Insectes, Coleopteres). Activite insecticide, action sur les cycles epidermocuticulaires durant la metamorphose et sur la longevite des adultes. Thesis, Dijon.

Soltani, N., Delbecque, J. P. and Delachambre, J. 1983. Penetration and insecticidal activity of Diflubenzuron in *Tenebrio molitor* pupae, Pestic. Sci. 14: 615–622.

Soltani, N., Delbecque, J. P., Delachambre, J. and Mauchamp, B. 1984. Inhibition of ecdysteroid increase by Diflubenzuron in *Tenebrio molitor* pupae and compensation of Diflubenzuron effect on cuticle secretion by 20-hydroxyecdysone. *Int. J. Inv. Rep. Dev.* 7: 323–332.

Van Eck, W. H. 1979. Mode of action of two benzoyl ureas as inhibitors of chitin synthesis in insects. Insect Biochem. 9: 295–300.

Wellinga, K., Mulder, T. and Van Daalen, J. J. 1973a. Synthesis and Laboratory evaluation of 1-(2,6-Disubstituted benzoyl)-3-phenylureas, a new class of insecticides. 1. 1-(2,6-Dichlorobenzoyl)-3-phenylureas. J. Agr. Food Chem. 21: 348–354.

Wellinga, K., Mulder, R. and Van Daalen, J. J. 1973b. Synthesis and laboratory evaluation of 1-(2,6-Disubstituted benzoyl)-3-phenylureas, a new class of insecticides. II. Influence of the acyl moiety on insecticidal activity. J. Agr. Food Chem. 21: 993–998.

Wright, J. E. and Harris, R. L. 1976. Ovicidal activity of Thompson-Hayward TH-6040 in the stable fly and horn fly after surface contact by adults. J. Eco. Entomol. 69: 728–730.

Wright, J. E. and Spates, G. E. 1976. Reproductive activity of the insect growth regulator TH-6040 against the stable fly and the house fly: effects on the hatchability. J. Econ. Entomol. 69: 365–368.

Yu, S. J. and Terriers, L. C. 1975. Activities of hormone metabolizing enzymes in house flies treated with some substituted urea growth regulators. Life Sci. 17: 619–626.

Yu, S. J. and Terriers, L. C. 1977. Ecdysone metabolism by soluble enzymes from three species of Diptera and its inhibition by the insect growth regulator TH 6040. Pestic. Biochem. Physiol. 7: 48–55.

6. Structure-activity relationships of benzoylphenyl ureas

Takahiro Haga, Tadaaki Toki, Tohru Koyanagi and Ryuzo Nishiyama

1 INTRODUCTION

Insecticides which can selectively disturb the development of immature insects have become of ever increasing interest. Benzoylphenyl ureas are known to interfere with the formation and/or deposition of cuticle chitin during larval molts, and to exhibit selective toxicity against some orders of insects in the larval stages. Owing to these promising characteristics, more benzoylphenyl urea derivatives are becoming commercialized, or are under development since the introduction of diflubenzuron (Figure 1).

We have synthesized a number of benzoyl (pyridyloxyphenyl) urea derivatives (*1*), and selected *N*-(4-(3-chloro-5-trifluoromethyl-2-pyridyloxy)-3,5-dichlorophenyl)-*N'*-(2,6-difluorobenzoyl) urea (chlorfluazuron, also IKI-7899) as a candidate compound to be developed (Nishiyama et al., 1979).

$$\text{(1)}$$

During the synthesis and screening program of (*1*), we have carried out the studies on the quantitative structure-activity relationship (QSAR) for the derivatives following the Hansch-Fujita approach, and found these studies very helpful to predict or rationalize the substituent patterns of active compound. There have been a few QSAR studies of the benzoylphenyl urea derivatives. Before describing our results, we would summarize the results of these studies.

2 QSAR STUDIES ON BENZOYLPHENYL UREAS

Verloop and coworkers (1977) first carried out the extensive structure-activity studies with several hundred derivatives. Instead of the conven-

Wright, J. E. and Retnakaran, A. (Eds), Chitin and Benzoylphenyl ureas. ISBN 978-94-010-8638-7.
© *1987, Dr W. Junk Publishers, Dordrecht.*

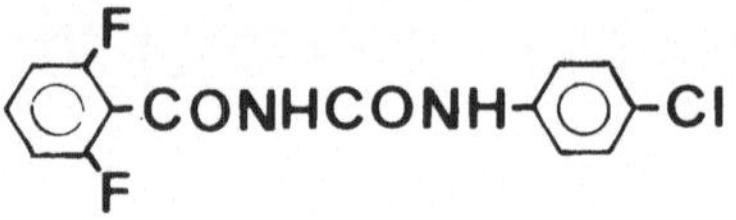

Diflubenzuron
PH-6040

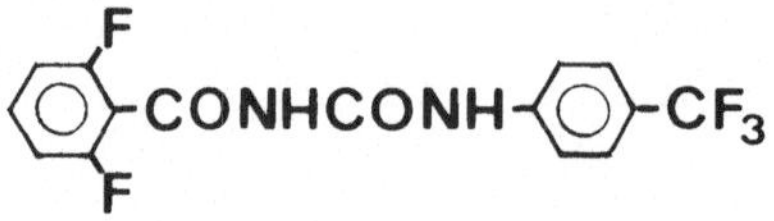

Penfluron
PH-6044

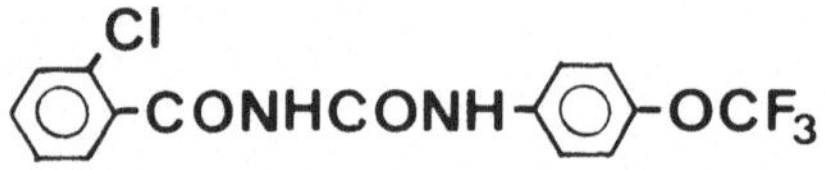

Triflumuron
BAY-SIR-8514

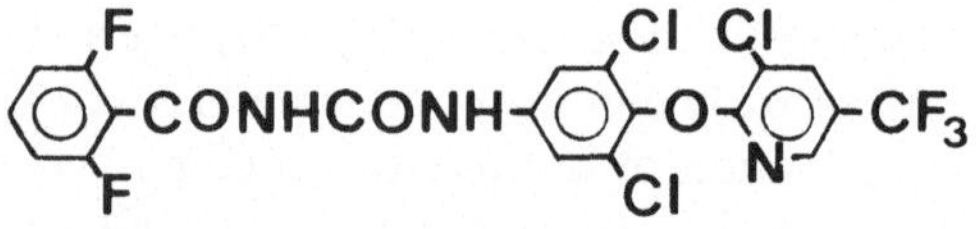

Chlorfluazuron
IKI-7899

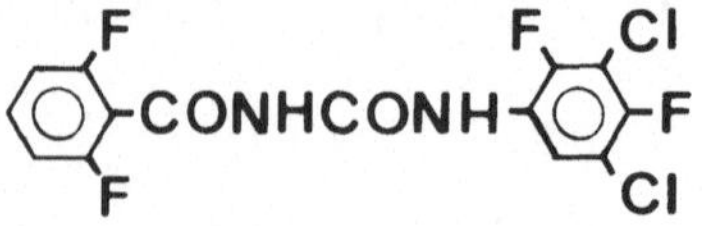

CME-134

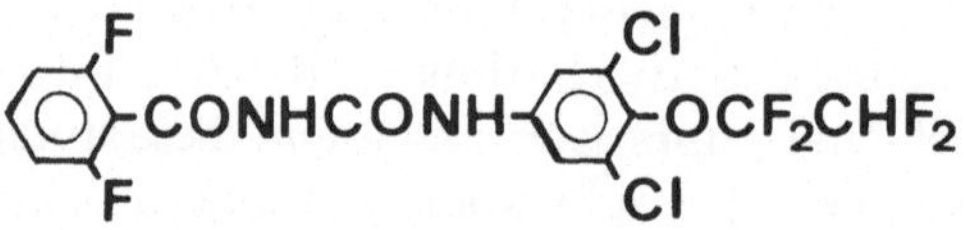

XRD-473

Figure 1. Benzoylphenyl urea insecticides, commercialized and under development.

tional steric parameters such as Taft E_s constants, they developed new steric substituent parameters, "STERIMOL parameters", and applied them to the QSAR analyses.

$$(2)$$

$$pED_{50} = 1.10\,\pi + 2.35\,\sigma - 0.40L - 0.27B_4 + 1.40D_1$$
$$- 0.70D_2 + 0.84$$

$$n = 48,\ r = 0.909,\ s = 0.408 \qquad \text{(Eq. 1)}$$

In tests against the cabbage white butterfly, *Pieris brassicae* and under non-synergistic conditions, they obtained Eq. 1, where in this and the other equations in this paper, n is the number of compounds, r is the correlation coefficient, and s is the standard deviation, pED_{50} is the logarithm of the reciprocal of the concentration required for a 50% kill of the larvae, L is the length of the substituent (R_3), B_4 is the width of STERIMOL parameter of the substituent (R_3), D_1 is an indicator variable (equals 1, when R_1 = F, and equals 0, when R_1 = Cl), D_2 is an indicator variable for the presence of the N-Methyl group (equals 0, when R_2 = H, and equals 1, when R_2 = Me). From this equation, it can be concluded that the substituent (R_3) at the *para* position of aniline ring must be hydrophobic, electron-withdrawing, short, and thin. And it is also suggested that preferable R_1 and R_2 are F and H, respectively.

In tests against larvae of the seed corn maggott, *Hylemya platura*, Yu and Kuhr (1976) obtained Eq. 2 and Eq. 3 in the presence of piperonyl butoxide (P.B.).

$$(3)$$

$$pLC_{50(synergized)} = -0.632 + 2.328\,\sigma$$
$$n = 7,\ r = 0.837,\ s = 0.490 \qquad \text{(Eq. 2)}$$

$$(4)$$

$$pLC_{50\,(synergized)} = 0.310 + 1.7673\,\sigma$$
$$n = 9,\ r = 0.729,\ s = 0.591 \qquad \text{(Eq. 3)}$$

These equations indicate that electron-withdrawing substituents are favorable for substituents on the aniline ring, and that fluorine is better than chlorine as the substituent on the benzoyl ring. They also recognized that in some cases larvicidal activity was synergized by P.B. and suggested the possibility that benzoylphenyl ureas are partially metabolized by the mixed-function oxidases of the larvae.

Casida and Hajjar (1979) studied the effects of benzoylphenyl ureas on milkweed bug, *Oncopeltus fasciatus* (Dallas) nymphs, and obtained a good correlation between their toxicity to fifth-instar nymphs and their potency as *in vitro* inhibitors of post-ecdysial chitin synthesis, which is expressed in Eq. 4 (assays at 3×10^{-6}M) and Eq. 5 (assays at 1×10^{-6}M), respectively.

$$pLD_{50} = -0.274 + 0.0234 \times (\text{inhibition}\%)$$
$$n = 11, r^2 = 0.574 \tag{Eq. 4}$$

$$pLD_{50} = 0.414 + 0.0350 \times (\text{inhibition}\%)$$
$$n = 8, r^2 = 0.781 \tag{Eq. 5}$$

These equations suggest that the primary mode of action of benzoylphenyl ureas may involve direct inhibition of chitin synthesis within the integument. However, they did not obtain any significant correlations as to the structure-activity relationships for these benzoylphenyl ureas.

Grosscurt and Tipker (1980) quantitatively examined the correlations as to the larvicidal and embryocidal activities against the house fly, *Musca domestica*. A significant correlation was found between embryocidal activity after adult feeding and embryocidal activity after injection in combination with the

(5)

hydrophobic constant

π

embryocidal activity after adult feeding (days) $= 0.33$ embryocidal activity after injection (days)

$+ 0.26 \pi - 0.12 \pi^2 + 0.04$

$$n = 24, r = 0.753, s = 0.509 \tag{Eq. 6}$$

A similar relationship also exists between the larvicidal activity and the embryocidal activity after injection as

$$\begin{matrix} \text{Larvicidal} \\ \text{activity} \\ (-\log \text{conc.}) \end{matrix} = 0.22 \begin{matrix} \text{Embryocidal activity} \\ \text{after injection} \\ (\text{days}) \end{matrix}$$

$$+ 0.62\,\pi - 0.21\,\pi^2 + 0.85\text{D} - 1.32$$

$$n = 23,\ r = 0.847,\ s = 0.480 \qquad \text{(Eq. 7)}$$

shown in Eq. 7, where D is an indicator variable in the presence of 2,6-F_2 group. From these two equations, it is suggested that embryocidal activity after injection is the basic activity even in the larvicidal activity by feeding. Presence of the hydrophobic parameter (π) term in these equations implies the importance of the transport process of the compound through the gut wall and the egg.

Fujita and coworkers (1984) examined the larvicidal activity against nondiapause larvae of rice stem borer, *Chilo suppressalis* Walker under conditions with and without piperonyl butoxide (P.B.). In the presence of P.B., they obtained Eq. 8, where $\Delta B_5\ (= B_5\,(X) - B_5\,(H))$ is the difference in the

$$(6)$$

$$\text{pLD}_{50}(\text{P.B.}) = 0.748\,\sigma_p - 0.398\,\Delta B_5 + 1.695\,\Sigma\pi$$

$$- 0.179\,(\Sigma\pi)^2 - 1.172\,\text{I} + 5.690$$

$$n = 29,\ r = 0.913,\ s = 0.303 \quad \text{(Eq. 8)}$$

maximum width of the substituent (X) in the STERIMOL parameter. I is the indicator variable, which equals 0 when $Y = F$, and equal 1 when $Y = Cl$, and $\Sigma\pi$ is the sum of the hydrophobic parameters for substituents on the aniline ring (X) and benzoyl ring (Y). Eq. 8 indicates that the electron-withdrawing and "thin" substituents on the aniline ring are preferable for the high insecticidal activity. Furthermore, it is suggested that the total hydrophobicity of the substituents on the aniline and benzoyl ring also relates to the activity, and that the optimum $\Sigma\pi$ value is about 5. Their studies also idicated that the oxidative metabolism of the compound in the larval body is very important in determining the activity.

In summary, conventional QSAR studies on benzoyl urea reveal the following results: first, the electron-withdrawing, hydrophobic, and not-bulky substituents on the aniline ring enhance the larvicidal activity; secondly, for some compounds the oxidative metabolism is very significant in the larval body. In the following chapter we will describe the results of the QSAR analyses on benzoyl (pyridyloxyphenyl) ureas.

3 QSAR STUDIES ON THE BENZOYL (PYRIDYLOXYPHENYL) UREAS

3.1 Physicochemical parameters of pyridyloxy group*

We have synthesized a number of pyridyloxy-substituted benzoyl-phenyl ureas without any information available on the physiochemical characteristics (e.g. hydrophobicity, electron-withdrawing power, and steric size) of the pyridyloxy group. In order to carry out the QSAR studies, we measured the hydrophobic parameter and the electronic parameter, the values of which are shown in Table 1 (Haga et al., 1984).

$$\text{R}_1\text{-benzene} - \text{CONHCONH} - \text{phenyl} - \text{R}_2 \quad \text{R}_1 \tag{5}$$

The hydrophobic parameter was obtained by high performance liquid chromatography (HPLC) method and expressed in $\log \kappa'_{pyridyloxy}$ (logarithm of the capacity factor) (Block et al., 1974) or π_{calcd}. Hydrophobic parameter of the pyridyloxy groups changes in a rather wide range as the substituent on the pyridine ring varies. And as the substituent becomes lipophilic, the hydrophobicity of the pyridyloxy group also increases. It is also shown that pyridyloxy group is generally less lipophilic, compared with the phenoxy group ($\pi_{C_6H_5O} = 2.08$) (Hansch and Leo, 1979).

Table 1. Physicochemical parameters for pyridyloxy group

R_3	$\log \kappa'_{pyridyloxy}$	π_{calcd}	σ_p
5-CF$_3$	0.358	1.120	0.025
3,5-Cl, CF$_3$	0.582	1.777	0.039
3,5-Cl$_2$	0.565	1.727	0.018
H	0.032	0.164	-0.051
5-Cl	0.317	1.000	-0.037
5-Br	0.371	1.158	-0.299
3,5-Br$_2$	0.658	2.000	0.004
5-I	0.455	1.405	-0.299
3,5-CF$_3$, Cl	0.616	1.877	0.025
3,5-Cl, CO$_2$Et	0.581	1.774	0.025
3,5-Cl, CN	0.184	0.610	0.060
3,5-NO$_2$, CF$_3$	0.271	0.865	0.087

* Details of the experiment will be published in the separate paper.

The electronic parameter was determined from ^{13}C-NMR shift and represented as σ_p (Spiesecke et al., 1961). As to the electronic parameter (σ_p) of the pyridyloxy group, variation in σ_p values is very small (from -0.299 to 0.087). These small values suggest that the pyridyloxy group is not a strongly electron-withdrawing group, even if the strongly attracting group (such as CF_3 or NO_2) is substituted on the pyridine ring.

As to the steric parameter, it can be easily understood that pyridyloxy group belongs to the rather bulky group. Thus, from the viewpoint of the conventional QSAR studies, it is estimated that the pyridyloxy group cannot be a favourable substituent for exhibiting the good insecticidal activity. However, actually pyridyloxy-substituted benzoylphenyl ureas show excellent larvicidal and embryocidal activity (Haga et al., 1982). In the following section, we would discuss the reason why the pyridyloxy-substituted benzoylphenyl ureas exhibit prominent insecticidal activity.

3.2 QSAR studies on the substituents on the aniline ring of benzoyl(pyridyloxyphenyl) urea

In order to rationalize the role of the pyridyloxy group we synthesized a set of benzoylphenyl urea derivatives (7), and measured the larvicidal activity against the common cotworm, *Spodoptera litura*.

$$\qquad (7)$$

In Table 2, log $\kappa'_{molecule}$ value, the logarithm of the capacity factor (κ') of each compound, and the pLC_{90} (ppm) value, the logarithm of the reciprocal of the concentration at which ninety percent of larvae were killed are shown.

Table 2. Larvicidal activities against *Spodoptera litura*

R_3	Log$\kappa'_{molecule}$	pLC_{90} (ppm)
5-CF_3	0.298	-1.63
3,5-Cl, CF_3	0.524	0
3,5-Cl_2	0.484	-0.58
H	-0.0436	-2.82
5-Cl	0.232	-1.59
5-Br	0.288	-1.62
3,5-Br_2	0.578	-0.301
3,5-CH_3, Br	0.505	-0.799
5-I	0.347	-1.58
3,5-CF_3, Cl	0.513	0
3,5-Cl, CO_2Et	0.486	-1.11
3,5-NO_2, CF_3	0.224	-2.11

118

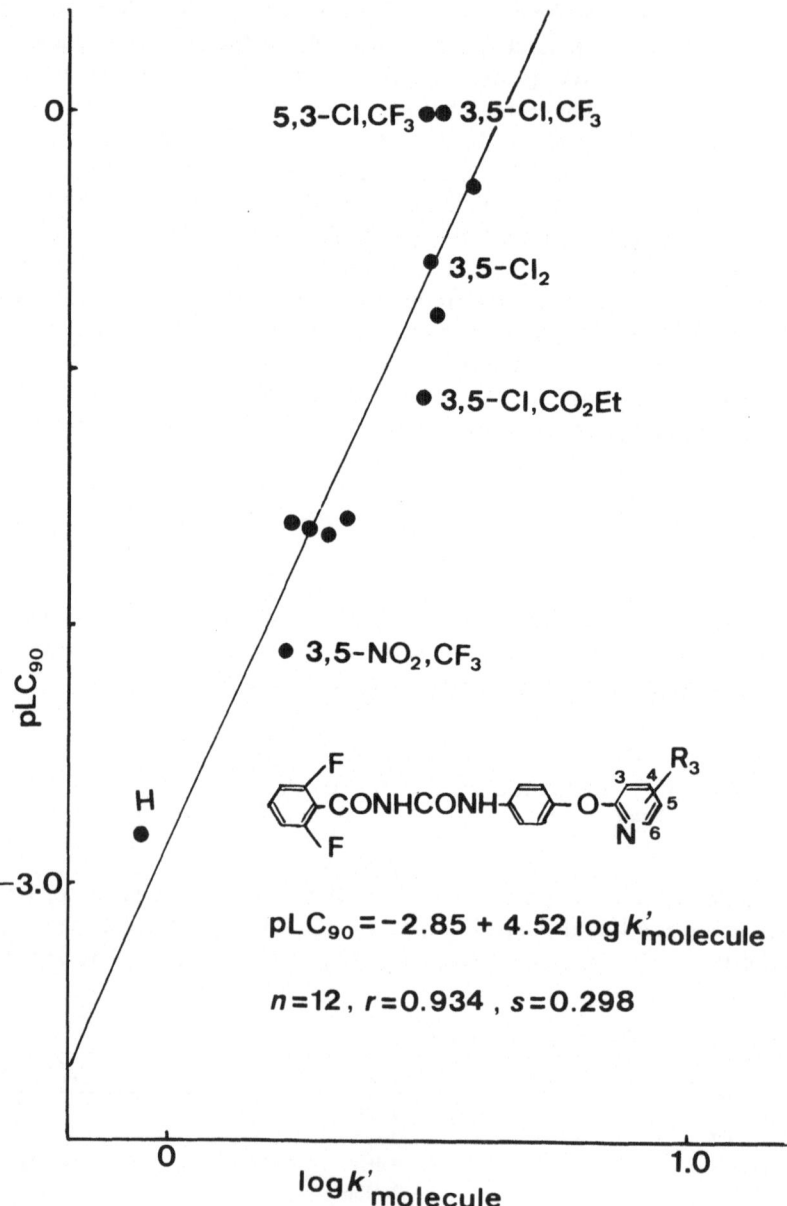

Figure 2. Larvicidal activity of benzoyl (pyridyloxyphenyl) – ureas against *Spodoptera litura*.

Eq. 9 was derived in good correlation by using the values of $\log \kappa'_{pyridyloxy}$ (logarithm of the capacity factor for the pyridyloxyphenyl group), and Eq. 10 was also formulated satisfactorily for the values of $\log \kappa'_{molecule}$ (logarithm of the capacity factor for the whole molecule) (Figure 2).

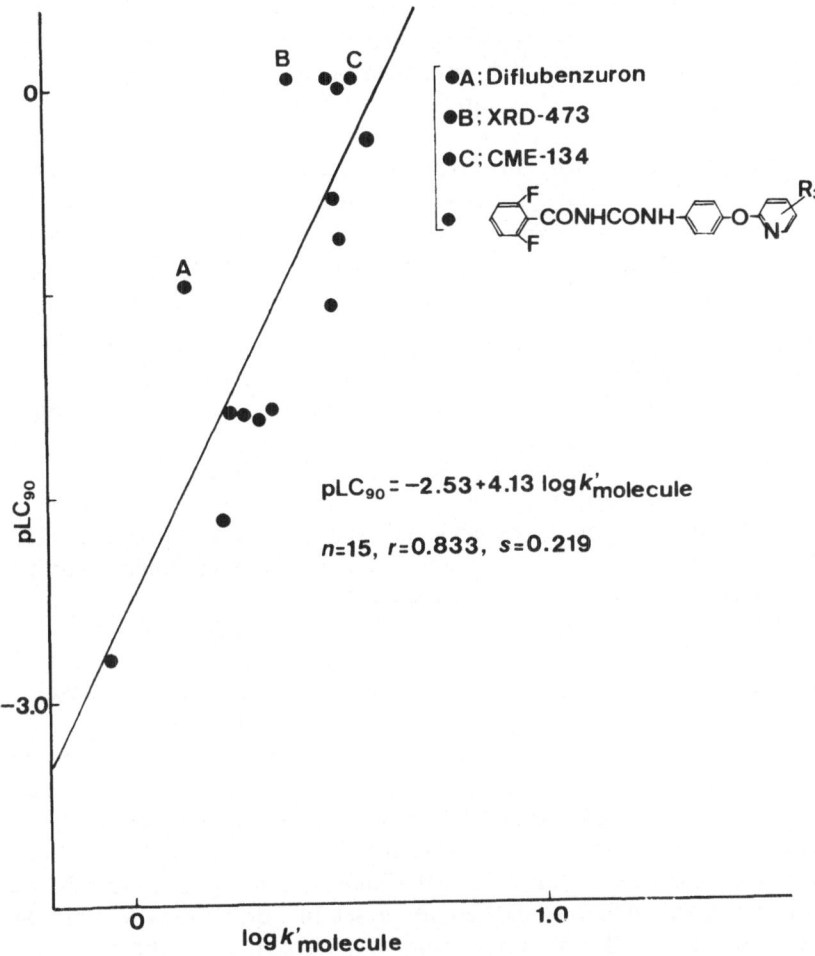

Figure 3. Larvicidal activity of benzoylphenyl ureas against *Spodoptera litura.*

$$pLC_{90} = -2.94 + 4.07\log \kappa'_{pyridyloxy}$$
$$n = 12, r = 0.907, s = 0.352 \qquad \text{(Eq. 9)}$$
$$pLC_{90} = -2.85 + 4.52\log \kappa'_{molecule}$$
$$n = 12, r = 0.934, s = 0.298 \qquad \text{(Eq. 10)}$$

Both Eq. 9 and Eq. 10 indicate that the larvicidal activity against *Spodoptera litura* is increased as the molecule becomes hydrophobic. Furthermore, since the values of the slope in both equations are almost equal, it is suggested that pyridyloxy group is responsible for the transport process of the compound. In such a case, substitution of the hydrophobic group (e.g. CF_3 or Cl) on the pyridine ring is favorable to increase the hydrophobicity of the whole molecule. Thus, the compound, in which trifluoromethyl and chloro group are substituted, exhibits the highest larvicidal activity among a set of compounds (7).

Other series of benzoyl ureas (diflubenzuron, CME-134, and XRD-473) without pyridyloxy substituents were subjected, together with the compounds (7), to the study of correlation for the larvicidal activity against *Spodoptera litura* to derive the correlation equation (Eq. 11) by using log $\kappa'_{molecule}$ (Figure 3).

$$pLC_{90} = -2.53 + 4.13 \log \kappa'_{molecule}$$

$$n = 15, r = 0.833, s = 0.219 \tag{Eq. 11}$$

Eq. 11 shows that the larvicidal activity against *Spodoptera litura* is fairly controlled by the hydrophobicity of the whole molecule. It is suggested at the same time that as far as larvicidal activity is concerned, there is little difference in the mechanism or rate of degradative metabolism among the compounds which are shown in Figure 3.

3.3 *Effects of substituents other than pyridyloxy group on the aniline ring*

As previously mentioned, the study of the effect of substituted pyridyloxy group in a set of compounds (7) on larvicidal activity against *Spodoptera litura* indicates that benzoyl (pyridyloxyphenyl) urea having chlorine atom and trifluoromethyl group on the positions of 3 and 5 of the pyridine ring, respectively shows the highest activity. Taking this result into account, SAR was studied for a set of compounds (8), in which substituents on the pyridine ring were fixed to 3-chloro- and 5-trifluoromethyl- groups and the *ortho* and/or *meta* substituents (R_2) on the aniline ring were changed.

$$(8)$$

The position and structure of the substituents (R_2), logarithm of the capacity factor for the whole molecules, larvicidal activities (pLC_{90}) (ppm) against *Spodoptera litura* and *Musca domestica*, and embryocidal activities (pLD_{50}) (mg/kg) against adult of *Musca domestica* are listed together in Table 3. The hydrophobicity of whole molecule showed the tendency to

Table 3. Larvicidal andembryocidal activities

R_2	$\log \kappa'_{molecule}$	pLC$_{90}$ (ppm) (Larvicidal) Spodoptera litura	pLC$_{90}$ (ppm) (Larvicidal) Musca domestica	pLD$_{50}$ (mg/kg) (Embryocidal) Musca domestica
H	0.524	0	0.20	-1.25
2-Cl	0.670	0	0.20	-1.37
2-Me	0.607	0	0.20	-1.10
2,5-Me	0.727	0	0.20	-1.31
3,5-Me, CO$_2$Me	0.582	0	0.20	-1.35
3,5-Cl$_2$	0.894	0	0.20	-0.37
diflubenzuron	0.143	-1.0	0.20	-1.27
XRD-473	0.393	0	0.20	-1.05
CME-134	0.538	0	0.20	-1.0

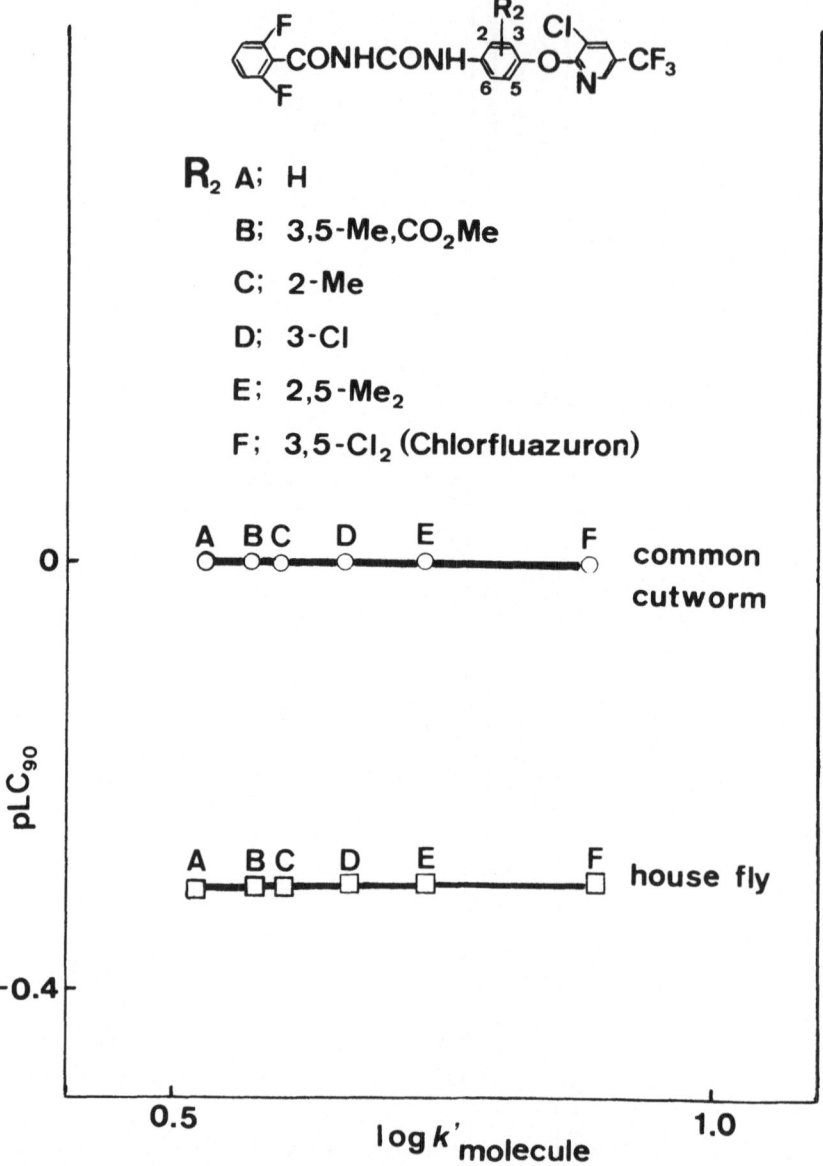

Figure 4. Larvicidal activity of benzoyl(pyridyloxyphenyl) – ureas against *Spodoptera litura* and *Musca domestica*.

change by the introduction of substituents on the aniline ring, and the 3,5-dichloro-substituted compound (chlorfluazuron) showed the highest log $\kappa'_{molecule}$ value. On the other hand, as to the larvicidal activities against *Spodoptera litura* and *Musca domestica*, no difference in activity was found among these compounds. These relationships can be drawn as in Figure 4. This result is in marked contrast to that obtained for the set of compounds (7) which show a good first order linear relationship, thus, it is suggested that there may be other factors which control the larvicidal

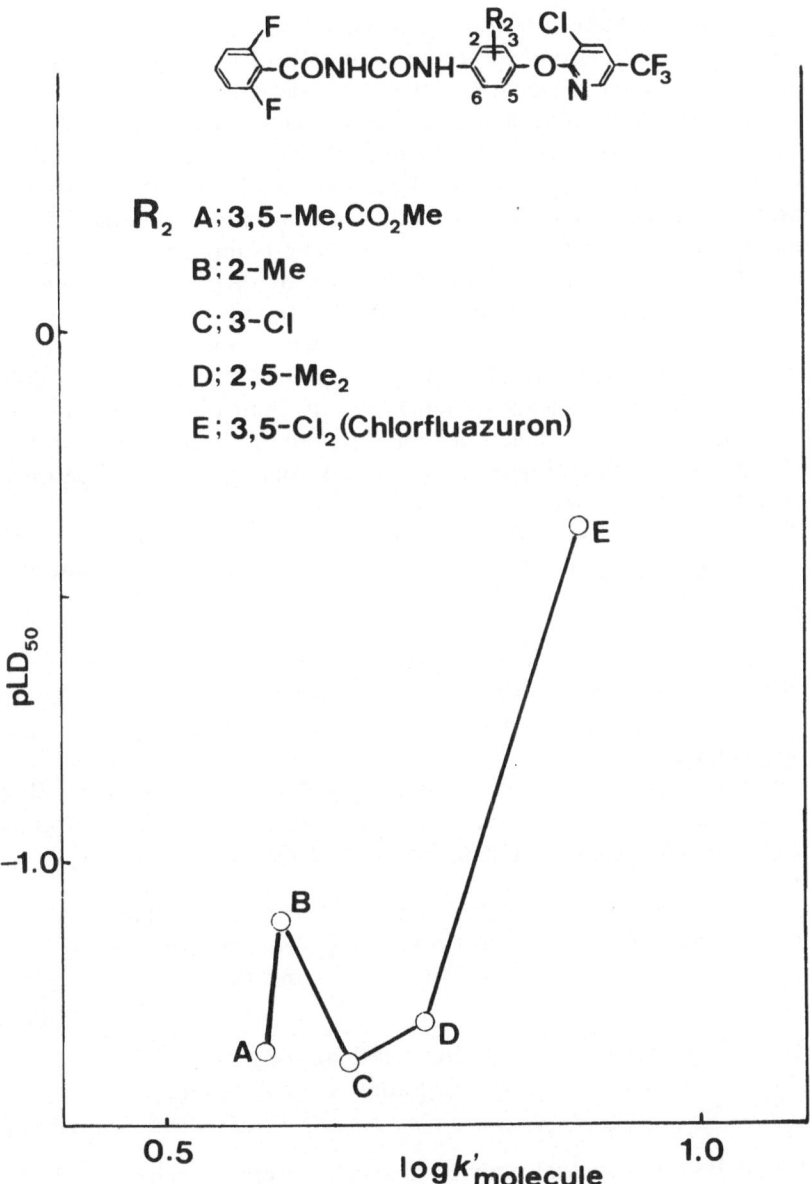

Figure 5. Embryocidal activity of benzoyl(pyridyloxyphenyl) – ureas against *Musca domestica.*

activity, other than the hydrophobicity of the whole molecule. As for the embryocidal effect against the adult of *Musca domestica*, 3,5-dichloro-substituted-compound (chlorfluazuron) showed the highest activity, and no linear relationship between activity and the hydrophobic parameter was obtained as shown in Figure 5.

Since the mechanism of the embryocidal action by benzoylphenyl urea derivatives is understood to be the inhibition of chitin formation in the embryo, analogous to the larvicidal action, linear relationship between the activity and the hydrophobic parameter should have existed as in set of compounds (7). Although we tried to analyze the relation of Figure 5 by introducing other physicochemical parameters (Hammett σ, Taft steric parameter E_s, etc.), reasonable correlation could not be obtained. Consequently, the assumption acceptable for these compounds (7), "there is little difference in the mechanism or rate of degradative metabolism among these compounds," is not applicable to (8).

Actually, Neumann et al. (1983) found that, in the body of the insects such as Egyptian cotton leafworm, *Spodoptera littoralis* to which chlorfluazuron shows significantly higher activity than diflubenzuron, chlorfluazuron has a much longer half life than diflubenzuron, which is metabolized in a short time (Figure 6). We found from preliminary experiments that although the compound whose aniline ring is substituted with only 3,5-Cl,CF_3-pyridyloxy group increases its larvicidal and embryocidal activity against *Spodoptera litura* and *Musca domestica* by adding a synergist (piperonyl butoxide), no synergism was observed for chlorfluazuron.

From these facts it can be said that by introducing substituents on the *ortho* and/or *meta* position of the aniline ring, the process of degradative metabolism of the insecticide is greatly affected, thereby reflecting the insecticidal activity.

By taking the above estimation into consideration, the behavior observed with respect to the larvicidal and embryocidal activity of compounds (8) is reasonably explained as in Eq. 12 and Eq. 13.

larvicidal activity $\propto$ decreased intrinsic activity $\times$ increased stability to degradative metabolism

(Eq. 12)

embryocidal activity $\propto$ increased stability to degradative metabolism

(Eq. 13)

The intrinsic insecticidal activity gradually decreases when the hydrophobicity of the insecticide molecule is increased by the introduction of substituents on the aniline ring. However, the rate of degradative meta-

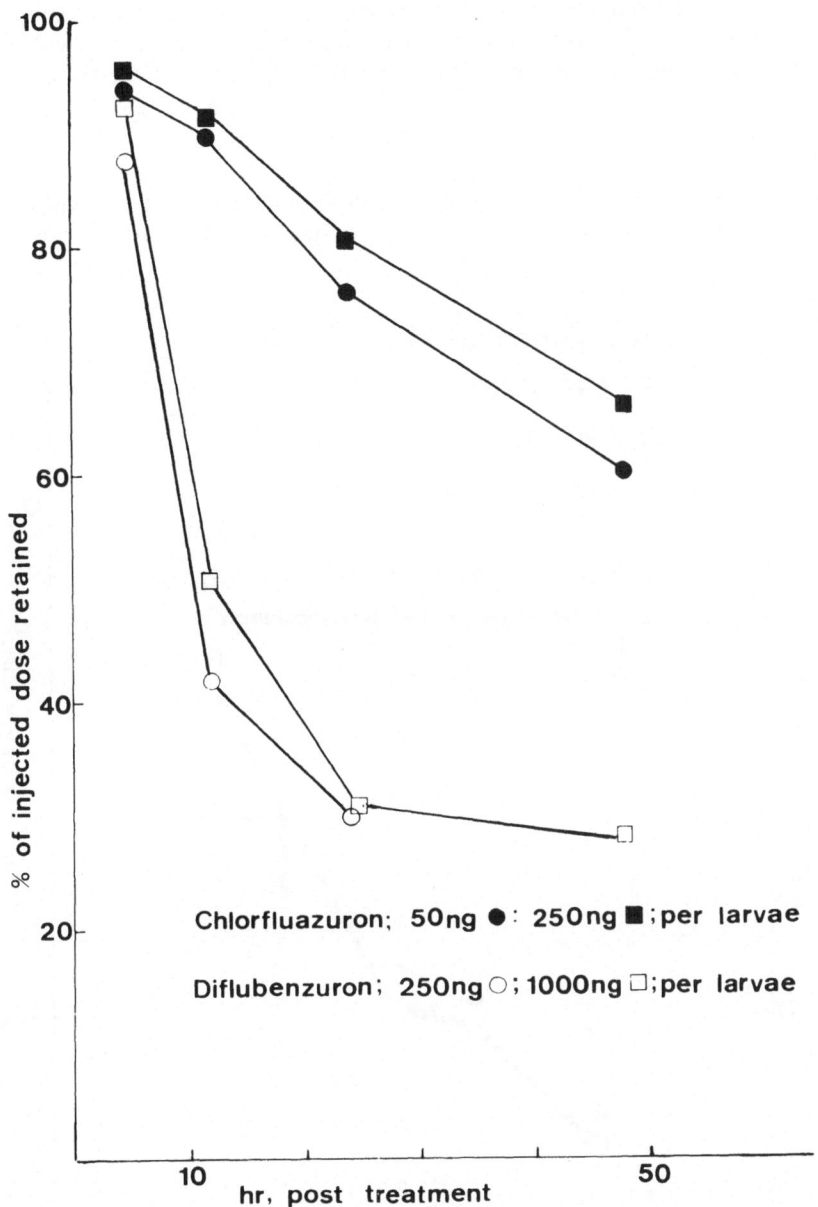

Figure 6. Elimination of chlorfluazuron and diflubenzuron from larvae of *Spodoptera littoralis.*

bolism is lowered by introducing substituents. Two factors, intrinsic activity and stability to degradative metabolism, compensate for each other, and key the overall activity constant at the excellent level (Eq. 12). In order to exhibit the embryocidal effect, insecticides should move through a longer process, "transportation to the embryo" in comparison with the larvicidal action. Therefore, it can be considered that since the stability in the stage of degradative metabolism becomes the largest factor to control the embryocidal activity, chlorfluazuron shows the highest activity (Eq. 13).

Figure 7 shows the larvicidal and embryocidal activity of chlorfluazuron against *Musca domestica*, in comparison with other series of

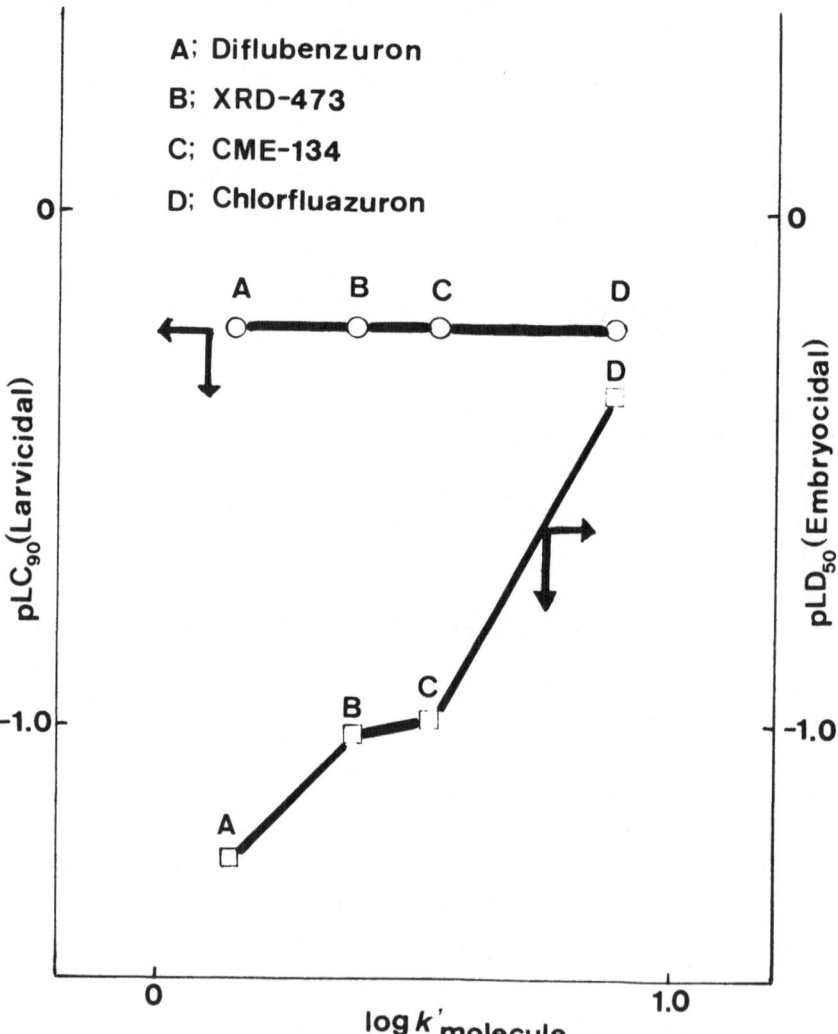

Figure 7. Larvicidal and embryocidal activities of chlorfluazuron and benzoylphenyl ureas against *Musca domestica*.

benzoylphenyl urea derivatives. Here, the strength of the embryocidal effect of chlorfluazuron is overwhelming, and it apparently suggests the stability of chlorfluazuron in insect bodies.

3.3 *Effects of substituents on benzoyl ring*

Effects of the substituents on benzoyl ring of benzoylphenyl urea on insecticidal activity against larvae of the yellow fever mosquito, *Aëdes aegypti* and *Pieris brassicae* were summarized by van Daalen et al. (1973). According to the report, 2,6-dihalosubstituted compounds showed the strongest activity. Above all, 2,6-difluoro derivative was the strongest. And if the 2- and/or 6- positions were substituted by groups other than halogen, no insecticidal activity was observed.

We have studied the QSAR for larvicidal activity against *Spodoptera litura* with a set of compounds (9), in which substituents on benzoyl ring of chlorfluazuron were changed.

$$(9)$$

Variety and position of the substituents, and pLC_{90} (log $1/LC_{90}$) values of the corresponding compounds are listed in Table 4. The regression analysis afforded a correlation equation (Eq. 14) related to Taft steric parameter E_s and indicator variable I (equals 1 or 2, when one or two substituents exist on *ortho* positions, respectively).

Table 4. SAR for substituents on benzoyl ring against *Spodoptera litura*

R_1	pLC_{90} (ppm)
H	0.301
2-F	1.0
2-Cl	1.0
2-Br	0.699
2-CH$_3$	1.0
2-NO$_2$	0.155
2,6-F$_2$	1.30
2,6-Cl, F	1.0
2,6-CH$_3$O	0.398
3-F	0.097
3-Cl	−2.91
4-Cl	−2.68
4-CH$_3$	−1.79
2,4,6-F$_3$	0.301

$$pLC_{90} = 0.593 + 0.233E_s^{ortho} + 3.15E_s^{meta}$$
$$+ 2.42E_s^{para} + 0.359 I$$
$$n = 14, r = 0.931, s = 0.597 \tag{Eq. 14}$$

According to Eq. 14, when substituents other than hydrogen exist on *meta* or *para* position, the insecticidal activity decreases. However, as to the *ortho* position the existence of substituents, unless they are bulky, is favorable to increase the activity.

It is noticeable contrary to the result by van Daalen et al. that compounds having substituents other than halogen such as methyl or methoxy are actually as active as halosubstituted compounds. This finding seems to give a hint to the drug design for any new benzoylphenyl urea insecticides.

4 SUMMARY

A number of benzoyl (pyridyloxphenyl) ureas were synthesized, and the larvicidal and embryocidal activities were tested against common cutworm, *Spodoptera litura*, and house fly, *Musca domestica*.

Correlation of the activities by means of physicochemical parameters reveals that not only the hydrophobicity of the whole molecule, but the susceptibility to the degradative metabolism is the controlling factor of the insecticidal activity. As to the embryocidal activity, the latter factor was estimated to be more important. The excellent insecticidal activity of chlorfluazuron (IKI-7899) seems to be due to the slow detoxification in the insect body.

5 REFERENCES

Block, J. H., Henry, D., Anderson, J. L., Carlson, G. R. 1974. J. Med. Chem., 17: 28.
Casida, J. E., Hajjar, N. P. 1979. Structure-activity relationships of benzoylphenyl ureas as toxicants and chitin synthesis inhibitors in Oncopeltus fasciatus. Pestic. Biochem. Physiol. 11: 33.
Fujita, T., Nakagawa, Y., Kitahara, K., Nishioka, T., Iwamura, H. 1984. Quantitative structure-activity studies of benzoylphenyl urea larvicides. I. Effect of substituents at Aniline Moiety against *Chilo suppressalis* Walker. Pestic. Biochem. Physiol. 21: 309.
Grosscurt, A. C., Tipker, J. 1980. Ovicidal and larvicidal structure-activity relationships of benzoylureas on the house fly (*Musca domestica*). Pestic. Biochem. Physiol. 13: 249.
Haga, T., Toki, T., Koyanagi, T., Nishiyama, R. 1982. In: Abstracts of 5th Int. Cong. of Pesticide Chemistry (IUPAC). IId-7: 1982.
Haga, T., Toki, T., Koyanagi, T., Nishiyama, R. 1984. In: Abstract of 17th Int. Cong. Entomology. S.4.1.6.
Hansch, C., Leo, A. J. 1979. In: Substituent Constants for Correlation Analysis in Chemistry and Biology. Wiley, New York, p. 65.
Neumann, R., Guyer, W. 1983. In: Abstracts of 10th Int. Cong. Plant Protection. p. 445.
Nishiyama, R., Fujikawa, K., Nasu, R., Toki, T., Yamamoto, T. U.S. Patent 4 173 737.

Spiesecke, H., Schneider, W. G. 1961. J. Chem. Phys. 35: 722, 731.

van Daalen, J. J., Wellinga, K., Mudler, R. 1973. Synthesis and laboratory evaluation of 1-(2,6-disubstituted benzoyl)-3-phenylureas, a new class of insecticides. II. Influence of the Acyl Moiety on insecticidal activity. J. Agric. Food Chem. 21: 993.

Verloop, A., Ferrell, C. D. 1977. In: Pesticide Chemistry in the 20th Century. vol. 37, pp. 237–270, ACS Symposium Series, American Chemical Society. Washington, D.C.

Yu, C. C., Kuhr, R. J. 1976. Synthesis and insecticidal activity of substituted 1-phenyl-3-benzoylureas and 1-phenyl-3-benzoyl-2-thioureas. J. Agric. Food Chem. 24: 134.

7. Toxicity of two benzoylphenyl ureas against insecticide resistant mealworms

I. Ishaaya and S. Yablonski

1 INTRODUCTION

The discovery of the insecticidal activity of benzoylphenyl ureas pointed to new pathways in insect control strategies. The first commercial compound, diflubenzuron, disrupts the molting process, thereby serving as an insect growth regulator (Mulder and Gijswijt, 1973; Verloop and Ferrell, 1977; Marks et al., 1982). Studies of the mode of action of diflubenzuron [1-(4-chlorophenyl)-3-(2,6-difluorobenzoyl)urea] revealed that the compound alters cuticle composition in insects, especially that of chitin (Ishaaya and Casida, 1974; Post et al., 1974; Sowa and Marks, 1975), thereby affecting the elasticity and firmness of the endocuticle (Grosscurt, 1978a; Grosscurt and Anderson, 1980). As the dietary diflubenzuron concentration is increased, the amount of cuticle chitin in the house fly larvae is progressively reduced without any appreciable effect on the cuticle protein level; thus, the protein chitin ratio increases from 3.4 in normal larvae to 14.3 in 2.5-ppm-treated larvae. This alteration could well affect the biophysical and biochemical characters of the glycoprotein, the main constituent of the endocuticle (Ishaaya and Casida, 1974). The reduced level of chitin in the cuticle seems to result from inhibition of biochemical processes leading to chitin formation (Post et al., 1974; Hajjar and Casida, 1979; Van Eck. 1979). It is not yet clear whether inhibition of chitin synthetase is the primary pathway for the reduced level of chitin, since diflubenzuron treatment did not inhibit either *in vitro* or *in vivo* the chitin synthetase activity obtained from the larval gut of flour beetle, *Tribolium*, larvae (Cohen and Casida, 1980). General disturbances in carbohydrate metabolism, as expressed by inhibition *in vivo* of invertase, trehalase and amylase, were observed in diflubenzuron-treated *Tribolium castaneum* larvae (Ishaaya and Ascher, 1977). This may result from a chain effect originating primarily from inhibition of chitin synthesis.

Diflubenzuron acts on larvae of various orders of insects (Grosscurt, 1978b) and exhibits in some insect species a relatively strong ovicidal

Wright, J. E. and Retnakaran, A. (Eds), Chitin and Benzoylphenyl ureas. ISBN 978-94-010-8638-7.
© *1987, Dr W. Junk Publishers, Dordrecht.*

activity (Ascher and Nemny, 1974; Holst, 1975; Ascher et al., 1979), fecundity suppression (Wright and Harris, 1976; Arambourg et al., 1977; Sarasua and Santiago-Alvarez, 1983) and contact toxicity (Ascher and Nemny, 1976). Treated larvae are usually fed normally until the apolytic stage and may skip in some cases two or three molts. In addition, older larvae are in general more tolerant than the younger ones (Neal, 1974; Rathburn and Boike, 1975; Ascher and Nemny, 1976; Busvine et al., 1976), so that some crop damage may still occur, depending on the susceptibilty of the species and on the time of application. Hence, some economically important cotton pests, such as *Spodoptera littoralis, Heliothis virescens* and *Trichoplusia ni*, cannot be controlled efficiently under field conditions (Grosscurt, 1978b). Accordingly, a search for more potent benzoylphenyl urea derivatives was continued. BAY SIR 8514 [1-(4-trifluoromethoxy-phenyl)-3-(2-chlorobenzoyl)urea] was found to exhibit 2 to 3 fold higher toxicity on house fly females (Chang, 1979) and on larvae of *S. littoralis* (Ascher et al., 1979) and *T. castaneum* (Ishaaya et al., 1981).

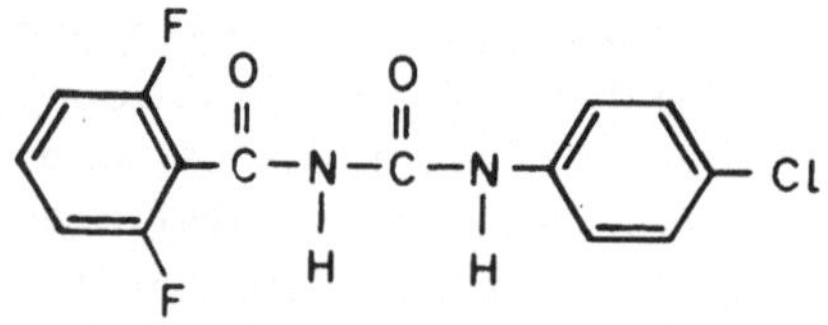

DIFLUBENZURON

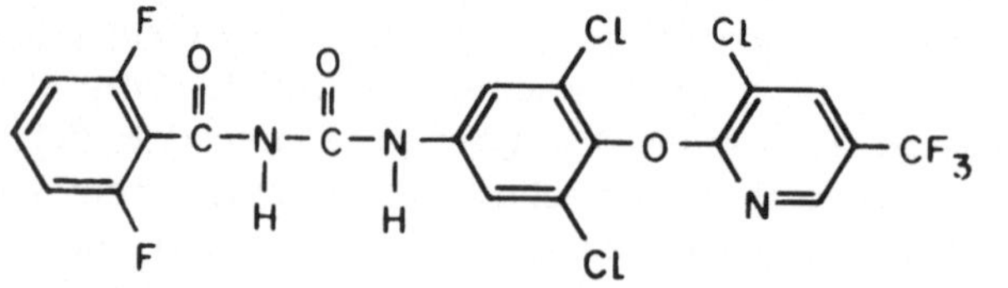

IKI - 7899
CGA 112,913
PP 145

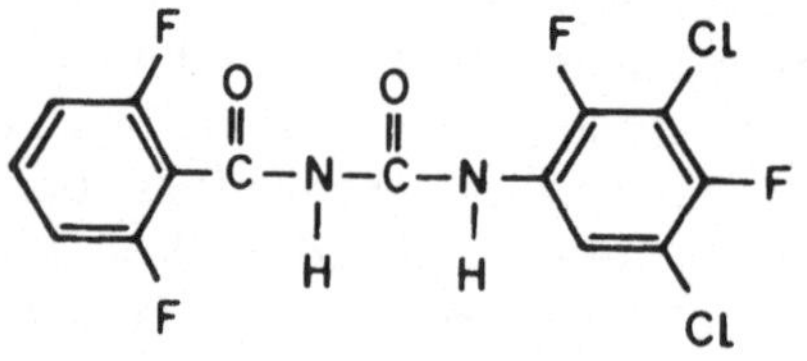

CME - 134

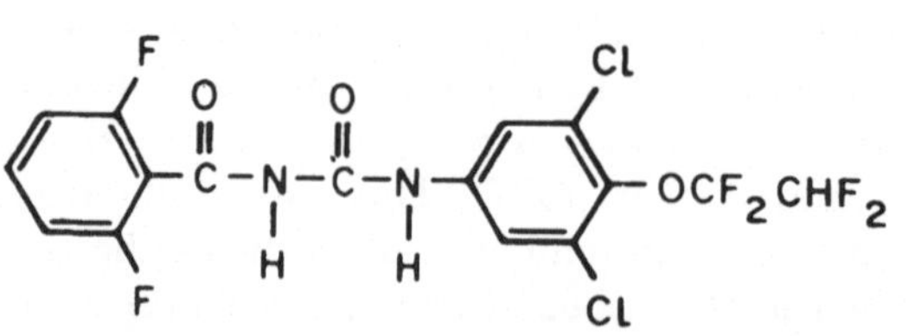

XRD - 473

Figure 1. Structure of the new acylureas in comparison with diflubenzuron.

Three novel benzoylphenyl ureas have been developed recently by Ishihara Sangyo Kaisha Ltd. chlorfluazuron, (IKI-7899), Celamerk GmbH & Company (CME 134) and Dow Chemicals (XRD-473), and were reported to exhibit a much higher potency than diflubenzuron on cotton leafworms such as *S. littoralis, S. exigua, Prodenia litura* and *Heliothis* spp. (Haga et al., 1982; Becher et al., 1983; Neumann and Guyer, 1983; Sbragia et al., 1983; Ascher and Nemny, 1984) (Figure 1). This research evaluates the toxicity and biochemical aspects of chlorfluazuron in comparison with those of diflubenzuron on malathion-resistant and -susceptible strains of *T. castaneum*.

2 BIOCHEMICAL AND TOXICOLOGICAL ASPECTS OF IKI-7899

Chlorfluazuron, IKI-7899 [N-2,6-difluorobenzoyl-N-4-(3-chloro-5-trifluoromethyl-pyridin-2-yloxy)-3,5-dichlorophenyl-urea], (also called CGA 112,913), one of the recent chitin synthesis inhibitors, exhibits excellent insecticidal and embryocidal activity against cotton lepidopterous pests such as *Heliothis* and *Spodoptera* species (Haga et al., 1982; Neumann and Guyer, 1983). Biochemical assays using glucose incorporation into chitin, as well as toxicological studies at early stages after treatment, reveal that the intrinsic activity of chlorfluazuron is weaker than that of diflubenzuron. On the other hand, chlorfluazuron possesses greater persistence in *S. littoralis*, thereby blocking chitin synthesis more efficiently than does diflubenzuron. This may explain the high toxicity of chlorfluazuron as compared with diflubenzuron on this pest (Neumann and Guyer, 1983). Our studies in the laboratory revealed that chlorfluazuron is 6- and 13-fold more toxic than diflubenzuron at LC_{50} and LC_{95}, respectively, on an undefined laboratory strain of *T. castaneum* larvae, and about 5-fold more toxic on topically treated *S. littoralis* larvae (Ishaaya, 1984; Ishaaya et al., 1984). Preliminary results indicate that chlorfluazuron inhibits chitin formation *in vivo* in house fly larvae, similar to the action of diflubenzuron, and this correlates with their larval toxicity (Ishaaya, unpublished results).

3 TOXICITY AND BIOCHEMICAL ASPECTS OF CHLORFLUAZURON AND DIFLUBENZURON ON DIFFERENT STRAINS OF TRIBOLIUM CASTANEUM

The greater potency of chlorfluazuron in comparison with that of diflubenzuron on *S. littoralis* seems to result from a lower metabolism and subsequently a much higher stability in the insect (Neumann and Guyer,

1983; Guyer and Neumann, 1984). Several reports indicate that diflubenzuron is metabolized by hydrolytic and oxidative reactions in mammalian and insect species (Metcalf et al., 1975; Verloop and Ferrell, 1977; Ivie, 1978; Ivie and Wright, 1978). Hence two well defined strains of *Tribolium castaneum* were selected for determining the potency of these two acylurea derivatives; a malathion-susceptible strain (bb) with low oxidative and hydrolytic activities, and a malathion-resistant strain, with relatively high mixed-function oxidase, esterase and epoxide hydrase activities (CTC-12) (Cohen, 1981; Wool et al., 1982). A third strain (Kano), with a relatively high-in-carboxyesterase activity but low-in-total esterase activity (Wool et al., 1982), was also used in these assays.

The rearing procedure and biological assays for *T. castaneum* were similar to those described previously (Ishaaya and Ascher, 1977; Ishaaya et al., 1983), using wheat flour containing 50 g dried yeast kg^{-1} under standardised laboratory conditions of 28°C and 70% relative humidity. For bioassays, the diet (10 g) was mixed with acetone solution (10 ml) containing the test compound, or with acetone (10 ml) alone as the control. The diet was distributed in 2-g portions in test vials (2 cm in diameter) following thorough mixing and solvent evaporation. Ten to 12 first-instar (0-3-h-old white in color) or fourth-instar (1.00 $\pm$ 0.05 mg) larvae were placed in each vial and held at 28°C for determination of cumulative larval mortality, larval weight gain, pupation and emergence. Larval weight in the first-instar assays was determined 14 days after the start of the experiment, and in the fourth-instar assays 4 days after treatment. Data are the mean and standard error values of 15–20 replicates of 12 larvae each. For comparison, log dosage-percentage mortality curves on a probit scale were used to determine LC_{50} and LC_{95} (concentrations needed for 50 and 95% larval mortality, respectively) or IC_{50} and IC_{95} (concentrations needed for 50 and 95% larval weight gain inhibition respectively) values for chlorfluazuron and diflubenzuron.

The CTC-12 and the Kano strains originated in the Pest Infestation Control Laboratory in Slough, England and the bb strain in the U.S.; they were identified biochemically and genetically by Cohen (1981) and Wool et al. (1982). According to this determination, the activity of the cytochrome P-450 and of epoxide hydrase in CTC-12 was four-fold higher than that in the bb strain, and the activity of the carboxyesterase in the Kano strain was 2.5-fold higher than that in the bb strain, while the total esterase activity in the Kano strain was about 60% than that of the bb strain (Wool et al., 1982).

In assays carried out with first instar larvae of *T. castaneum*, the effect of chlorfluazuron on larval mortality (Figure 2), larval weight gain (Figure 3) and pupation and emergence (Table 1) was much greater than that of diflubenzuron in both the malathion-susceptible (bb) and the malathion-resistant strain with the high MFO and epoxide hydrase activities (CTC-

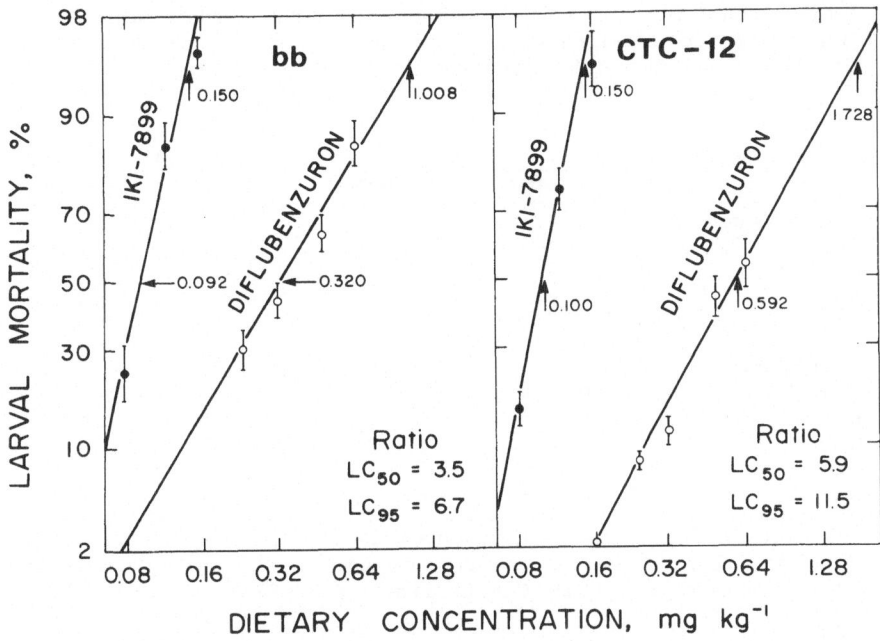

Figure 2. Log-dosage percentage mortality curves on a probit scale for dietary applied diflubenzuron and chlorfluazuron on a malathion-susceptible strain (bb) and a malathion-resistant strain with high MFO and epoxide, hydrase activities (CTC-12). Ten to fifteen replicates of 15 first-instar, 0–3-h-old larvae were kept on treated diet until pupation and emergence. Cumulative larval mortality was determined. Data are mean ± SE values.

12). According to LC_{50} and LC_{95} values, the potency of chlorfluazuron was similar in both strains, while that of diflubenzuron was much lower in the CTC-12 strain (Figures 2 & 3). The LC_{50} ratios of diflubenzuron: chlorfluazuron for the bb and CTC-12 strains were 3.5 and 5.9, respectively, and the LC_{95} ratios were 6.7 and 11.5, respectively (Figure 2). Similarly, the effect of chlorfluazuron on larval weight gain was the same in both bb and CTC-12 strains, while that of diflubenzuron was much lower in the CTC-12 strain, as expressed by much higher IC_{50} and IC_{95} values (Figure 3). The lower potency of diflubenzuron in the CTC-12 strain seems to result from a higher metabolism due to greater oxidase and esterase activities (Wool et al., 1982). These findings concur with those obtained by Neumann and Guyer (1983), who showed that diflubenzuron was more susceptible than chlorfluazuron to detoxification in *S. littoralis* larvae.

In assays carried out with fourth-instar larvae of *T. castaneum*, of both the bb and CTC-12 strains, chlorfluazuron was considerably more potent than diflubenzuron (Table 2). The LC_{50} value of chlorfluazuron for cumulative larval and pupal mortality was about 0.3 mg kg^{-1} for both the bb and the CTC-12 strains, while that of diflubenzuron was greater than

Table 1. Effect of chlorfluazuron and diflubenzuron on pupation and emergence of first-instar of *Tribolium castaneum* larvae of the bb strain with low MFO activity, and of the CTC-12 strain with relatively high MFO and epoxide hydrase activities.

Compound and dietary concentrations, mg kg^{-1}	Data relative to control, %[a]			
	bb		CTC-12	
	Pupation	Emergence	Pupation	Emergence
Chlorofluazuron				
0.08	76 ± 7	76 ± 7	76 ± 3	74 ± 3
0.12	14 ± 3	12 ± 3	18 ± 2	14 ± 2
0.16	4 ± 1	2 ± 1	5 ± 2	4 ± 2
0.24	0	0	0	0
Diflubenzuron				
0.16	99 ± 1	94 ± 3	—	—
0.24	69 ± 5	69 ± 5	92 ± 1	92 ± 1
0.32	56 ± 5	51 ± 7	88 ± 2	88 ± 2
0.48	—	—	56 ± 6	56 ± 6
0.64	15 ± 4	6 ± 3	46 ± 7	46 ± 7

[a]Data are mean and SE values of 10–15 replicates of 15 first-instar 0–3-h-old larvae each, kept under standardised conditions on treated diet until pupation and emergence. Pupation and emergence of the untreated bb and CTC-12 strains were over 96%.

Table 2. Effect of chlorfluazuron and diflubenzuron on weight gain, pupation and emergence of fourth-instar larvae of bb and CTC-12 strains of *Tribolium castaneum*

Compound and dietary concentrations, mg kg^{-1}	Data relative to control, %[a]					
	bb			CTC-12		
	ΔW[b]	P	E	ΔW	P	E
Chlorofluazuron						
0.1	104 ± 4c	97 ± 2c	93 ± 2c	102 ± 3c	99 ± 1c	99 ± 1c
0.2	103 ± 5c	87 ± 4d	80 ± 6d	96 ± 2c	82 ± 6d	66 ± 8d
0.4	104 ± 4c	84 ± 4d	21 ± 4e	97 ± 3c	68 ± 6e	36 ± 6e
0.8	95 ± 5c	34 ± 8e	3 ± 2f	99 ± 2c	31 ± 9f	7 ± 3f
Diflubenzuron						
1.0	106 ± 5c	93 ± 3c	92 ± 3c	95 ± 3c	95 ± 3c	94 ± 3c
2.0	99 ± 5c	90 ± 5c	88 ± 5cd	91 ± 3c	95 ± 2c	95 ± 2c
3.0	96 ± 5c	91 ± 4c	91 ± 4cd	88 ± 2c	89 ± 3cd	87 ± 3d
6.0	96 ± 5c	84 ± 4c	81 ± 4d	90 ± 2c	82 ± 2d	79 ± 2c

[a]Data are the mean and SE values of 15–20 replicates of 12 larvae each. Fourth-instar larvae (1.00 ± 0.005 mg) were introduced into the medium and maintained until adult emergence was complete. The average larval weight gain (ΔW) for the untreated bb and CTC-12 strains was 0.88 ± 0.05 and 1.42 ± 0.05 mg, respectively, and pupation (P) and emergence (E) values for both strains were over 94%.
[b]Larval weight gains (ΔW) relative to control were determined 4 days after treatment.
[c,d,e,f]Represent data which differ significantly from each other at P = 0.05 within the same group.

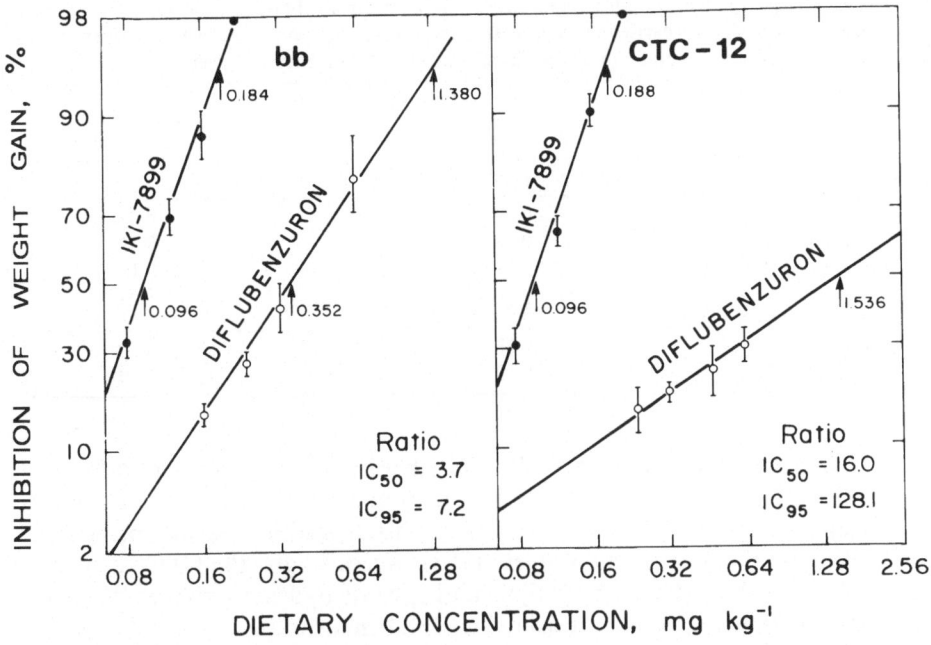

Figure 3. Log-dosage percentage weight gain inhibition curves on a probit scale for dietary applied diflubenzuron and chlorofluazuron on a malathion-susceptible strain (bb) and a malathion-resistant strain with high MFO and epoxide hydrase activities (CTC-12). Assay and replicates are as in Fig 2. Larval weight gain was determined 14 days after start of the experiment.

6 mg kg^{-1}. Larval weight gain 4 days after treatment was not affected, as in the case of first-instar larvae (Figure 2), by either chlorfluazuron or diflubenzuron (Table 2); hence, larval weight in *T. castaneum* is affected by benzoylphenyl ureas only after prolonged feeding. At 0.4 mg kg^{-1} chlorfluazuron 84 and 68% of the bb and CTC-12 strain larvae pupated respectively, but only 21 and 36% of the adults emerged (Table 2). These results indicate that chlorfluazuron exerts its effect also during the pupal stage.

The LC$_{50}$ values of chlorfluazuron and diflubenzuron representing cumulative larval and pupal mortality, in assays using first- and fourth-instar larvae of various strains of *T. castaneum*, are summarized in Table 3. With all the test strains and assays, the LC$_{50}$ values of chlorfluazuron were lower than those of diflubenzuron. The potency of chlorfluazuron on the various strains of *T. castaneum*, using either first- or fourth-instar larvae, was similar. The LC$_{50}$ values in the case of first-instar larvae ranged between 0.09 and 0.11 mg kg^{-1}, and those of fourth-instar larvae were

Table 3. LC$_{50}$ values of chlorfluazuron and diflubenzuron obtained with bb, CTC-12 and Kano strains. Assays carried out with first-instar larvae kept on treated diet until emergence. LC$_{50}$ values representing cumulative larval and pupal mortality are given.

Tribolium strains	LC$_{50}$ values (mg kg^{-1}) of	
	chlorofluazuron	diflubenzuron
	First instar larvae	
bb	0.09	0.32
CTC-12	0.10	0.59
Kano	0.11	0.16
	Fourth-instar larvae	
bb	0.28	> 10
CTC-12	0.30	> 10

0.28–0.30 mg kg^{-1} (Table 3). On the other hand, the potency of diflubenzuron differs greatly from one strain to another, which seems to result from the susceptibility of diflubenzuron to degrading enzymes present in an insect (Neumann and Guyer, 1983). Accordingly, the CTC-12 strain, with its high oxidative, hydrolytic and epoxide hydrase activities (Wool et al., 1982), was the most resistant to diflubenzuron.

The high potency and prolonged stability of chlorfluazuron in malathion -susceptible and -resistant strains of *Tribolium*, and in other agricultural pests such as *Spodoptera* and *Heliothis* species (Haga et al., 1982; Neumann and Guyer, 1983; Ishaaya et al., 1984), along with its low mammalian toxicity (Neumann and Guyer, 1983), make this compound a potential insecticide for controlling important agricultural pests, and especially those resistant to organophosphorus compounds.

4 ACKNOWLEDGEMENTS

The authors thank Mrs. Zmira Mendelson for her skillful technical assistance, Dr. D. Wool of the University of Tel Aviv for providing us with various identified strains of *Tribolium castaneum*, and Dr. K. Urano, of Ishihara Sangyo Kaisha Ltd., for the supply of chlorofluazuron. This paper constitutes contribution No. 1265-E, 1984 series, from the Agricultural Research Organization, The Volcani Center, Bet Dagan, Israel.

5 REFERENCES

Arambourg, Y., Pralavorio, R. and Dolbeau, C. 1977. Premieres observations sur l'action du diflubenzuron (PH 6040) sur la fecondité, la longevité et la viabilité des oeufs de *Ceratitis capitata* Wied. (Dipt. Trypetidae). Rev. Zool. Agric. Pathol. Veg. 76: 118–126.

Ascher, K. R. S. and Nemny, N. E. 1974. The ovicidal effect of PH 60–40 [1-(4-chlorophenyl)-3-(2,6-difluorobenzoyl)-urea] in *Spodoptera littoralis* Boisd. Phytoparasitica 2: 131–133.

Ascher, K. R. S. and Nemny, N. E. 1976. Contact activity of diflubenzuron against *Spodoptera littoralis* larvae. Pestic. Sci. 7: 447–452.

Ascher, K. R. S. and Nemny, N. E. 1984. The effect of CME 134 on *Spodoptera littoralis* eggs and larvae. Phytoparasitica 12: 13–27.

Ascher, K. R. S., Nemny, N. E., Eliyahu, M. and Ishaaya, I. 1979. The effect of BAY SIR 8514 on *Spodoptera littoralis* (Boisduval) eggs and larvae. Phytoparasitica 7: 177–184.

Becher, H. M., Becher, P., Prokic-Immel, R. and Wirtz, W. 1983. CME, A new chitin synthesis inhibiting insecticide. 10th Int. Congr. Pl. Protection, Brighton, pp. 408–415.

Busvine, J. R., Rongsriyam, Y. and Bruno, D. 1976. Effect of some insect development inhibitors on mosquito larvae. Pestic. Sci. 7: 153–160.

Chang, S. C. 1979. Laboratory evaluation of diflubenzuron, penfluron, and BAY SIR 8514 as female sterilants against the house fly. J. Econ. Ent. 72: 479–481.

Cohen, E. 1981. Epoxide hydrase activity in the flour beetle *Tribolium castaneum* (Coleoptera, Tenebrionidae). Comp. Biochem. Physiol. 69B: 29–34.

Cohen, E. and Casida, J. E. 1980. Inhibition of *Tribolium* gut chitin synthetase. Pestic. Biochem. Physiol. 13: 129–136.

Grosscurt, A. C. 1978a. Effect of diflubenzuron on mechanical penetrability, chitin formation, and structure of the elytra of *Leptinotarsa decemlineata*. J. Insect Physiol. 24: 827–831.

Grosscurt, A. C. 1978b. Diflubenzuron: some aspects of its ovicidal and larvicidal mode of action and an evaluation of its practical possibilities. Pestic. Sci. 9: 373–386.

Grosscurt, A. C. and Anderson, S. O. 1980. Effect of diflubenzuron on some chemical and mechanical properties of the elytra of *Leptinotarsa decemlineata*. Proc. K. ned. Akad. Wet. 83C: 143–150.

Guyer, W. and Neumann, R. 1984. Activity and fate in Lepidoptera of CGA 112913 compared with diflubenzuron. XVII Int. Congr. Entomology, Hamburg, Abstract Volume, p. 129.

Hajjar, N. P. and Casida, J. E. 1979. Structure activity relationships of benzoylphenyl ureas as toxicants and chitin synthesis inhibitors in *Oncopeltus fasciatus*. Pestic. Biochem. Physiol 11: 33–45.

Haga, T., Tobi, T., Koyanagi, T. and Nishiyama, R. 1982. Structure activity relationships of a series of benzoyl-pyridyloxyphenyl-urea derivatives. Abstracts, 5th Int. Congr. Pesticide Chemistry (IUPAC), Kyoto, p. IId-7.

Holst, H. 1975. Die fertilitätsbeeinflussende Wirkung des neuen Insektizids PDD 60–40 bei *Epilachna varivestis* Muls. (Col. Coccinellidae), und *Leptinotarsa decemlineata* Say (Col. Chrysomelidae). Z. PflKrankh. PflSchutz 82: 1–7.

Ishaaya, I. 1984. Effect of a new benzoylphenyl urea derivative, IKI-7899, on growth and development of *Tribolium castaneum* and *Musca domestica vicina*. XVII Int. Congr. Entomology, Hamburg, Abstract Volume, p. 128.

Ishaaya, I. and Ascher, K. R. S. 1977. Effect of diflubenzuron on growth and carbohydrate hydrolases of *Tribolium castaneum*. Phytoparasitica 5: 149–158.

Ishaaya, I., Ascher, K. R. S. and Yablonski, S. 1981. The effect of BAY SIR 8514, diflubenzuron, and Herculus 24108 on growth and development of *Tribolium castaneum*. Phytoparasitica 9: 207–209.

Ishaaya, I. and Casida, J. E. 1974. Dietary TH 6040 alters composition and enzyme activity of housefly larval cuticle. Pestic. Biochem. Physiol. 4: 484–490.

Ishaaya, I., Elsner, A., Ascher, K. R. S. and Casida, J. E. 1983. Synthetic pyrethroids: Toxicity and synergism on dietary exposure of *Tribolium castaneum* (Herbst.) larvae. Pestic. Sci. 14: 367–372.

Ishaaya, I., Nemny, N. E. and Ascher, K. R. S. 1984. The effect of IKI-7899, a new chitin synthesis inhibitor, on larvae of *Tribolium castaneum* and *Spodoptera littoralis*. Phytoparasitica 12: 193–197.

Ivie, G. W. 1978. Fate of diflubenzuron in cattle and sheep. J. Agric. Fd. Chem. 26: 81–89.

Ivie, G. W. and Wright, J. E. 1978. Fate of diflubenzuron in the stable fly and house fly. J. Agric. Fd. Chem. 26: 90–94.

Marks, E. P., Leighton, T. and Leighton, F. 1982. Modes of action of chitin synthesis inhibitors. In: "Insecticide Mode of Action" (J. R. Coasts, Ed.) Chapter 10, pp. 281–313, Academic Press, Inc., New York, NY.

Metcalf, R. L., Lu, P. Y. and Bowlus, S. 1975. Degradation and environmental fate of 1-(2,6-difluorobenzoyl)-3-(4-chlorophenyl) urea. J. Agric. Fd. Chem. 23: 359–364.

Mulder, R. and Gijswijt, M. T. 1973. The laboratory evaluation of two promising new insecticides which interfere with cuticle deposition. Pestic. Sci. 4: 737–745.

Neal, J. W. 1974. Alfalfa weevil control with the unique growth disruptor TH 6040 in small plot tests. J. Econ. Ent. 67: 300–301.

Neumann, R. and Guyer, W. 1983. A new chitin synthesis inhibitor CGA 112 913: its biochemical mode of action as compared to diflubenzuron. 10th Int. Congr. Plant Protection, Brighton, pp. 445–451.

Post, L. C., de Jong, B. J. and Vincent, W. R. 1974. 1-(2,6-Disubstituted benzoyl)-3-phenylurea insecticides: inhibitors of chitin synthesis. Pestic. Biochem. Physiol. 4: 473–483.

Rathburn, C. B. and Boike, A. H. 1975. Laboratory and small scale field tests of altosid and Dimilin for the control of *Aedes taeniorhynchus* and *Culex nigripalpus* larvae. Mosq. News 35: 540–546.

Sarasua, M. J. and Santiago-Alvarez, C. 1983. Effect of diflubenzuron on the fecundity of *Ceratitis capitata*. Entomologia Exp. Appl. 33: 223–225.

Sbragia, R. J., Bisabri-Ershadi, B. and Rigterink, R. H. 1983. XRD-473, a new acylurea insecticide effective against *Heliothis*. 10th Int. Congr. Plant Protection, Brighton, pp. 417–424.

Sowa, B. A. and Marks, E. P. 1975. An *in vitro* system for the quantitative measurement of chitin synthesis in the cockroach: inhibition by TH 6040 and polyoxin D. Insect Biochem. 5: 855–859.

Van Eck, W. H. 1979. Mode of action of two benzoylphenyl ureas as inhibitors of chitin synthesis in insects. Insect Biochem. 9: 295–300.

Verloop. A. and Ferrell, C. D. 1977. Benzoylphenyl ureas - A new group of larvicides interfering with chitin deposition. In "Pesticide Chemistry in the 20th Century" (J. R. Plimmer, Ed.), ACS Symp. Ser. No. 37, pp. 237–270, American Chemical Society, Washington, D. C.

Wool, D., Noiman, S., Manheim, D. and Cohen, E. 1982. Malathion resistance in *Tribolium* strains and their hybrids: inheritance patterns and possible enzymatic mechanisms (Coleoptera, Tenebrionidae). Biochem. Genet. 20: 621–636.

Wright, J. E. and Harris, R. L. 1976. Ovicidal activity of Thompson-Hayward TH 6040 in the stable fly and horn fly after surface contact by adults. J. Econ. Ent. 69: 728–730.

8. Environmental Fate and Properties of 1-(4-chlorophenyl)-3-(2,6-difluorobenzoyl)urea (Diflubenzuron, DIMILIN®)

Gary M. Booth, Daniel C. Alder, Milton L. Lee, Melvin W. Carter, Robert C. Whitmore and Robert E. Seegmiller

1.0 INTRODUCTION

Diflubenzuron (DFB) is an insect growth regulating chemical that interferes with cuticle deposition (Mulder and Gijswijt, 1973) by inhibiting the synthesis of cuticular chitin (Deul, DeJong, and Kortenbach, 1978). It is the active ingredient of DIMILIN®* which is used to control such insects as mosquitoes, gypsy moths, cotton and soybean pests, and VIGILANTE®** has potential use in fly control (Wright, 1978). A variety of alternative names for this chemical have been used in the literature. These include: IUPAC, 1-(4-chlorophenyl)-3-(2,6-difluorobenzoyl) urea; CA index, N-[[(4-chlorophenyl)amino]carbonyl]-2,6-difluorobenzamide; Wiswesser line-formula notation, GR DMVMVR BF FF; code names PH 60-40 (generally used outside the United States), TH-6040 (United States), DU 112307 (Philips-Duphar, former research code), ENT-29054 (USDA), and OMS-1804 (WHO).

Clearly, this compound has the potential of being widely used in both the aquatic and terrestrial environment.

For the past 10 years, we have been investigating the environmental fate and toxic effects of DFB under both laboratory and field conditions. The purpose of this paper is to report on the environmental behavior of this chemical in terms of particle size development, fate in water and forest litter, soil adsorption/desorption behavior, metabolism in a simulated activated sludge system, effects on selected non-target organisms, effects on upland game bird reproduction, teratogenic and biochemical effects in chick embryos, and potential perturbations on mammalian testosterone, body weight, and food consumption. Extensive detail on the methods, particularly those of the physiological and metabolic experiments, are

---

* Registered trademark of Duphar B. V. Netherlands.
** Registered trademark of American Cyanamid.

Wright, J. E. and Retnakaran, A. (Eds), Chitin and Benzoylphenyl ureas. ISBN 978-94-010-8638-7.
© *1987, Dr W. Junk Publishers, Dordrecht.*

provided as a resource for future reference since these studies have not appeared previously in the literature. We believe these experiments demonstrate the bioavailability of DFB under field use conditions in some of the most critical parts of our ecosystem.

2.0 MATERIALS AND METHODS

2.1 *Importance of particle size*

Glass microscope slides were dipped in a soil/water slurry and allowed to air-dry. Acetone solutions of DFB and aqueous dispersions of air-milled DIMILIN® W-25 were pipeted onto the soil-coated slides as well as on uncoated glass slides and allowed to dry. Diflubenzuron crystals were then photographed using a compound microscope.

2.2 *Fate in water and forest litter*

A pond in Salt Lake County, Utah, was treated with 3 applications of DIMILIN® W-25 at a rate of 0.25 lbs ai/A using a hand-operated spray applicator. Water samples were taken at pre-treatment, 1 hour, 1 day, 3 days, 7 days, and 14 days post-treatment for each of the 3 applications and analyzed using published high pressure liquid chromatography (HPLC) procedures (Rabenort et al., 1978; Booth and Ferrell, 1977). Recently (Summer 1985), a 500 acre tract of forest land in West Virginia was treated for gypsy moth with one application of DIMILIN® W-25 at a rate of 0.13 lbs ai/A using a helicopter equipped with a spray boom. Forest litter samples were taken at pre-treatment, 1 day, 4 days, 10 days, and 21 days post-treatment. Residue analysis was done using a modified HPLC procedure of Rabenort et al. (1978) where a methylene chloride:hexane (1:1) and methylene chloride:methanol (99:1) column extraction were used in place of acetone and additionally a mixed column of silica gel, neutral alumina, and florisil was used. The latter project is under the direction of West Virginia University, via College of Agriculture and Forestry.

2.3 *Soil adsorption/desorption behavior*

2.3.1 *Chemicals*

Air milled ^{14}C-DIMILIN® (S.A. $= 1.25 \times 10^5$ dpm/μg) was supplied by Thompson-Hayward Chemical Company, Kansas City, KA. The radioactive purity of DIMILIN® was greater than 95% as determined by thin layer chromatography (tlc). Unlabelled DIMILIN® was used with the radioactive sample to make the appropriate concentration in the soil/water matrix.

2.3.2 *Soil types*

Four agricultural soils and four sediments from selected mosquito habitats were used in the study. The physical-chemical properties of the eight soil types are shown in Table 1.

2.3.3 *Experimental design*

To determine the adsorptive properties of DIMILIN®, a total of 50 mL of distilled water was added to 25 g of each soil type in separate 200 mL erlenmeyer flasks and shaken for 10 minutes on a Gyratory Shaker held at 300 rpm.

After 10 minutes, each flask was dosed with 1 uCi of ^{14}C-DIMILIN® plus an appropriate amount of unlabelled DIMILIN® to obtain approximately 0.5 ppm or 2.0 ppm. Agitation on the shaker was resumed for 60 minutes taking 3 mL samples of the soil/water matrix at 5, 10, 20, and 60 minutes. The 3 mL samples of the soil/water were then centrifuged at 2000 rpm for 5 minutes and then 1 mL of the supernatant water was drawn off and counted in 5 mL of Aquasol® to determine the unadsorbed DIMILIN®.

To determine the desorptive properties of the DIMILIN®, the soil/water matrix was suction-filtered through a buchner funnel and the filtered soil was placed into a 200 mL erlenmeyer flask. A total of 50 mL of fresh distilled water was added to the soil and mixed on a Gyratory Shaker for 60 minutes while taking 3 mL aliquots of soil/water at 5, 10, 20, and 60 minute intervals. The 3 mL soil/water aliquots were then centrifuged at 2000 rpm and 1 mL of the supernatant was pipetted into 5 mL of Aquasol® and counted as described above. All treatments were done in triplicate for each soil.

In both the adsorptive and desorptive experiments, the total radiocarbon in the soil was determined to assure that the ^{14}C inventory was accurate. The radiocarbon in the soil was determined by combusting aliquots of the soil in a Harvey Biological Material Oxidizer (BMO), trapping the ^{14}CO$_2$ in 10 mL of Oxyfluor®, and counting in a Delta 300 Liquid Spectrometer.

2.4 *Metabolism by a simulated activated sludge system*

2.4.1 *Chemicals*

Air-milled UL ^{14}C-DFB (Lot No. 1057-028, S.A. = 1.71×10^5 dpm/μg), UL ^{14}C-4-chlorophenylurea (CPU, Lot No. 46692-1, S.A. =

Table 1. Summary of physical/chemical properties of soil-types used in adsorption/desorption experiments.

Sample	pH	1/3 atm % H$_2$O	15 atm % H$_2$O	% H$_2$O available	% O.M.[a]	cec[b]	% sand	% clay	% silt	Texture class
Provo River	8.11	16.37	9.30	7.07	4.15	29.65	56.0	12.0	32.0	Sandy loam
Jordon River	7.75	19.02	7.74	11.28	0.97	24.76	51.0	22.0	27.0	Sandy clay loam
Golf course	8.12	22.01	10.01	12.00	1.32	28.71	55.0	20.0	25.0	Sandy loam
Utah Lake	8.48	40.12	22.77	17.35	4.80	38.70	19.0	42.0	39.0	Silty clay loam
Rice Clay	5.80	42.70	24.10	18.60	2.55	41.80	14.90	66.1	19.0	Clay
Troup sand	6.10	5.70	1.12	4.60	0.56	5.70	84.40	3.6	12.0	Loamy sand
Olivier Silt Loam	8.40	26.86	4.98	—	0.81	4.94	14.70	10.8	74.5	Silt loam
Commerce Loam	5.50	25.82	11.33	—	1.95	9.73	49.40	24.8	25.8	Sandy clay loam

[a] % O.M. = % organic matter.
[b] cec = cation exchange capacity.

1.13 × 10⁴ dpm/μg), technical grade DFB (Lot No. PP416) and unlabelled standards of DFB, CPU, 4-chloroaniline (CA) 2,6-difluorobenzoic acid (DFBA), and 2,6-difluorobenzamide (DFBAM), were obtained from Thompson-Hayward Chemical Company, Kansas City, KA. All of the standards were ⩾ 94% pure. The organic solvents used in the study were reagent or nanograde purity.

2.4.2 SEP-PAK cartridges

SEP-PAK C_{18} cartridges were obtained from Waters Associates, Inc., Milford, MA. These disposable cartridges were used to extract radiocarbon residues from the aqueous phase of the simulated activated sludge system.

2.4.3 Scintillation equipment

The total residues in the bacterial floc (bacteria + waste products) were determined by combusting appropriate samples in a BMO using 10 mL of Oxyfluor® to trap the $^{14}CO_2$. Total $^{14}CO_2$ evolved from each flask as a natural product of metabolism was also trapped in Oxyfluor®. Organic extracts were counted in 10 mL of Aquasol® with a Delta 300 or an Isocap 300 Liquid Scintillation System.

2.4.4 Experimental design

The activated sludge used in this metabolism study was taken from a previous study. A copy of the exact protocol used in generating the activated sludge is available from the authors. It should be noted that only activated sludge from control tanks and periodic (or shock) test tanks were used in the present study. Use of the term periodic or shock test refers to the fact that the test system was exposed to a DFB-treated synthetic sewage mixture for a 5 hour period; after which all periodic test system influent reservoirs were emptied, cleaned, and refilled with DFB-free synthetic sewage for each remaining feeding of the respective dosing period. This treatment-sequence was repeated every 3 days through 17 days with simultaneous increases in the dosage every 3 days. The dosages ranged from 1.0 mg/L on day 1 to 100 mg/L on day 17.

On day 19 in a previous study, activated sludge from the control (non-treated) and shock-dose tanks were removed, washed, and then used in the metabolism study. The activated sludge was washed by taking 250 mL aliquots of the sludge, centrifuging at 2500 rpm for 10 minutes, removing the supernatant, resuspending the floc in enough synthetic sewage to make 100 mL, and recentrifuging to remove the supernatant. This washing procedure was then repeated two more times. After removing the supernatant from the third rinse, the floc was resuspended in enough synthetic sewage to make 100 mL. This washed fraction was then used in the study.

The original non-treated sludge samples are hereafter referred to as "unacclimated" samples while the shock-dose samples are referred to as "acclimated" samples meaning that the sludge from the previous study was "acclimated" (or exposed) to increasing shock-doses of DFB. The settleability of the "unacclimated" sludge was 21 mL/100 mL and for the "acclimated" sludge it was was 22 mL/100 mL. The total suspended solids (TSS) for the "unacclimated" and "acclimated" samples were determined to be 1615 mg/L and 1869 mg/L, respectively. A total of 20 mL of the "unacclimated" activated sludge was placed into each of twenty 125 mL suction flasks and 20 mL of the "acclimated" activated sludge was placed into each of twenty 125 mL suction flasks. Of the twenty "unacclimated" samples, 8 were spiked with 12.89 μg of ^{14}C-DFB plus 7.11 μg of unlabelled DFB for a final concentration of 1 ppm and a revised S.A. = 1.10 × 10^5 dpm/ug. Eight of the "unacclimated" samples were spiked with 12.89 ug of ^{14}C-diflubenzuron and 1987.11 ug of unlabelled DFB for a final concentration of 100 ppm and a revised S.A. = 1.10 × 10^3 dpm/ug. Each of the remaining four "unacclimated" samples were spiked with only acetone (254 ul) and served as controls. This procedure was repeated for the twenty "acclimated" samples.

The top of each flask was covered with a rubber septum while a 9 cm piece of rubber tubing with a clamp was fitted on the side arm. Two replicate DFB-treated flasks and 1 control (untreated) flask were analyzed at 1, 4, 24, and 48 hours post-treatment for both the "acclimated" and "unacclimated" samples. The parameters that were measured included total residues (TR), total extractable residues (TER), and total bound residues (TBR). Each flask was processed in an identical manner at each sampling period. To trap the ^{14}CO$_2$, a hypodermic syringe, backed with a stream of nitrogen, was inserted through the rubber septum while simultaneously releasing the clamp and the nitrogen was bubbled through ca. 80 mL of Oxyfluor®. Replicate 10 mL samples were then counted in a liquid scintillation counter (LSC) and the total radiocarbon determined. The activated sludge from each flask was centrifuged at 2500 rpm for 10 minutes to separate the floc from the aqueous portion. The average aqueous volume from the forty flasks was 42 ± 12 mL while the average wet-weight of the floc samples was 1.55 ± 0.16 g. Variation in aqueous volumes was due to rinse-water added in the rinsing of each 125 mL flask. The TR of the aqueous sample was determined by counting replicate 1 mL samples in 10 mL of Aquasol® in a LSC. A portion of the floc sample was combusted in a BMO and counted in 10 mL of Oxyfluor® in a LSC. A portion of each remaining aqueous sample was extracted with C$_{18}$ SEP-PAK cartridges. Following activation of the cartridges and the addition of the sample (ca. 100,000 dpm), the sequence of volume rinses included 20 mL of acetone, 10 mL of hexane and 10 mL of MeOH. Each fraction was counted for TR. Since the acetone fraction contained the majority of radiocarbon, this extract was analyzed for metabolites by tlc, auto-radiography, and mass spectrometry (MS).

The remaining floc in each sample was extracted by the following procedure:

1. The floc sample was placed in an Omni-mixer canister along with 40 mL of acentonitrile which filled the canister ca. half-full.

2. The floc-acentonitrile matrix was homogenized for 1 minute at ca. 75% capacity of the Omni-mixer.

3. The homogenate was poured into a 500 mL round-bottom flask. The canister and blades were rinsed 4 × with acetonitrile which was added to the 500 mL flask. Additional acetonitrile was added to the 500 mL flask to bring the volume to 150 mL.

4. A total of 15 mL of distilled water was also added to the 500 mL flask and mixed with the floc-acetonitrile matrix.

5. The mixture was then gently refluxed for 30 minutes. Refluxing was judged to begin when the mixture was boiling and a condensation ring had formed on the reflux condensor. When refluxing was completed, the mixture was allowed to cool and the reflux condensor was rinsed with acetonitrile which was added to the total volume.

6. The mixture was then filtered via aspiration through a Buchner Funnel with Whatman No. 4 filter paper. The filter paper and cake were rinsed with additional acetonitrile which was added to the filtrate.

7. The filtrate was evaporated to dryness in a Roto-vap at 47°C. The residue was then dissolved in 20 mL of acetone and the TR was quantified by LSC and the metabolite distribution was determined using two-dimensional tlc and confirmation of the residues with MS.

8. The TR of the acetonitrile-water fraction, which was evaporated off, was also determined by LSC.

9. The original vessel that contained the floc sample and the canister and whipping-blades were rinsed 2 × with 5 mL portions of acetone and these were pooled and the TR determined.

10. The filter paper and filter cake were air-dried and combusted to determine the TBR.

2.4.5 Analytical methods

2.4.5.1 *Thin layer chromatography and autoradiography*. — The floc and aqueous extracts were spotted on EM silica gel 60 F-254 (0.25 mm gel thickness) aluminum-backing tlc plates and eluted first in benzene:ethanol:acetic acid (93:7:2 v/v) and second in benzene:ethanol (9:1 v/v). The elution time was ca. 1.75 hours in each direction. Kodak X-OMAT film was used in making autoradiograms from the extracts on the tlc plates. Following the development of the autoradiograms, the silica gel containing the radiocarbon (usually 5 spots) was scraped from the plates and counted in 10 mL of Aquasol®. The autoradiogram-spots were labeled 00 (origin), 01 (CPU), 02 (DFBAM), 03 (CA), 04 (DFB), and 12 (DFBA). Two other unknown compounds (11 and 21) were occasionally found but they were generally only a few percent of the total extracted.

2.4.5.2 *Mass spectrometry*. — Mass spectrometry was used on selected samples to confirm the identifications tentatively assigned from the R_f values of the separated compounds on the tlc plates. Mass spectra of the extracted spots were made and compared to the spectra obtained from pure samples of DFB and other probable degradation products.

The mass spectra were obtained on a Hewlett-Packard 5982 mass spectrometer coupled to a 5934A data system.

The samples were introduced via the direct inlet probe (DIP) which was then heated from ambient to ca. 200°C to vaporize the sample. The mass spectrometer was operated in the electron impact (EI) mode at 70 ev and the ionization source pressure was less than 1×10^{-6} torr. The mass spectra of standards of DFB and four possible degradation products are shown in Figures 13A–F. Figure 13F shows the spectrum of CA taken from the EPA/NIH mass spectral data base (Heller and Milne, 1978) which is also shown in Figure 13E. A comparison of the reference mass spectrum to the mass spectrum obtained for this study confirms the reliability of the mass spectra. Although the fragmentation patterns of these compounds are relatively simple, each is distinctively unique. For increased sensitivity and elimination of impurity ions, selected ion monitoring (SIM) was used for quantitation at trace levels. In each case, the appropriate masses for the compound to be verified were scanned.

2.4.5.3 *Statistical methods*. — The general linear model and the Rummage Program (Bryce, 1980) were used for the analysis.

The codes used in the analyses included the following:

Total residue analysis:
1. C21 = Log_{10} CO_2 radiocarbon (ug)
2. C22 = Log_{10} FLOC radiocarbon (ug)
3. C23 = Log_{10} Aqueous radiocarbon (ug)

Metabolism analysis:
1. C25 = Log_{10} FLOC CPU radiocarbon (%)
2. C26 = Log_{10} FLOC CA radiocarbon (%)
3. C27 = Log_{10} FLOC DFB radiocarbon (%)
4. C28 = Log_{10} Aqueous CPU radiocarbon (%)
5. C30 = Log_{10} Aqueous DFB radiocarbon (%)

The mathematical model used was as follows:

$$Y_{ijkl} = u + D_i + S_j + DS_{ij} + T_k + DT_{ik} + ST_{jk} + DST_{ijk} + E_{ijkl}$$

where:

Y = dependent variable (TR CO_2, TR FLOC, . . . etc.),

u = overall mean,

D_i = effect of the ith dose (i = 1 (1 ppm), 2 (100 ppm)),

S_j = effect of the jth sludge condition (j = 1 (unacclimated), 2 (acclimated)),

DS_{ij} = interaction effect of the ith dose with the jth sludge condition,

T_k = effect of the kth time (k = 1 (1 hour), 2 (4 hours), 3 (24 hours), 4 (48 hours)),

DT_{ik} = interaction effect of the ith dose with the kth time,

ST_{jk} = interaction effect of the jth sludge condition with the kth time,

DST_{ijk} = interaction effect of the ith dose with the jth sludge condition, with the kth time,

E_{ijkl} = residual of the lth replicate in the ijk^{th} dose-sludge condition-time combination.

Analysis of variance (ANOV) was performed for the above model with accompanying F-tests for each effect and interaction and also including individual contrasts, means, and plots of means. The contrasts for dose, sludge-condition, and time respectively were 1 ppm with 100 ppm, "unacclimated" with "acclimated," and 1 hour with each of the other time periods. For the interaction effect, each of the respective contrast interactions were obtained with the analysis.

The ANOV was performed mainly to obtain comparisons of the "unacclimated" and "acclimated" sludge conditions and the interaction of these conditions with dose and time.

2.5 Non-target toxicity studies

2.5.1 Segmented worms and midges

A total of 8 applications of air-milled DIMILIN® W-25 was aerially (fixed-wing aircraft) applied at a rate of 0.13 lbs ai/A into Provo Bay. Additionally, DIMILIN® 1-G was also applied 8 × in a similar manner at a rate of 0.12 lbs ai/A into the Bay. The 8 spray periods were 7 June, 29 June, 13 July, 27 July, 10 August, 24 August, 7 September, and 21 September 1976. Segmented worms (Oligochaeta) and midges (Chironomidae) were sampled 24 times from 4 June to 27 August with standard benthic sampling methods to determine population trends.

2.5.2 *Fungal growth and CO_2 evolution*

2.5.2.1 *Chemicals.* — Air-milled DIMILIN® W-25 was used throughout these studies. In addition, the inert ingredient used in the commercial preparation of DIMILIN® was also studied. Both of these materials were supplied by Thompson-Hayward Chemical Company, Kansas City, KA.

The concentrations of the active ingredient were: 0, 1, 10, 50, and 100 ppm in most experiments. The inert samples contained 100 ppm.

2.5.2.2 *Fungi.* — The fungi used in the toxicity experiments were pure stains of: 1) *Rhizopus arrhizus*, 2) *Aspergillus niger*, 3) *Fusarium oxysporum*, 4) *Mycorrhizae* — *Rhizopogan vinicolor*, 5) *Pythium debaryanum*, and 6) *Trichoderma viride*.

Germination experiments were not done on *Pythium* or *Rhizopogan* since these organisms did not form spores under normal growth conditions.

2.5.2.3 *Mycelial growth.* — Fresh PDA fungal media was prepared and added to flasks containing an appropriate amount of DIMILIN® w-25 or inert ingredient to yield dilutions of 100 ppm, 50 ppm, 10 ppm, 1 ppm, 0 ppm DFB and also replicate flasks of 100 ppm inert ingredient. The media + DIMILIN® W-25 was autoclaved and then poured into sterile petri dish plates. When the agar hardened, a given fungus was aseptically transferred using a sterile cork borer from a pure culture to the middle of the appropriate plates. All of the plates were run in triplicate. Mycelial growth outward from the center of the plate was recorded at various time intervals depending on the type of fungus.

Additional toxicity data was collected using conventional DAM tubes, but the data showed identical results to the petri dish technique.

2.5.2.4 *Germination of spores.* — A plate of given fungi was grown to maximum growth. Approximately 25 mL of an appropriate media (glucose-phosphate) was poured onto the plate. A sterile glass rod was used to gently stir the mycelia to break off the sporangia from the sporangiophore. The solution was then poured through 8 layers of sterile cheese cloth into a sterile collecting flask. A sterile pasteur pipette was used to withdraw a sample of spores to determine the optical density. The spore solution was adjusted such that a given number of drops per volume used would give an adequate number of spores per liquid media.

All of the treatments were studied in triplicate and the percent germination was recorded at specified time intervals. The flasks were all kept at 25°C on shakers in a walk-in growth chamber.

2.5.2.5 *Mycelial weight.* — A constant concentration of spores from the germination solutions were inoculated into sterile flasks containing the appropriate media + DIMILIN® W-25. The exact media differed somewhat for each organism used. All flasks were incubated on shakers. After the organisms had grown to confluency ($\simeq$ 1 week), the solutions were filtered with sterile Whatman No. 1 filters. The mycelial growth was collected on the filter paper, wrapped in aluminum foil, lyopholized and weighed.

2.5.2.6 *CO$_2$ evolution from Rhizopus arrhizus*. — A spore solution of *Rhizopus* was inoculated on a sterile media plate and allowed 8–10 hours to germinate. A small volume (10 mL) of the spore solution was injected into the treatment plates and then placed into a "startrek" CO_2 monitoring apparatus to determine possible changes or inhibition in respiratory metabolism. The CO_2 was monitored for a total of 162 hours.

2.6 Effect on bobwhite quail reproduction

2.6.1 Experimental design

A total of 864 bobwhite quail was obtained from a commercial breeding farm and acclimatized for 44 days prior to the initiation of the study.

The birds were housed in 9 GQF battery breeding pens designed for bobwhite quail. The quail were housed 4 per cage (2 male, 2 female). Each cage was 30.4 × 60.9 cm. The cages were arranged 6 in a row with 4 rows per battery unit. Each battery unit thus consisted of 24 cages. Three battery units were placed into each of 3 separate rooms.

The treatments used in this dietary feeding study were Control (C), Control + Corn Oil (C$_0$), 2.5 ppm (T$_1$), 25 ppm (T$_2$), and 250 ppm (T$_3$). Pure (99.9%) air-milled DFB was used in the game bird feed. Because the treatments were of necessity assigned to a cage of 4 birds, the cage was the experimental as well as the observational unit. The 5 treatments were randomly assigned to the cages. The procedure described below was used for each of the 36 rows (9 battery units × 4 rows) but an independent randomization was performed for each row.

Because there were 5 treatments and 6 cages per row, one treatment was replicated (R) in each row. The replicated treatment was selected at random for each row. A random permutation of the first 6 numbers was thus made. The 5 treatments plus R were randomly assigned to one of the first 6 digits and this determined the assignment of the treatments to the cage.

For example, in battery 1 the replicate treatment was 2 or control + corn oil (Co). The random assignment of treatments to digits was 1 = C, 2 = Co, 3 = T$_3$, 4 = T$_1$, 5 = T$_2$, and 6 = R. The random permutation of 6 digits was 1, 3, 6, 5, 4, and 2. The assignment of treatments to the top layer in rank from left to right was thus: C, T_3, Co, T_2, T_1, and Co. This was repeated 35 more times.

The statistical model for the study was as follows:

$$
\begin{aligned}
Y_{ijklm} = {} & \mu + R_i = P_j + RP_{ij} + L_k + RL_{ik} + PL_{ji} \\
& + RPL_{ijk} + T_l + RT_{il} + PT_{jl} + RPT_{ijl} + LT_{kl} \\
& + RLT_{ikl} + PLT_{jkl} + RPLT_{ijkl} + C_{(ijkl)m} + W_n \\
& + RW_{in} + PW_{jn} + RPW_{ijn} + LW_{wn} + RLW_{ikn} \\
& + PLW_{jkn} + RPLW_{ijkn} + TW_{ln} + RTW_{iln} \\
& + PTW_{jen} + RPTW_{ijln} + LTW_{kln} + RLTW_{ikln} \\
& + PLTW_{jkln} + RPLTW_{ijkln} + CW_{(ijkl)mn}
\end{aligned}
$$

R_i = effect of ith room $i = 1, 2, 3$
P_j = effect of jth position $j = 1, 2, 3$ (Battery Unit)
L_k = effect of kth layer $k = 1, 2, 3, 4$ (Level or row)
T_l = effect of lth treatment $l = 1, .., 5$
$C_{(ijkl)m}$ = effect of mth cage in $ijkl^{th}$ cell $m = 1, 2$ (for one T per rack)
W_n = effect of nth week $n = 1, \ldots, 12$

The above model was appropriate for all reproductive parameters studied except for the eggshell thickness data. A revised model is described in the Results and Discussion for that parameter.

The rooms were kept at a constant temperature of 74°F, well ventilated and free of major disturbances. Rooms were lighted with Sylvania GroLux Fluorescent lamps, model #F96T12/GRO/HO giving a maximum wattage of 59 foot candles at any one level. The birds were placed on 17 hours light: 7 hours dark before and during the study to maximize egg production. The birds were fed and watered daily. The drop pans and water trays were cleaned bi-weekly, and the only other disturbances were a weekly weighing of feed and a daily gathering of eggs.

The total length of the study was 12 weeks from May 1, 1977 to July 23, 1977.

2.6.2 *Egg care*

The eggs were gathered daily, and identification of code treatment, cage and date was placed on the egg with a pencil. The eggs were placed immediately in a humidified cooler at a temperature of 55°F and a humidity of 75%. They were stored under these conditions for one to ten days, then placed weekly into a commercial (Humidaire Model 300A) circulation air-incubator.

Before the incubation period, all the eggs were examined for cracks and all uncracked eggs were then placed into the incubator. The incubator was kept at 99.5°F and a wet bulb temperature of 70–75°F. (This was in compliance with the manufacturer's guidelines for best results.)

The eggs were candled at 6 days and at $18\frac{1}{2}$ days. Any eggs which had not developed into a full embryo within the $18\frac{1}{2}$ days were considered infertile. During the $18\frac{1}{2}$ days the eggs were automatically rotated 90° every hour. At the conclusion of $18\frac{1}{2}$ days the fertile eggs were then placed by treatment into separate levels of a Huimidaire Model 300A hatcher. The temperature of the hatcher was 99.5°F and a wet bulb temperature of 75–80°F.

Between 21–22 days after beginning of the incubation period, the eggs hatched. After a 24-hour drying period, the new chicks were placed in separate brooder levels and observed for 14 days. The new chicks received a diet of Purina "Startena", a wild game bird feed, mixed with suggested amounts of "Liv", a high protein supplement for game birds. The brooders were inspected and cleared of all dead chicks and debris and the water replaced daily. The water was supplemented with a multiple vitamin, "Headstart," specially made for chicks and adult birds. At the end of the 14 days the birds were inspected for abnormalities.

All of the equipment, brooders, hatchers, etc. were cleaned with a disinfectant called "Hi-Lethal" and then rinsed well. In addition, with every new group put into the incubator, 1 mL of formaldehyde was placed inside the incubator on a towel; also 5 mL of formaldahyde was placed in the bottom of the hatcher with every new hatch.

The embryonated eggs that did not completely hatch were considered "no-hatch." The infertile and "no-hatch" shells were used for egg shell thickness examination and measurement.

2.6.3 *Feeding of the birds*

The birds were fed by using a $\frac{1}{2}$ cup plastic scoop and by carefully leveling the cup with a straight knife. This gave an accurate measurement of $\pm$ 2.0 grams. We calculated the amount of feed consumed per bird by weighing the feed at the beginning of each week, adding to this figure the amount of feed given for the week, minus the excess feed left in the feed tray at the end of the week. By dividing this amount by the number of birds in the cage we could get an accuracy of $\pm$ 3.5 grams. However, this did not include spillage or wastage. The feed level was kept less than half full, which helped prevent wastage and yet gave plenty of feed for the birds.

To prevent the birds in adjacent cages from eating other treatments by reaching into the adjoining cage feed tray, we placed wooden blocks into the feed tray with 3″ × 5″ cards stapled to the blocks.

Each feed tray was labeled with a colored tape and coded for a specific feed. All the feed canisters were similarly coded along with the plastic scoops used. This prevented any mistakes or contamination of one treatment with another.

2.6.4 *Feed preparation*

To prepare the different treatment levels of DFB in the normal feed, specific amounts of the chemical were dissolved in corn oil. The mixing was done with a commercial Hobart mixer with a capacity of 16 pounds of feed.

To prepare the 2.5 ppm, 25 ppm and 250 ppm, the following relationships were used:

2.5 ppm: 0.114796 g DFB, 906.9 g corn oil, 45 kg feed

25 ppm: 1.14796 g DFB, 906.9 g corn oil, 45 kg feed

250 ppm: 11.4796 g DFB, 906.9 g corn oil, 45 kg feed

The corn oil control was prepared by putting the same portion of the corn oil with the feed but without DFB. The other control feed was untreated "Layena."

The mixing of the feed was done on a bi-weekly schedule. The feed was weighed to an accuracy of $\pm$ 1 pound. The corn oil was measured with an accuracy of $\pm$ 0.25 grams and the DFB to an accuracy of $\pm$ 0.0005 grams. The mixed feeds were stored in separate air-tight, 20-gallon metal canisters.

2.6.5 *Egg shell thickness*

Eggs from each of the five treatments were randomly selected from the no-hatch and non-fertile eggs during each of the 12 weeks of the study. A total of 508 eggs were measured at three different positions on the egg for shell thickness. There was an average of 8.47 eggs per treatment per week with a minimum of 5 and a maximum of 15. The shell thicknesses were measured with a micrometer according to published procedures (Longcore et al., 1971).

2.6.6 *Parameters studied*

The following reproduction data were recorded for every one-week set of eggs:
1. total eggs laid
2. # cracked eggs
3. # eggs set
4. # fertile eggs
5. # hatched eggs
6. egg shell thicknesses

7.   feed consumption (g/bird/week)
8.   # deaths (Adults)
9.   # 14-day-old chicks surviving
10. # 14-day-old chicks surviving per unit time (day, week, and season).
11. % cracked
12. % fertile
13. % hatch
14. % 14-day-old chicks surviving
15. % 14-day-old chick survival/hen

2.6.7 *Statistical analysis*

The data were analyzed by general least squares procedures. This was accomplished by use of the RUMMAGE Program. A basic two-way analysis of variance was run on each of the following dependent variables: eggs laid, eggs cracked, eggs set, eggs fertile, eggs hatched, egg shell thickness, and feed consumed. The egg data were recorded as eggs/cage/week and the feed data as grams/bird/week.

2.7 *Teratogenic and biochemical effects on chick embryos*

2.7.1 *Teratology*

Chicken eggs from Lake Shore Egg Farm in Spanish Fork, Utah, were divided into three groups and incubated at 39°C and 30% relative humidity. Group-one eggs were injected with a solution containing 10 mg DFB (99.4%, air-milled) in 0.1 mL of peanut oil. This dispersion was sterilized in a steam bath for 20 minutes prior to injection. Group-two eggs were injected with 0.1 mL of peanut oil without DFB; group-three eggs were left uninjected. The first two groups of eggs were further divided into groups that were injected on days 1, 2, 3, 4, and 6 of incubation.

Injections were made with a tuberculin syringe passed through an opening in the shell in the region of the air sac. Solutions were injected near the embryonic coelom within the sinus terminalis in eggs in which these structures could be identified (days 3, 4, 5, and 6) or near the blastoderm in those eggs in which the sinus terminalis could not be identified (days 1 and 2). The injected material was placed as close as possible to blood vessels without rupturing them. After injecting, the openings in each egg were covered with tape and the eggs were returned to the incubator.

After ten days incubation, all eggs were recovered and the embryos were killed. Abnormalities and embryonic deaths were tabulated and the embryos were preserved in Carnoy's fixative (60% ethanol, 30% chloroform, 10% acetic acid).

2.7.2 *Chondroitin sulfate and hyaluronic acid methods*

Chicken eggs were injected with 10 mg of DFB (> 99.4% pure, air-milled) in 0.1 mL of peanut oil at 48, 24, and 2 hours prior to recovery of eight-day limb cartilage. Control eggs were injected with 0.1 mL of peanut oil prior to recovery.

For the *in vitro* data, cartilage was recovered and labeled at 38°C for 2 hours in Waymouth's tissue-culture medium containing 5 uci/mL $Na_2^{35}SO_4$ or [^{3}H]-glucosamine. Labeling was followed by the cartilage with four rinses of 1 mL saline solution. After desiccation in acetone and under-vaccum, individual tibias (cartilage) were weighed to the nearest ug with a Cahn M-10 electrobalance. The cartilages were pooled in groups of four tibias and homogenized at 4°C in 1 mL of distilled water. The homogenate was then sonicated for 15 seconds with a Biosonic IV sonicator (VWR Scientific) at the low setting.

The labeled sonicate was used as follows: An assay for glucuronic acid was performed on 0.5 mL of the sonicate (Bitter and Muir, 1962); 50 uL of sonicate was counted in a scintillation counter; 50 uL was precipitated with cetylpyridinium chloride (CPC) prior to counting the precipitate with the scintillation counter; and 50 uL was treated with chondroitinase, CPC precipitated, and then counted. Toluene containing PPO and POPOP was used as the scintillation fluid.

The *in ovo* experimental protocol was basically the same except that the radiolabeled [^{35}S] and [^{3}H]-glucosamine were injected into "whole" embryos three hours before excising the tibias.

2.7.3 *CPC treatment*

Samples of the sonicate (50 uL) were placed on Whatman 3 MM filter paper (2.3 cm diameter) and allowed to dry. The filter papers were placed in a beaker containing 50 mL of 1% CPC solution at room temperature. The beaker was agitated 60 minutes during which time one change in solution was made. The filter papers were then rinsed twice with 95% ethanol followed by one rinse with ethyl ether. Filters were then dried, placed in separate scintillation vials, and counted.

Enzyme treated homogenates were similarly treated with CPC to determine how much of the CPC precipitable fraction was enzyme degradable.

2.7.4 *Chondroitinase treatment*

Cartilage homogenates labeled with $Na_2^{35}SO_4$ and [^{3}H]-glucosamine were treated with Chondroitinase-ABC (Miles Laboratories, Inc.) by adding equal parts of homogenate to the enzyme solution. The enzyme was prepared at a concentration of two units/mL in 0.02 M Tris-HCL buffer, pH 7.2. The enzyme-homogenate solution was incubated for four hours at 38°C. The chondroitinase procedure was that of Yamagata *et al.* (1968).

2.7.5 *Hyaluronidase treatment*

Additional cartilage homogenates labeled with $Na_2^{35}SO_4$ or $[^3H]$-glucosamine were treated with streptomyces hyaluronidase (Calbiochem), an enzyme specific for hyaluronic acid. The enzyme was prepared at a concentration of 0.1 mg/mL in 0.01 M sodium acetate and 0.15 M NaCl buffer, pH 5.7. Equal quantities of homogenate and enzyme solution were incubated at 38°C for three hours. The hyaluronidase procedure was that of Solursh (1975).

2.7.6 *Glucuronic acid assay procedure*

A total of 3 mL of 0.025 M sodium tetraborate in concentrated sulfuric acid was placed in tubes and cooled to 4°C. The cartilage homogenate (0.5 mL) or uronic acid standard was layered onto the acid. The tubes were close with teflon lined caps and agitated, at first gently, then vigorously. The tubes were then heated for ten minutes in a boiling water bath and allowed to cool to room temperature. Carbazole reagent (0.125%) in absolute ethanol was then added to each tube (0.1 mL per tube). The tubes were agitated as before and heated in a boiling water bath for an additional 15 minutes, and cooled to room temperature. Optical density was read at 530 nm in a 1 cm cell.

2.7.7 *Embryonic transfer of DFB*

Eggs for this experiment were obtained from the same source, incubated, prepared for inspection and injected in the same way as were the eggs for the nonradioactive technical DIMILIN® teratological study.

Eggs were then divided into three groups: those to be injected with 10 mg of DFB in 0.1 mL peanut oil plus 2–3 uL ^{14}C-DFB (S.A. = 17.42 mCi/mMole), those to be injected with 0.1 mL of 0.9% saline plus ^{14}C-DFB, and those to be uninjected but recovered, sonicated, and extracted in the same way as the eggs from the other two groups. All eggs were injected at four days of incubation and were recovered six to fourteen days after incubation.

Eggs were recovered and the shells, yolks, and embryos were separated. In some eggs, the supporting membranes and egg white were also separated from shell, yolk and embryo.

Recovered embryos were weighed and washed five times in one mL of Locke's solution (9 mL total wash). All five washes were combined and counted in Aquasol® for radioactivity. The embryos were then sonicated for 15 seconds in 1–2 mL methanol. A 50 uL aliquot of the sonicate was counted for radioactivity. The remaining sonicate was centrifuged for 15 minutes at 60,000 RPM's and a 50 uL aliquot of the supernatant was counted for radioactivity.

Total volume of yolk was measured and then the yolk was mixed by gentle sonication until homogeneous. After sonication, a 50 uL aliquot of yolk was counted for radioactivity. The remaining yolk was extracted three times with methanol, the three extracts were combined and 1 mL of combined extract was counted for radioactivity. The remaining extract was concentrated to a few mL. By evaporating most of the methanol in a Rotovap. The extract was concentrated to less than one mL by passing nitrogen gas over it. This highly concentrated yolk extract was chromatographed with standards of DFB, CPU, DFBA, and CA with glass-backed chromatography plates (silica gel, EM, 0.25 mm).

Recovered membranes and shells were weighed and 0.15–0.3 g of membrane or shell were heated (60°C) in 1 mL of Protosol® and 3 drops of water. After heating for 24 hours, shells and membranes still in Protosol® and water were treated with 10% benzoyl peroxide, heated again at 60°C for 1 hour and placed in the dark for eight hours or longer. After this "dark" treatment, the shells and membranes, were placed in 10 mL of Aquasol® and couted for radioactivity.

2.8 Investigations on rat serum testosterone, weight, and food consumption

2.8.1 Rats

A total of 231 male Sprague-Dawley rats approximately 23 days old and weighing 50–60 g were obtained from Gibco Animal Resources Lab, P.O. Box 4220, Madison, WI. Each rat was weighed and randomly assigned as a day-0 control animal or as a treatment animal and placed in separate cages.

2.8.2 Chemical

DFB ($\geq 98\%$ pure air-milled) was obtained from Thompson-Hayward Chemical Co., Kansas City, KS.

2.8.3 Experimental design

DFB was mixed in corn oil and added to pulverized rat-block at concentrations of 0, 75, 150, 300, and 3000 ppm based on the weight of the food. The treated food was placed in earthen jars and the total food consumption was computed weekly in order to estimate the amount of DFB consumed. All animals were housed at the Bringham Young University Animal Facility with a light regimen of 12 h light, 12 h dark and a temperature of 22°C. The food and water were supplied ad libitum. On the day of sacrifice, each of the 2 animals per cage were weighed and total food consumption tallied. Since each cage was both the experimental and observational unit, the serum from each of the two rats was pooled and testosterone analysis was completed in duplicate for each sample.

2.8.4 *Radioimmunoassay (RIA) procedure*

The antigen, antibody, additional reagents, and general procedures for RIA testosterone analysis were, according to the protocol, supplied by Radioassay Systems Laboratories, Inc., Carson, CA.

2.8.5 *Statistical analysis*

The statistical model used in this study was:

$$Y_{ij} = \mu + T_i + D_j + TD_{ij} + \varepsilon_{ij}$$

where:

μ = overall mean
T_i = effects of the ith treatment; $i = 1, 2, 3, 4, 5$
D_j = effect of the jth day; $j = 1, 2, 3, 4, 5$
TD_{ij} = effect of the interaction of the ith treatment with the jth day of sacrifice
ε_{ij} = experimental error for the ij^{th} experimental unit

An analysis of variance (ANOV), covariance, a square-root transformation of the ANOV, and a log transformation of the ANOV data were performed on all of the data using the RUMMAGE Program.

3.0 RESULTS AND DISCUSSION

3.1 *Importance of particle size*

Earlier work has shown that the size of DFB particles can influence the rate of soil degradation (Verloop et al., 1975, Verloop and Ferrell, 1976) as well as insect toxicity (Mulder and Gijswijt, 1973). In general, when the mean particle size is 10 microns or larger, the soil halflife of this compound increases up to $16 \times$. Acetone solutions of DFB clearly cause remarkable crystal growth (Figure 1B) while air-milled aqueous-applied aliquots of DIMILIN® W-25 on soil slides shows the particle size remains in the 1–5 micron size (Figure 1A). Investigators should be cautious and use only air-milled samples of DFB in environmental fate studies especially if they are using complex matrices such as soil. Note in Figure 1B (arrow) that the soil particles seemed to be "centers" for crystal growth. Reports on DFB showing lack of soil degradation (Metcalf et al., 1975) most likely is related to crystal growth associated with applications of acetonic solutions to soil.

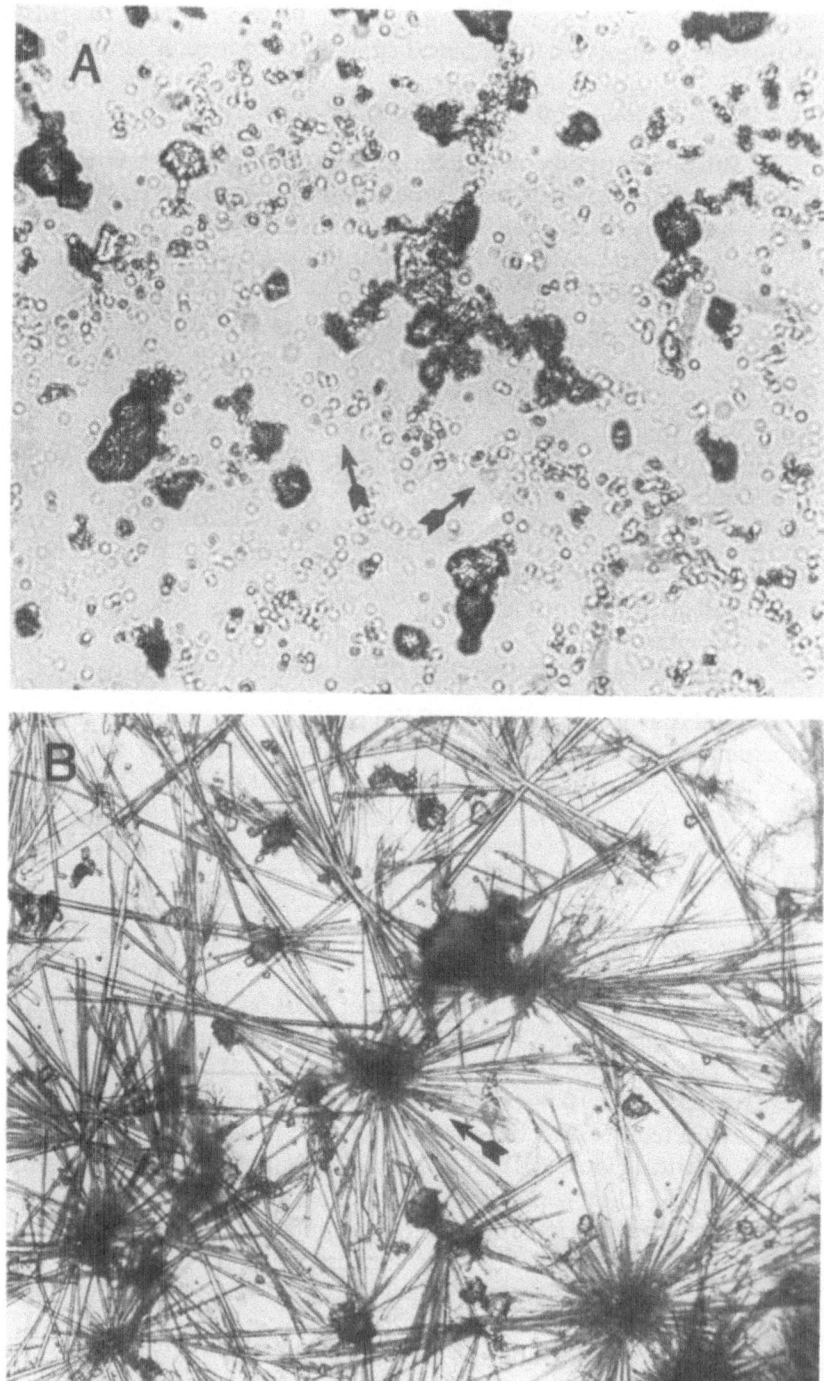

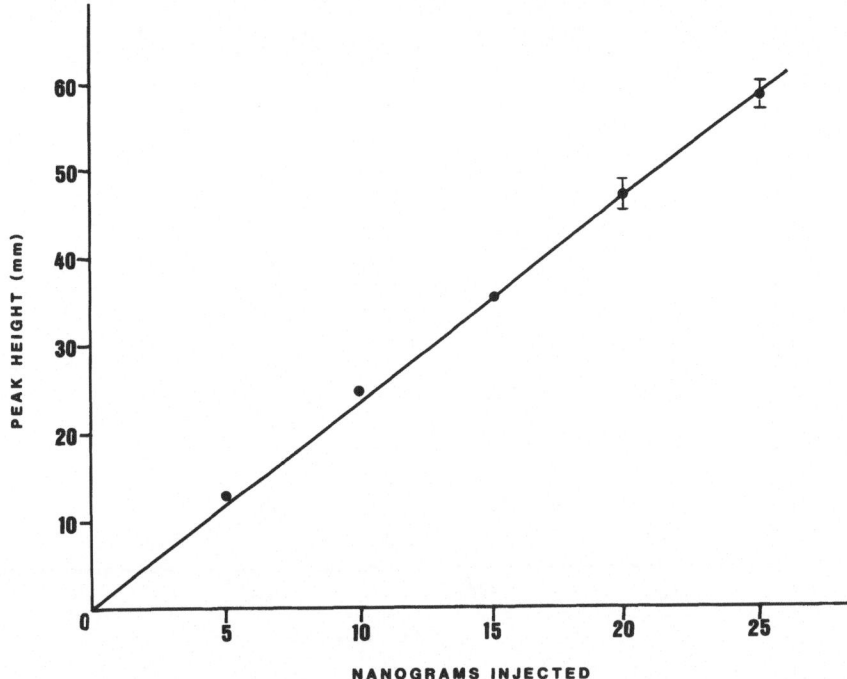

Figure 2. Diflubenzuron standard curve using a Tracor high pressure liquid chromatograph and a C_{18} reverse phase column.

This penomenon could be common to all benzyoylphenyl ureas that have a high melting point and low water solubility and thus a high energy of crystallization. This may mean that the complete dissolution of the larger particles would be very low (Verloop and Ferrell, 1977) resulting in greater environmental stability. In other words, the rate of dissolution of particles could be correlated linearly with their surface area.

3.2 *Fate in water and forest litter*

The sensitivity of the HPLC residue method from our laboratory is generally 5 ng for pure standards and seems to be linear up to at least 25 ng (Figure 2). This method has worked well for evaluating DFB levels from environmental samples. In 1976, our laboratory was asked to review all of the published and unpublished information on the environmental fate of

Figure 1. (A) Air-milled DIMILIN° W-25 crystals on soil particles (arrows) following application in water. (B) Diflubenzuron crystals on soil particles (arrows) following application in acetone.

162

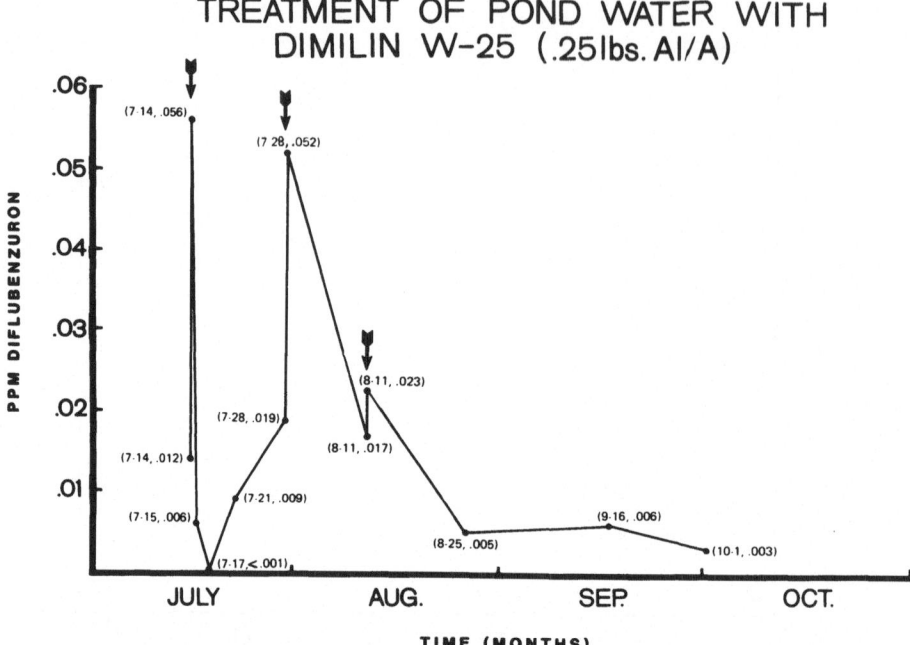

TREATMENT OF POND WATER WITH
DIMILIN W-25 (.25lbs. AI/A)

Figure 3. Diflubenzuron residues measured over time from pond water treated with three applications of 0.25 lbs ai/A. Arrows indicate application dates while numbers in parentheses indicate month-day ppm; for example, (7–14, 0.012) is a sample taken on July 14th which had 0.012 ppm diflubenzuron.

DFB. Over 20 papers on the fate in water (Booth, 1976) showed virtually no detectable residues (less than 10 ppb) after several days. And this is consistent with what we have found in Utah with repeated applications at 0.04 lbs ai/A (Booth and Ferrell, 1977) or 0.25 lbs ai/A (Figure 3, arrows show application dates) to ponds. DFB simply does not persist in the water column under field conditions.

Insecticide residues in the forest litter are a concern to wildlife biologists because of the potential contact with birds, mammals, and other forest-inhabiting non-target organisms. We have been cooperating this past summer (1985) in a West Virginia University sponsored study on the impact of DIMILIN® on the forest community. One of the objectives was to evaluate the residue profile in forest litter. Figure 4B shows a typical chromatogram of the forest litter (day 4 post-treatment) compared with a 25 ng standard (Figure 4A), a 0.05 ppm fortified litter sample (Figure 4C) and a control litter sample (Figure 4D). These data clearly show that DFB is found at less than 0.05 ppm 4 days following treatment and that our modified residue analysis method (Christman, 1985) works well on these

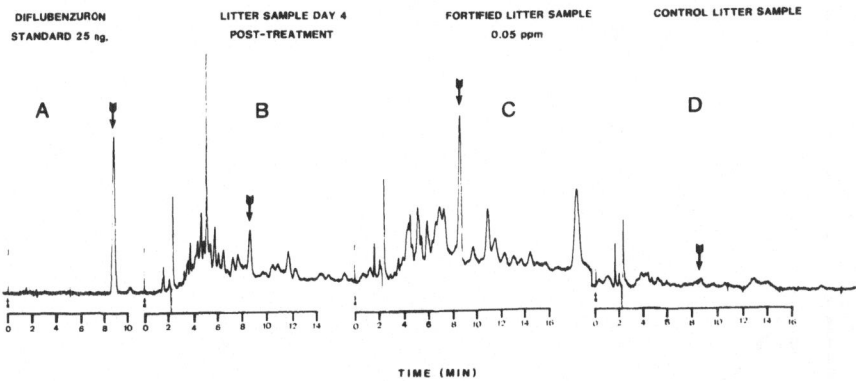

Figure 4. High pressure liquid chromatograms of selected samples. Small arrows indicate injection points, while large arrows indicate diflubenzuron peaks. (A) Diflubenzuron 25 ng standard. (B) Litter sample day 4 post-treatment. (C) 0.05 ppm fortified litter sample. (D) Control litter sample.

samples. All of the forest litter samples analyzed to date (days 1, 4, 10, and 21 days post-treatment) show residues of less than 0.05 ppm indicating that DFB apparently does not persist in this type of matrix.

3.3 *Soil adsorption/desorption behavior*

These experiments summarize the adsorption and desorption behavior of DIMILIN® to 8 soil types.

DIMILIN® clearly adsorbs rapidly to all eight soil types (Figures 5–8). Within the first 5 minutes of the adsorption phase, the amount remaining in the water ranges from 1.46–8.21% of the original dose. It appears that an equilibrium between the soil and water is established within the first 5 minutes on every soil type except for the Olivier Silt Loam (Figures 8A–B) where the water actually picks up radiocarbon with time. Also, no concentration dependence was observed for any of the soils.

Very limited desorption occurred with the eight soil types, and in fact the rice clay showed almost no desorption (Figures 7A and 7B). After 60 minutes, the range of concentration in the water was 0.77–19.53% with the Olivier Silt Loam showing the greatest desorption (Figures 8A and 8B). However, except for the Olivier Silt Loam, the desorption was consistently less than 6% throughout the 60-minute incubation period. No relationship between concentration of the DIMILIN® and desorption was observed with any of the soil types except, again, for the Olivier Silt Loam (Figures 8A and 8B).

164

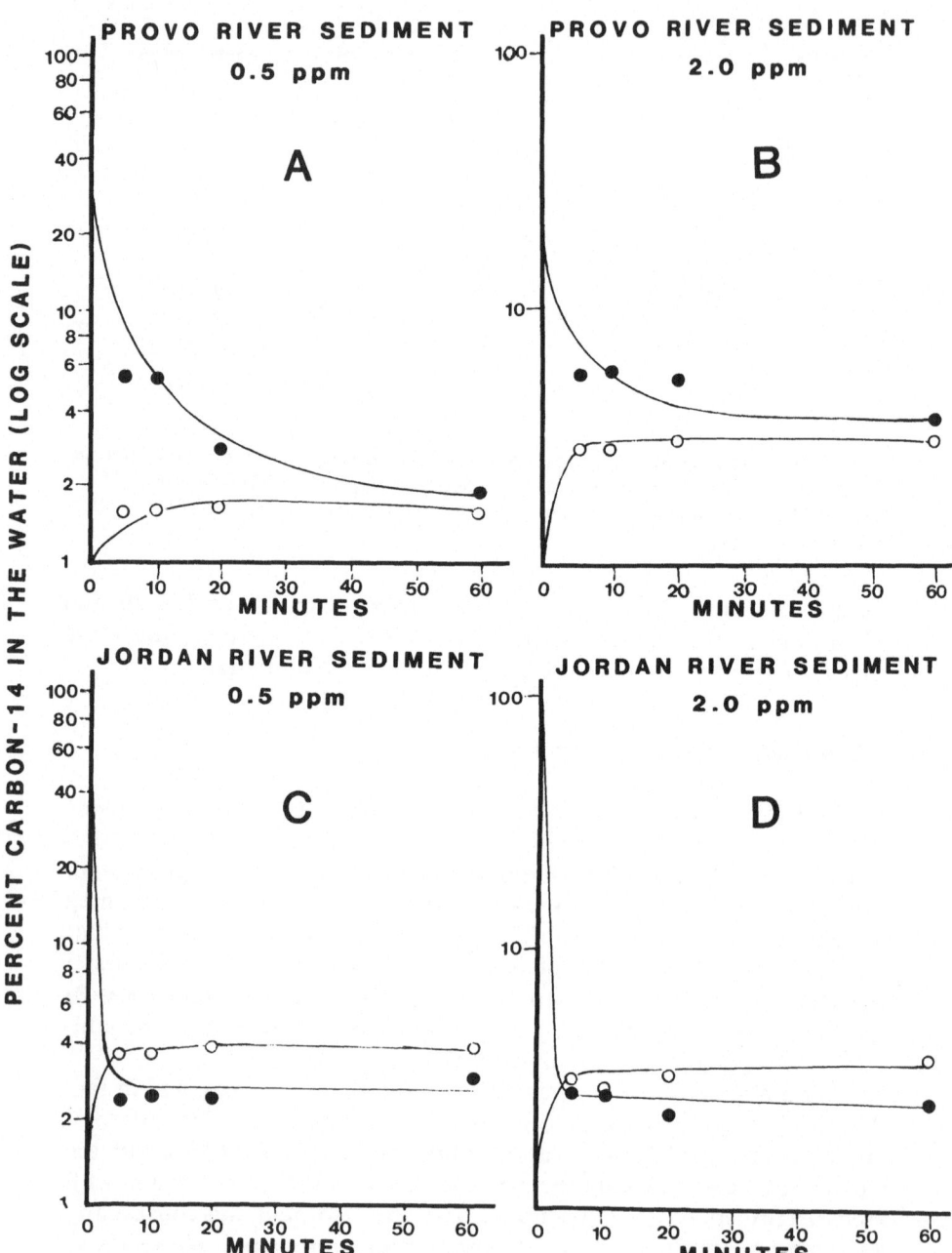

Figure 5. Adsorption (solid circles) and desorption (open circles) of DIMILIN® from selected soil samples as measured by the percent of DIMILIN® in the water. (A) Provo River sediment using 0.5 ppm DIMILIN® W-25. (B) Provo River sediment using 2.0 ppm DIMILIN® W-25. (C) Jordon River sediment using 0.5 ppm DIMILIN® W-25. (D) Jordan River sediment using 2.0 ppm DIMILIN® W-25.

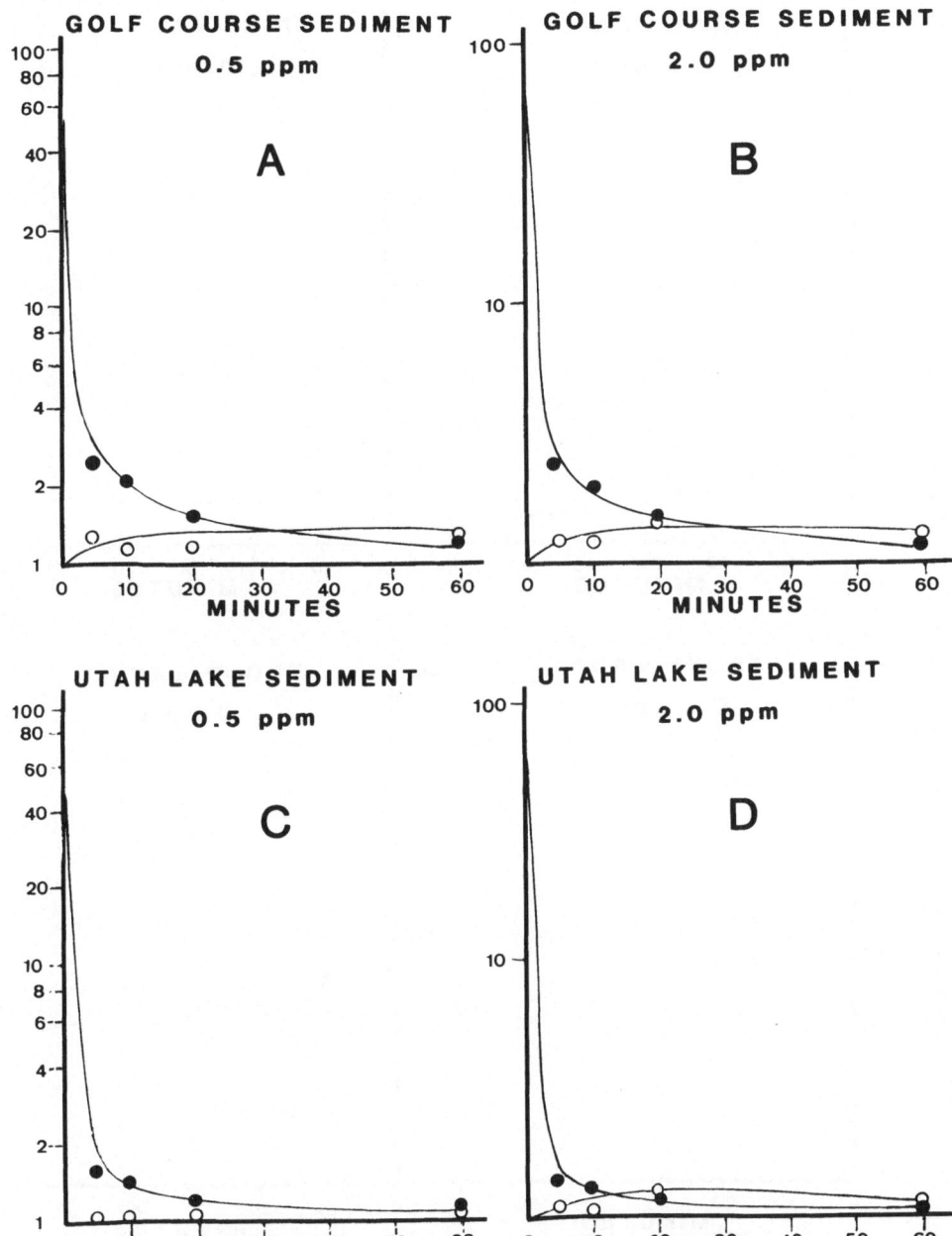

Figure 6. Adsorption (solid circles) and desorption (open circles) of DIMILIN® from selected soil samples as measured by the percent of DIMILIN® in the water. (A) Golf course sediment using 0.5 ppm DIMILIN® W-25. (B) Golf course sediment using 2.0 ppm DIMILIN® W-25. (C) Utah Lake sediment using 0.5 ppm DIMILIN® W-25. (D) Utah Lake sediment using 2.0 ppm DIMILIN® W-25.

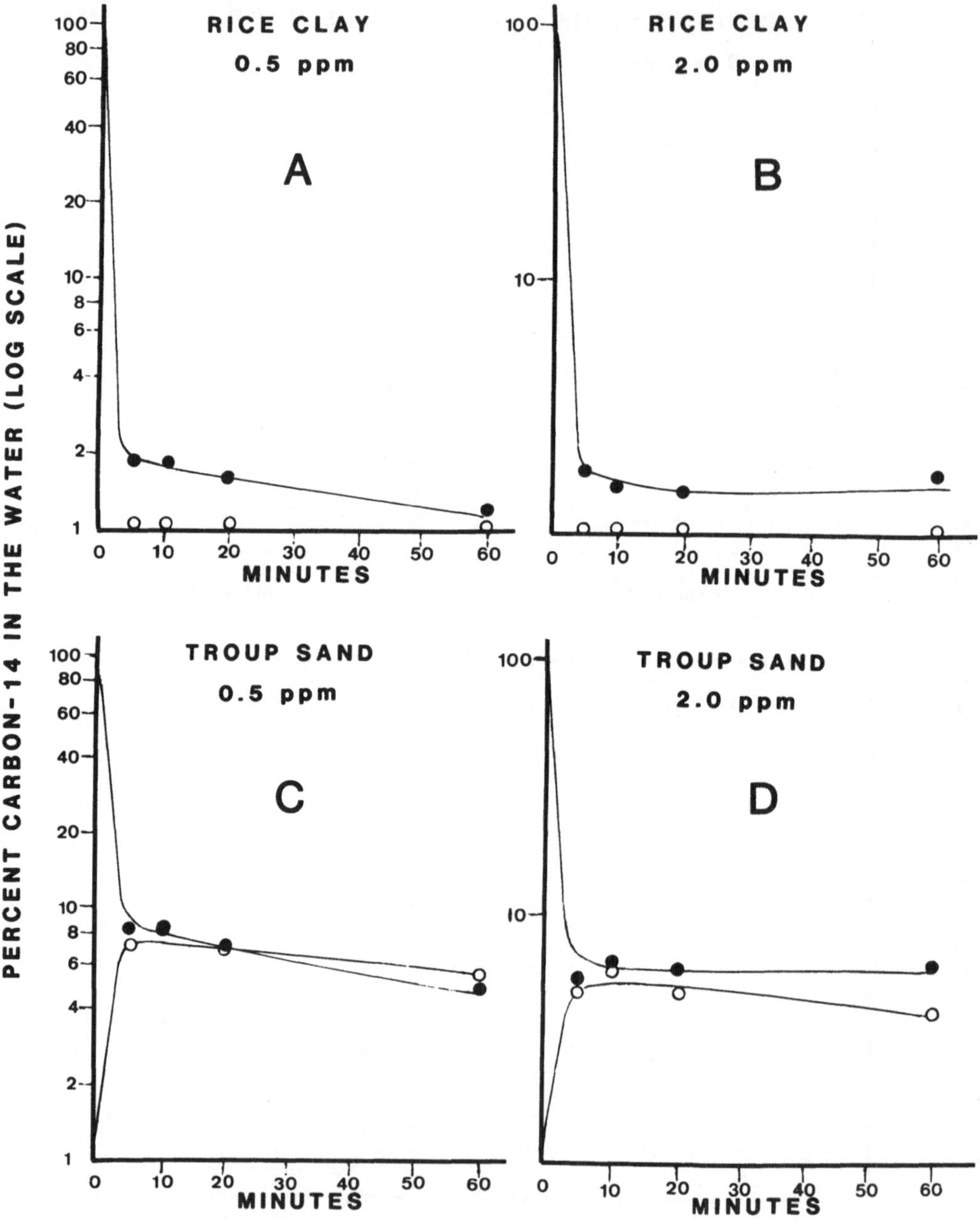

Figure 7. Adsorption (solid circles) and desorption (open circles) of DIMILIN® from selected soil samples as measured by the percent of DIMILIN® in the water. (A) Rice clay using 0.5 ppm DIMILIN® W-25. (B) Rice clay using 2.0 ppm DIMILIN® W-25. (C) Troup sand using 0.5 ppm DIMILIN® W-25. (D) Troup sand using 2.0 ppm DIMILIN® W-25.

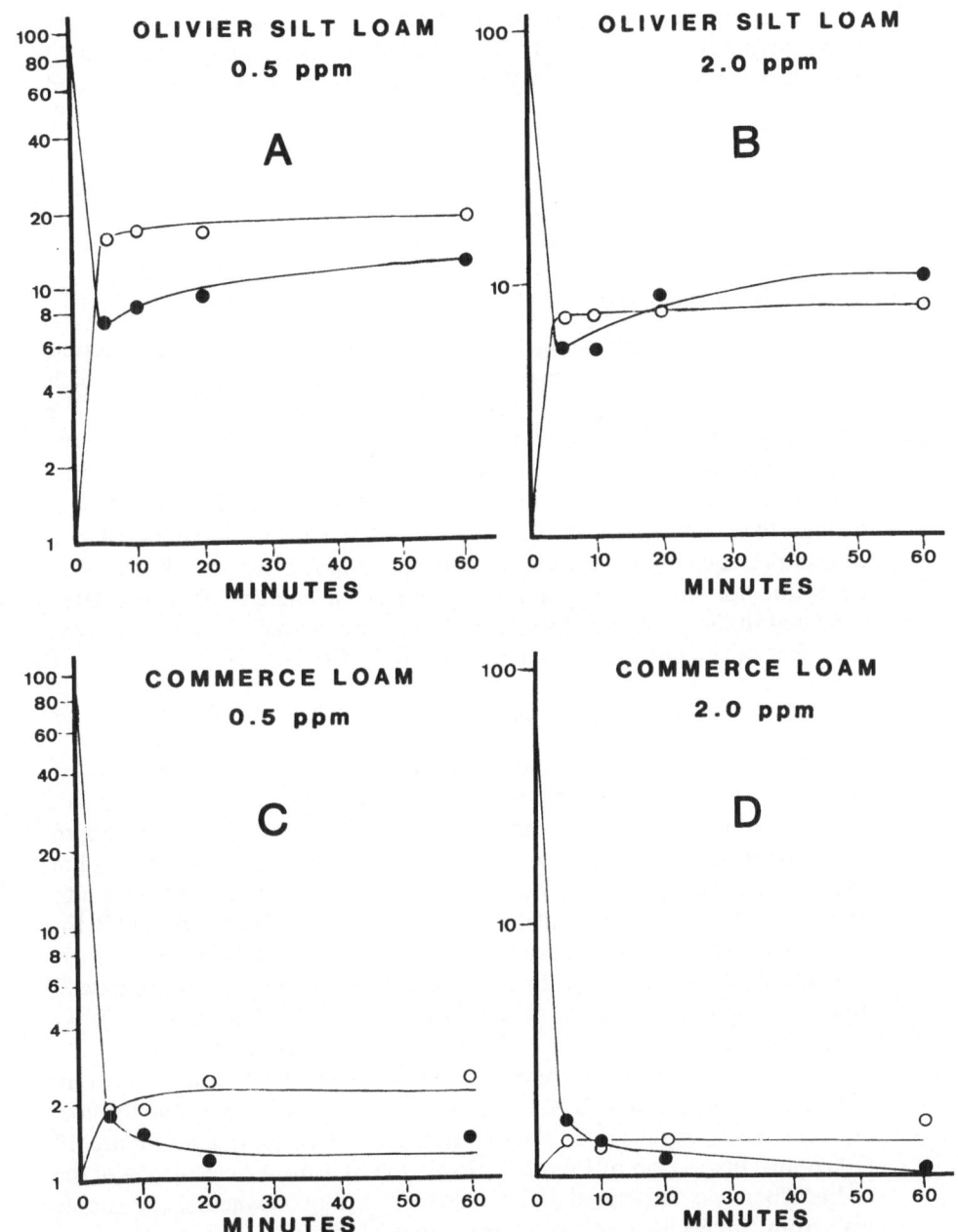

Figure 8. Adsorption (solid circles) and desorption (open circles) of DIMILIN® from selected soil samples as measured by the percent of DIMILIN® in the water. (A) Olivier silt loam using 0.5 ppm DIMILIN® W-25. (B) Olivier silt loam using 2.0 ppm DIMILIN® W-25. (C) Commerce loam using 0.5 ppm DIMILIN® W-25. (D) Commerce loam using 2.0 ppm DIMILIN® W-25.

The rapid adsorption precluded the calculation of any kinetic relationships. Summarizing, DIMILIN® was found to adsorb very rapidly to all eight soil types (greater than 87% of the initial amount), with limited desorption from all soil types except for the Olivier Silt Loam which apparently desorbs DIMILIN® with time.

3.4 *Metabolism by a simulated activated sludge system*

3.4.1 *General comments*

The term "activated sludge" is derived from wastewater being mixed with air or oxygen for a length of time to develop a brown floc which consists principally of microorganisms and their waste products. The activated sludge process provides the environment necessary to keep the microorganisms under controlled conditions so they can remove organic material from the watewater and thereby produce an effluent-quality high enough that beneficial uses of receiving waters will not be hindered.

In a previous report unpublished from our laboratory we showed that DFB did not inhibit any of the operating parameters of a simulated activated sludge process. The intent of the present study was to determine the effect of the activated sludge matrix on DFB.

3.4.2 *TR in Oxyfluor® (as $^{14}CO_2$), floc, and aqueous*

Table 2 summarizes the radiocarbon inventory for the entire study. Generally, the recovery of the radiocarbon for each experiment was good. For example, over time the average recovery for the 1 ppm "unacclimated" study was $\bar{x} = 21.42 \pm 1.37$ ug and for the 1 ppm "acclimated" study was $\bar{x} = 22.20 \pm 1.41$ ug. Since 20 ug was placed into each flask, this represents a mean percent recovery of 107% and 111% for the "unacclimated" and "acclimated" sludge-conditions respectivley. A similar calculation over time was made for the 100 ppm "unacclimated" and "acclimated" portion of the study and the values were $\bar{x} = 2280 \pm 64$ ug (114%) and $\bar{x} = 2278 \pm 102$ (114%) respectively.

The % radiocarbon (of the applied dose) in each fraction is shown in the parentheses in Table 2. A plot of these % values against time for floc, aqueous, and Oxyfluor® is shown in Figure 9 (1 ppm) and Figure 10 (100 ppm). Inspection of Figure 9 shows that at 1 hour ca. 57–75% of the radiocarbon was associated with the floc depending on whether the sample was from "unacclimated" or "acclimated" sludge. Hence, there was a tremendous affinity of DFB for bacterial Floc consistent with other experiments which show that bacteria use DFB as a carbon and nitrogen source (Booth and Ferrell, 1977). But there was a significant disappearance of

DFB over time which generally follows first order kinetics. The most rapid elimination ($t_{1/2}$ = 1.53 hours, k = 4.53 × 10^{-1} hours^{-1}) occurred in the 1 ppm "unacclimated floc commensurate with a significant increase in aqueous radiocarbon (compare Figure 9, curve 1 and 3). Thus, in this sample the disappearance of floc-radiocarbon generally follows first order kinetics with an r^2 = 0.96. But, in the "acclimated" sample, the $t_{1/2}$ = 1451.67 hours or k = 4.77 × 10^{-2} hours^{-1} with a much smaller increase in aqueous radiocarbon and an r^2 = 0.60 (compare Figure 9, curves 2 and

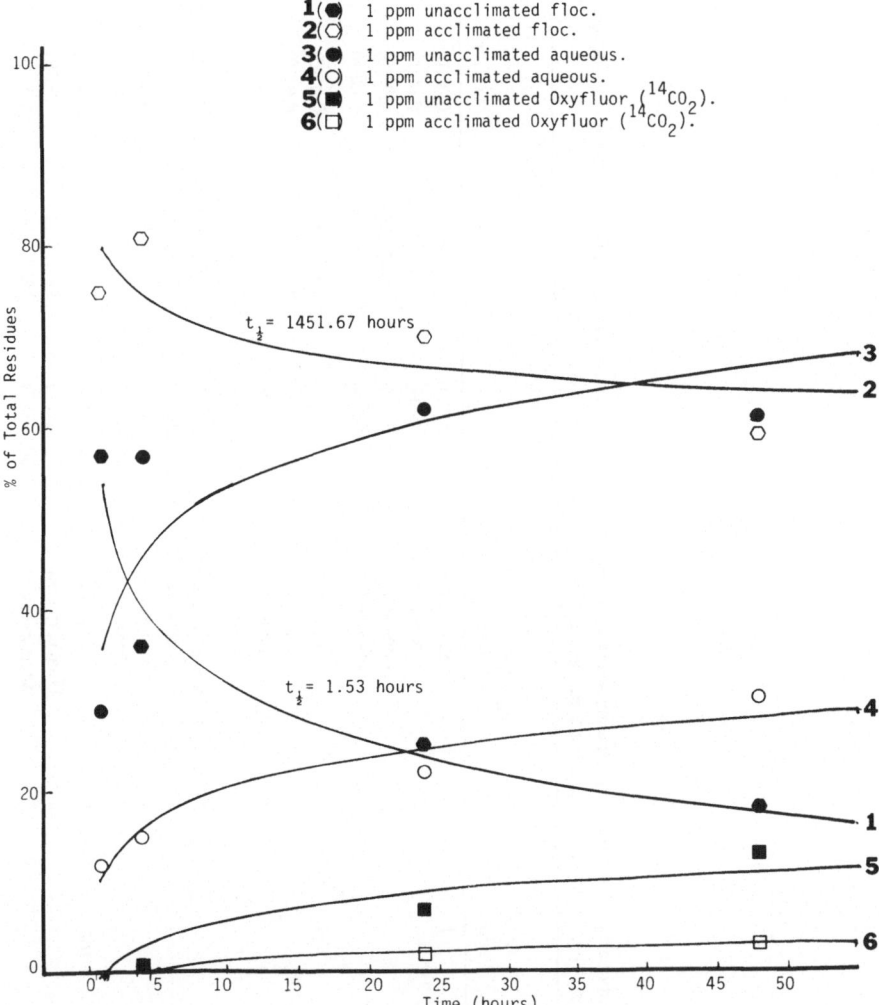

Figure 9. Total residue profile for the 1 ppm study in the floc, aqueous, and Oxyfluor® fractions. Appropriate half lives are shown for the floc samples.

Table 2. Radiocarbon inventory from the activated sludge metabolism study using liquid scintillation methods. Values are total micrograms while numbers in parenthesis refer to the percent of the total radiocarbon applied to each flask.

Time (hours)	DFB (ppm)	Rep.	Sludge Treatment[a]	Micrograms				
				Floc[b]	Aqueous	Oxyfluor	Other[c]	Total
1	1	1	Unacclimated	9.22(50)	6.46(35)	0.00(0)	2.82(15)	18.50
		2	Unacclimated	14.56(64)	4.91(22)	0.00(0)	3.12(14)	22.59
		1	Acclimated	17.30(79)	2.51(11)	0.00(0)	2.02(9)	21.84
		2	Acclimated	14.97(70)	2.84(13)	0.00(0)	3.67(17)	21.49
	100	1	Unacclimated	1914.91(84)	303.59(13)	0.05(0)	47.82(2)	2266.37
		2	Unacclimated	1913.70(84)	276.13(12)	0.04(0)	96.91(4)	2286.78
		1	Acclimated	1939.33(86)	237.50(12)	0.05(0)	41.19(2)	2254.08
		2	Acclimated	1941.38(86)	298.39(13)	0.02(0)	23.66(1)	2263.45
4	1	1	Unacclimated	8.22(37)	12.31(55)	0.12(1)	1.75(8)	22.41
		2	Unacclimated	7.69(35)	12.63(58)	0.13(1)	1.35(6)	21.79
		*1	Acclimated	21.84(85)	2.98(12)	0.00(0)	0.74(3)	25.55
		2	Acclimated	17.30(77)	3.98(18)	0.05(0)	1.02(5)	22.36
	100	1	Unacclimated	1698.03(76)	398.90(18)	0.91(0)	149.40(7)	2247.24
		2	Unacclimated	1879.32(83)	343.79(15)	0.95(0)	47.89(2)	2271.95
		1	Acclimated	2030.97(87)	250.27(11)	1.06(0)	52.45(2)	2334.75

		2	Acclimated	1817.94(87)	205.41(10)	1.21(0)	76.54(4)	2101.11
24	1	1	Unacclimated	4.95(24)	13.09(63)	1.46(7)	1.38(7)	20.88
		2	Unacclimated	5.31(25)	13.11(61)	1.49(7)	1.52(7)	21.42
		1	Acclimated	15.58(70)	4.59(21)	0.33(1)	1.67(8)	22.17
		2	Acclimated	14.85(70)	4.92(23)	0.34(2)	1.24(6)	21.35
	100	1	Unacclimated	1812.88(81)	182.12(8)	8.18(0)	233.46(10)	2236.64
		2	Unacclimated	1913.30(85)	225.12(10)	12.10(1)	103.86(5)	2254.38
		1	Acclimated	2047.02(86)	187.04(8)	8.33(0)	137.68(6)	2380.07
		2	Acclimated	2037.60(89)	133.85(6)	6.75(0)	111.72(5)	2289.92
48	1	1	Unacclimated	3.74(18)	12.80(62)	2.72(13)	1.24(6)	20.50
		2	Unacclimated	3.99(18)	13.25(59)	2.71(12)	2.52(11)	22.47
		1	Acclimated	12.33(57)	6.35(30)	0.60(3)	2.21(10)	21.49
		2	Acclimated	12.90(60)	6.47(30)	0.63(3)	1.34(6)	21.34
	100	1	Unacclimated	1720.87(77)	349.23(16)	14.44(1)	159.56(7)	2244.10
		2	Unacclimated	1599.96(66)	456.57(19)	7.51(0)	368.27(15)	2432.32
		1	Acclimated	1854.63(85)	158.42(7)	11.46(1)	159.70(7)	2184.21
		2	Acclimated	2102.34(87)	138.36(6)	12.43(1)	159.71(7)	2412.84

[a] Unacclimated = simulated activated sludge which had never been exposed to diflubenzuron prior to this study.

[a] Acclimated = simulated activated sludge which had been exposed to periodic (shock) doses of diflubenzuron prior to this study.

[b] Floc = microorganisms plus their waste products.

[c] Other = flask rinses.

172

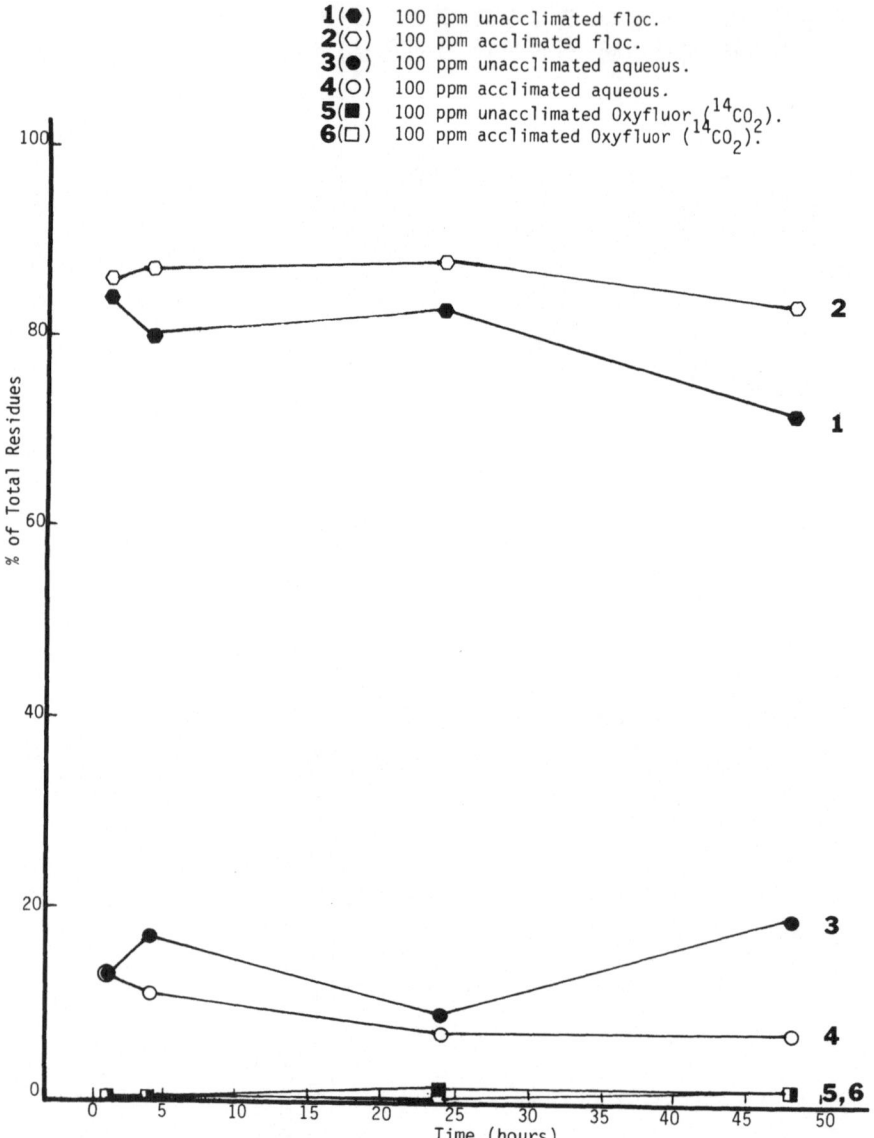

Figure 10. Total residue profile for the 100 ppm study in the floc, aqueous, and Oxyfluor®
fractions.

4). At 48 hours there was ca. 13% of the TR as $^{14}CO_2$ in the "unac-climated" sludge and ca. 3% in the "acclimated" samples (Figure 9).

The TR data for the 100 ppm studies are plotted in Figure 10. These data changed very little over time and therefore were not subjected to kinetic analysis. Apparently, when megadoses of DFB are presented to sewage bacteria, there is a rapid up-take but very little elimination over time. The $^{14}CO_2$ levels were essentially 0 regardless of sludge-condition while the aqueous TR averaged ca. 10% through time. Realistically, however, residue levels in water sewage plants will never approach 100 ppm.

The TR ANOV F-tests for CO_2, aqueous and floc analysis clearly indicates that for the CO_2 data only the main effects of days (D), samples (S), and treatments (T) and D × S interaction were significant. The D × S significant interaction indicates that the difference in the CO_2 level between 1 and 100 ppm in the "unacclimated" sludge was higher than the equivalent difference in the "acclimated" sample. The significance of these being different is indicated by the D × S interaction. All other inter-actions were non-significant. Analysis of the TR aqueous data shows that the main effects and all interactions were significant. The important item is the D × S interaction. In practical terms this interaction indicates that the difference in TR between "unacclimated" and "acclimated" sludge at 1 ppm was significantly higher than the same difference at 100 ppm. The data show that the main effects and all interactions were significant. In general, the difference in floc TR in "unacclimated" and "acclimated" samples at 1 ppm was significantly higher than that at 100 ppm. In other words, the elimination of TR in the 1 ppm floc "unacclimated" samples was faster over time compared to the comparable "acclimated" samples, while in the 100 ppm samples there was little change in either sludge-condition.

3.4.3 *Extraction and metabolism*

Table 3 summarizes the extraction of DFB-associated residues from the floc and aqueous samples. The average aqueous and floc extraction across time and dose was $\bar{x} = 66.53 \pm 13.42\%$ and $95.31 \pm 5.5\%$ respectivley.

The disappearance of DFB from the aqueous environment appeared to follow first order kinetics with a $t_{1/2} = 4.36$ hours ($k = 1.59 \times 10^{-1}$ hours^{-1}) for the 1 ppm "unacclimated" samples, $t_{1/2} = 14.52$ hours ($k = 4.77 \times 10^{-2}$ hours^{-1}) for the 1 ppm "acclimated" samples, $t_{1/2} = 181.69$ hours ($k = 3.81 \times 10^{-3}$ hours^{-1}) for the 100 ppm "unacclimated" sam-ples, and $t_{1/2} = 99.20$ hours ($k = 6.99 \times 10^{-3}$ hours^{-1}) for the 100 ppm "acclimated" samples. In each case there was a comparable increase in CPU depending on the rate of DFB disappearance (Figure 11). The relationship of DFB to CPU was carefully examined since CPU was the only major metabolite found in the aqueous fraction (Table 4). The data in Figure 11 clearly shows that the parent compound disappears faster in 1 ppm "unacclimated" samples than in 1 ppm "acclimated" samples which in turn was faster than both the 100 ppm samples; but, there apparently was little difference in the disappearance of DFB in the 100 ppm "unac-climated" and "acclimated" samples. These visual observations were confirmed statistically.

174

Table 3. Organic-solvent extraction of diflubenzuron-associated residues from the floc and Aqueous fractions.

Time (hours)	DFB (ppm)	Rep.	Sludge Treatment[a]	Sample[b]	Micrograms					% Extraction[d]
					TR[e]	TER	TBR	Total[c]	TER Lost	
1	1	1	Unacclimated	Floc	6.26	4.21	1.47	5.68	0.58	77
				Aqueous	0.96	0.83	0.16	0.99	0	83
		2	Unacclimated	Floc	8.18	8.74	0.37	9.11	0	100
				Aqueous	0.91	0.91	0.16	1.07	0	82
		1	Acclimated	Floc	11.48	11.06	0.34	11.40	0.08	97
				Aqueous	0.86	0.81	0.09	0.90	0	90
		2	Acclimated	Floc	11.95	10.12	0.35	10.47	1.48	97
				Aqueous	–	–	–	–	–	–
	100	1	Unacclimated	Floc	1419.79	1152.08	63.23	1215.31	204.48	96
				Aqueous	75.79	51.42	22.77	74.19	1.6	70
		2	Unacclimated	Floc	1093.33	1077.53	50.33	1127.86	34.53	95
				Aqueous	86.72	65.33	20.74	86.07	0.65	76
		1	Acclimated	Floc	1326.75	1291.39	44.05	1335.44	0	100
				Aqueous	86.92	43.72	38.03	81.75	5.17	56
		2	Acclimated	Floc	1077.20	1186.95	73.38	1260.33	0	100
				Aqueous	85.59	42.42	44.77	87.19	0	48
4	1	1	Unacclimated	Floc	6.36	4.82	0.39	5.21	1.15	94
				Aqueous	1.09	0.81	0.34	1.15	0	69
		2	Unacclimated	Floc	5.37	4.30	0.36	4.66	0.71	93
				Aqueous	1.12	0.81	0.32	1.13	0	71
		1	Acclimated	Floc	15.02	14.12	0.41	14.53	0.49	97
				Aqueous	0.87	0.49	0.42	0.91	0	52
		2	Acclimated	Floc	14.31	10.98	0.32	11.30	3.01	98
				Aqueous	0.95	0.70	0.26	0.96	0	73
	100	1	Unacclimated	Floc	1191.11	1020.38	62.50	1082.88	108.23	95

	2	Unacclimated	Aqueous	100.32	62.88	38.17	101.05	0	62
			Floc	1215.09	1060.50	109.84	1170.34	44.75	91
	1	Acclimated	Aqueous	86.13	53.35	33.65	87.00	0	61
			Floc	1305.45	1117.71	35.17	1152.88	152.57	97
	2	Acclimated	Aqueous	89.41	53.93	34.37	88.30	1.11	62
			Floc	1358.38	1083.60	29.37	1112.97	245.41	98
24	1	Unacclimated	Aqueous	57.82	38.84	32.87	71.71	0	43
			Floc	4.21	2.48	0.50	2.98	1.23	88
	2	Unacclimated	Aqueous	1.04	0.79	0.18	0.97	0.07	83
			Floc	4.33	2.60	0.54	3.14	1.19	88
	1	Acclimated	Aqueous	1.22	0.98	0.28	1.26	0	77
			Floc	9.23	7.24	0.74	7.98	1.25	92
	2	Acclimated	Aqueous	0.98	0.74	0.27	1.01	0	72
			Floc	9.65	9.05	0.73	9.78	0	100
100	1	Unacclimated	Aqueous	0.87	0.70	0.28	0.98	0	68
			Floc	1358.16	1176.93	58.35	1235.28	122.88	96
	2	Unacclimated	Aqueous	79.88	70.73	22.83	93.56	0	71
			Floc	1108.98	909.46	101.99	1011.45	97.53	91
	1	Acclimated	Aqueous	70.27	57.22	21.95	79.17	0	69
			Floc	1199.13	1118.76	29.21	1147.97	51.16	98
	2	Acclimated	Aqueous	88.01	61.03	23.90	84.93	3.08	73
			Floc	1244.77	809.99	29.98	839.97	404.80	98
48	1	Unacclimated	Aqueous	86.29	63.25	37.74	100.99	0	56
			Floc	2.71	1.47	0.57	2.04	0.67	79
	2	Unacclimated	Aqueous	1.12	0.90	0.20	1.10	0.02	82
			Floc	3.85	1.68	0.84	2.52	1.33	78
	1	Acclimated	Aqueous	1.07	0.91	0.19	1.10	0	82
			Floc	8.16	8.05	0.58	8.63	0	93
	2	Acclimated	Aqueous	1.10	0.65	0.36	1.01	0.09	67
			Floc	12.69	8.62	0.57	9.19	3.50	96

Table 3. Continued.

Time (hours)	DFB (ppm)	Rep.	Sludge Treatment[a]	Sample[b]	Microgram TR[e]	TER	TBR	Total[c]	TER Lost	% Extraction[d]
48	100	1	Unacclimated	Aqueous	0.97	0.85	0.36	1.21	0	63
				Floc	1494.92	1117.10	75.30	1192.40	302.52	95
				Aqueous	98.76	53.71	52.75	107.46	0	47
		2	Unacclimated	Floc	1137.58	1008.96	75.46	1084.42	53.16	93
				Aqueous	108.30	69.51	53.77	123.28	0	50
		1	Acclimated	Floc	1168.29	1093.05	44.88	1137.93	30.36	96
				Aqueous	77.03	57.23	27.79	84.99	0	64
		2	Acclimated	Floc	1345.13	1158.21	33.90	1192.11	153.02	97
				Aqueous	83.37	60.67	28.91	89.58	0	65

[a] Unacclimated = simulated activated sludge which had never been exposed to diflubenzuron prior to this study.

[a] Acclimated = simulated activated sludge which had been exposed to periodic shock doses of diflubenzuron prior to this study.

[b] Each sample to be extracted was divided into Floc (bacteria plus waste products) and aqueous.

[c] Total = TER + TUR or TBR.

[d] % extraction = TR − (TUR or TBR)/TR × 100.

[e] TR = Total residues actually used in the extraction.

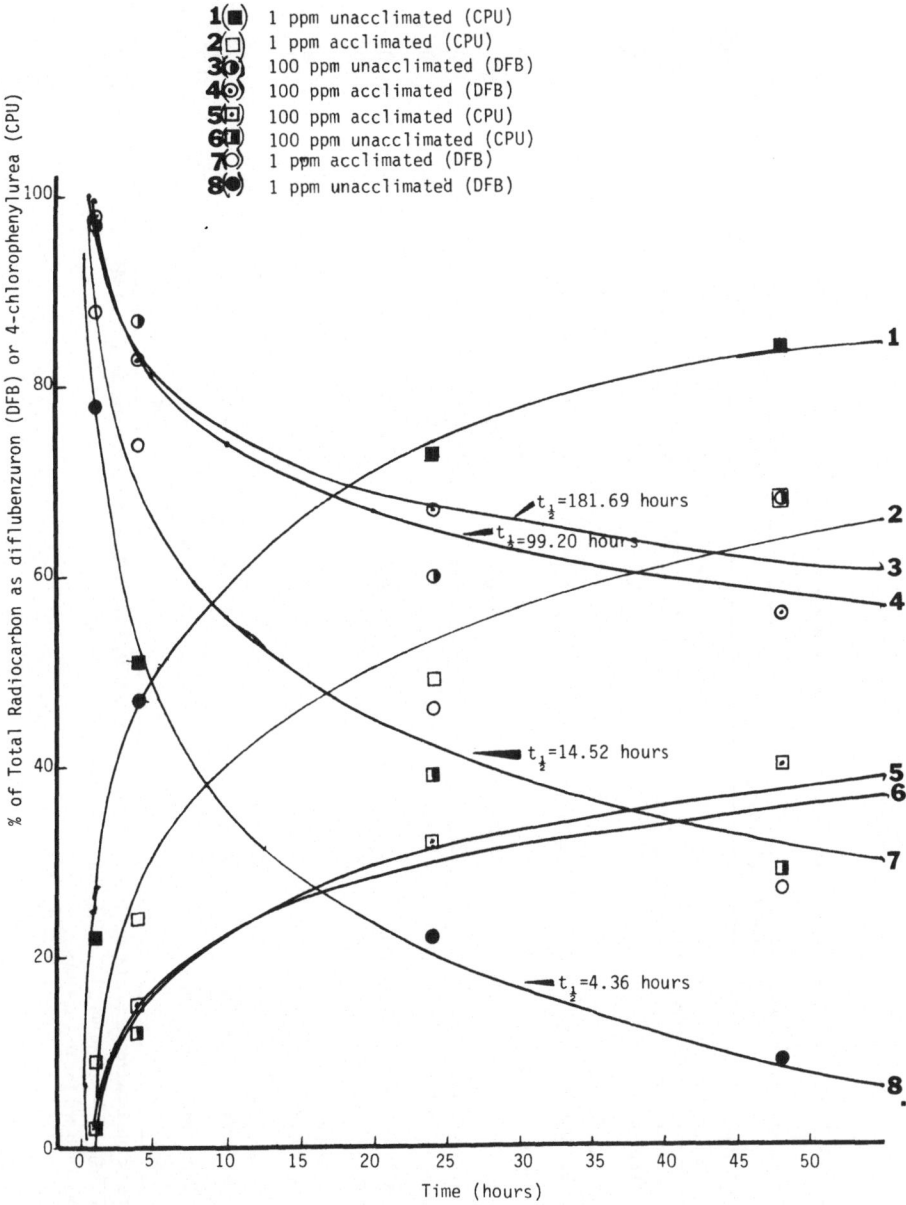

Figure 11. The percent of the total radiocarbon as diflubenzuron (DFB) or 4-chlorophenylurea (CPU) plotted against time for the four sludge treatments. The half lives for the parent compound (DFB) are shown for each treatment. The data are for the aqueous fraction of each study only.

Table 4. Mean percent (± S.D.) residue distribution in aqueous-extracts by time, diflubenzuron-treatment, and a sludge-treatment. Values are rounded to neareast whole number.

Time (hours)	DFB (ppm)	Sludge Treatment	Residue[a] (% of total)							
			00	01	02	03	04	11	12	21
1	1	Unacclimated	0	22 ± 1	0	1	78 ± 1	0	0	0
		Acclimated	0[b]	9[b]	0[b]	2[b]	88[b]	0	0	0
	100	Unacclimated	0	2	0	1	97	0	0	0
		Acclimated	0	2	0	1	98 ± 1	0	0	0
4	1	Unacclimated	0	52 ± 2	0	1	47 ± 2	1	1	0
		Acclimated	0	24 ± 23	0	2	74 ± 24	1 ± 1	0	0
	100	Unacclimated	0	12 ± 5	0	2 ± 1	87 ± 5	0	0	0
		Acclimated	1 ± 1	15 ± 6	0	2	83 ± 6	0	0	0
24	1	Unacclimated	1	73 ± 4	1	2 ± 1	22 ± 3	0	2 ± 1	0
		Acclimated	1 ± 1	49 ± 1	1	2	46 ± 3	0	1	0
	100	Unacclimated	0	39 ± 12	0	2	60 ± 12	0	0	0
		Acclimated	0	32 ± 2	0	2	67 ± 2	0	0	0
48	1	Unacclimated	2 ± 1	84 ± 1	2 ± 1	1	9 ± 1	2	0	0
		Acclimated	0	68 ± 1	1	1 ± 1	27 ± 1	2	1	1
	100	Unacclimated	0	29 ± 13	0	3 ± 1	68 ± 13	1 ± 1	0	1 ± 1
		Acclimated	0	40 ± 4	1 ± 1	2	56 ± 4	1	0	1 ± 1

[a] Residue codes from the tlc plates were: 00 = origin; 01 = 4-chlorophenylurea (CPU); 02 = 2,6-difluorobenzamide (DFBAM); 03 = 4-chloroaniline (CA); 04 = diflubenzuron (DFB); 11 = unknown; 12 = 2,6-difluorobenzoic acid (DFBA); 21 = unknown.

[b] Only one replicate was analyzed for this sample.

The only interactions that were significant in the Aqueous CPU data were the D × T and D × S. On the average, the 1 ppm CPU values were higher than the 100 ppm CPU values. Although the D × T interaction was significant, the 1 ppm values were consistently higher at every time interval. The difference in CPU content between "acclimated" and "unacclimated" samples was greater at 1 ppm than the equivalent difference at 100 ppm (Figure 12).

Table 5 summarizes the materials-balance in the floc extract. As Figure 12 shows, the major residue was DFB ($88.44 \pm 12.62\%$ averaged over T, D, and S) which remained consistently high in all treatments except in the 1 ppm "unacclimated" samples where DFB appears to decline. These observations were consistent with the presence of CPU in each sample. In other words, in the 1 ppm "unacclimated" floc samples, CPU tends to increase over time while CPU in all other treatment-samples remained very low (Figure 12, curves 1, 5, 2 and 6). Inspection of Figure 12 shows that the difference in residues at 1 ppm in the "unacclimated" and "acclimated" samples was greater than the same difference at 100 ppm. The significance of this difference is indicated by the D × S interaction. The 1 ppm "unacclimated" floc-DFB curve with a $t_{1/2} = 57.98$ hours ($k = 1.20 \times 10^{-2}$ hours^{-1}) best followed a linear relationship with an $r^2 = 0.55$. Except for the D × S interaction, all other interactions for the presence of CPU in the floc were non-significant.

The presence of CA in floc extracts averaged less than 5% and clearly showed no T, D, or S effects.

3.4.4 *Metabolite verification*

To provide absolute verification of the metabolite distribution, the spots from selected tlc plates were extracted and analyzed by mass spectrometry. By comparing the SIM spectra of selected "acclimated" and "unacclimated" floc samples and aqueous extracts to the standard spectra shown in Figure 13, it was clear that the correct metabolite identities were assigned. It is not surprising that trace levels of parent DFB were found in the origin spot of the tlc plates since tlc procedures are not always 100% efficient in resolving complex mixtures. Due to the low quantities of material in the unknown spots and to the complexity of the background matrix, it was not possible to obtain mass spectra that were useful for the successful identification of these spots.

3.5 *Non-target toxicity studies*

3.5.1 *Segmented worms and midges*

Based on population trends, Figure 14 clearly shows that neither oligochaetes or chironomids were affected by 6 applications of DIMILIN®

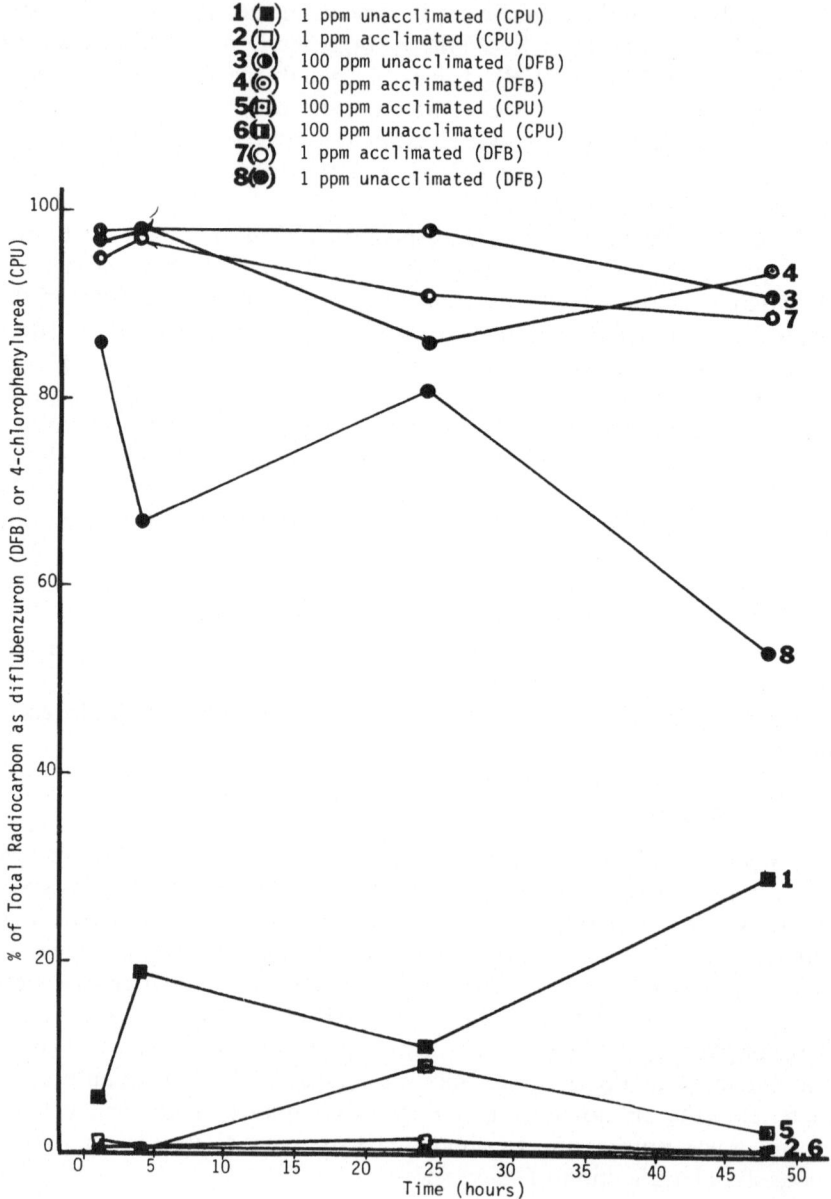

Figure 12. The percent of the total radiocarbon as diflubenzuron (DFB) or 4-chlorophenylurea (CPU) plotted against time for the four sludge treatments. The data are for the floc fraction of each study only.

Table 5. Mean percent ($\pm$ S.D.) residue distribution in floc-extracts by time, diflubenzuron-treatment, and sludge-treatment. Values are rounded to nearest whole number.

Time (hours)	DFB (ppm)	Sludge Treatment	Residue[a] (% of total)							
			00	01	02	03	04	11	12	21
1	1	Unacclimated	0	6 ± 1	1 ± 0	3 ± 1	86 ± 1	1 ± 1	3 ± 1	1 ± 1
		Acclimated	1 ± 1	1 ± 1	2 ± 1	3 ± 1	95 ± 2	0	0	0
	100	Unacclimated	0	0	1 ± 1	2	98 ± 1	0	0	0
		Acclimated	0	0	1 ± 1	2	97 ± 1	0	0	0
4	1	Unacclimated	1	19 ± 15	2 ± 1	3	67 ± 27	5 ± 6	5 ± 4	0
		Acclimated	1 ± 1	0	1 ± 1	3 ± 2	97 ± 2	0	0	0
	100	Unacclimated	0[b]	0[b]	1[b]	1[b]	98[b]	0[b]	0[b]	0[b]
		Acclimated	0	0	1 ± 1	2 ± 1	98 ± 1	0	0	0
24	1	Unacclimated	3 ± 1	11 ± 11	1 ± 1	5 ± 6	82 ± 18	0	0	0
		Acclimated	1 ± 1	1 ± 1	1	7 ± 6	91 ± 8	0	1	0
	100	Unacclimated	0	0	0	2 ± 1	98	0	0	0
		Acclimated	1 ± 1	9 ± 13	1	3 ± 1	86 ± 16	1 ± 1	1 ± 1	0
48	1	Unacclimated	4	29 ± 1	0	14 ± 1	53	0	0	0
		Acclimated	1 ± 1	2 ± 1	1 ± 1	5 ± 6	89 ± 8	0	2 ± 3	2 ± 2
	100	Unacclimated	0	0	3 ± 2	7 ± 6	91 ± 8	0	0	0
		Acclimated	0	2 ± 2	1 ± 1	2 ± 1	94 ± 6	1 ± 1	1 ± 1	0

[a] Residue codes from the tlc plates were: 00 = origin; 01 = 4-chlorophenylurea (CPU); 02 = 2,6-difluorobenzamide (DFBAM); 03 = 4-chloroaniline (CA); 04 = diflubenzuron (DFB); 11 = unknown; 12 = unknown; 21 = 2,6-difluorobenzoic acid (DFBA); 21 = unknown.

[b] Only one replicate was analyzed for this sample.

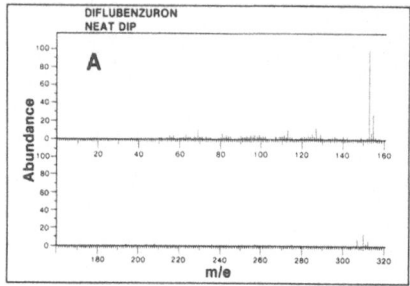

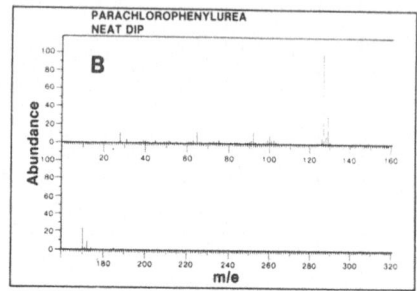

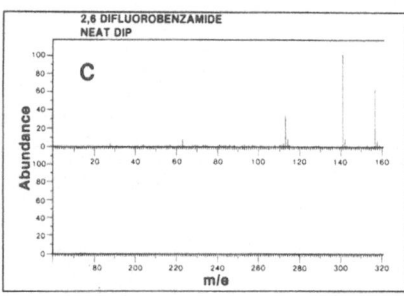

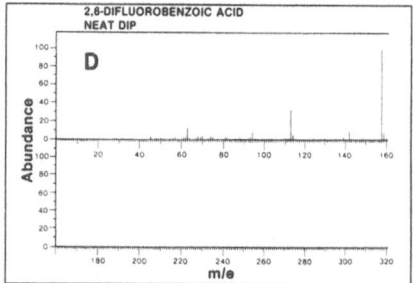

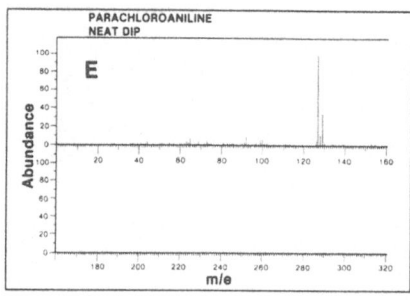

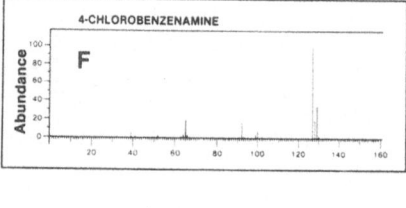

Figure 13. Mass spectra of purified standards used in the study. (A) Diflubenzuron. (B) Para-chlorophenylurea. (C) 2,6-difluorobenzoic acid. (E) Parachloroaniline. (F) Para-chloroaniline, also called 4-chlorobenzenamine (EPA/NIH standard).

W-25 and DIMILIN® 1-G. The average number of organisms per Ekman dredge sample was ca. 100 and 40 for segmented worms and midges, respectively. It was not surprising to find no affects since DFB residues in sediments have generally been less than 0.05 ppm (Booth and Ferrell, 1977), although there are reports of higher levels in sediments from certain parts of the United States (Booth, 1976). Even though DFB adsorbs rapidly to soil (Figures 5–8), microorganisms associated with soil have

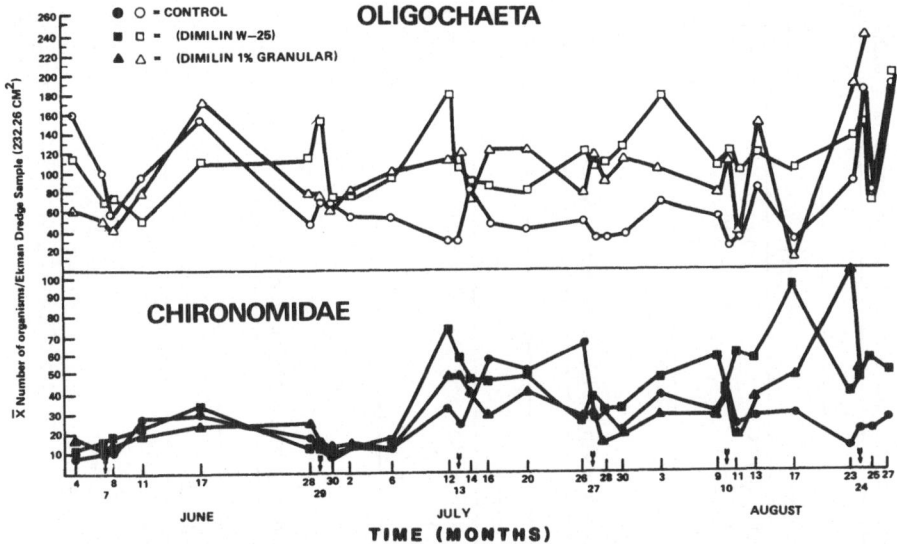

Figure 14. Population trends of segmented worms (Oligochaeta) and midges (Chironomidae) from Provo Bay, Utah Lake sampled over time during 6 applications of 0.13 lbs ai/A of DIMILIN® W-25 and 0.12 lbs ai/A.

been shown to metabolize it principally to CPU (Booth and Ferrell, 1977) contributing to low residue levels in sediments. Furthermore, *Pseudomonas* can use DFB as a sole carbon and nitrogen source (Booth and Ferrell, 1977) thereby reducing the available DFB in the environment.

3.5.2 *Fungal growth and CO_2 elimination*

3.5.2.1 *Mycelial growth.* — The mycelial growth of each fungus was measured in cm from the center of each inoculated plate. Table 6 shows the mycelial growth for *Rhizopus arrhizus*. There were no significant differences in the growth at 24, 33, and 48 hours after inoculation for any of the treatment levels. The growth appeared to be linear.

Table 7 shows the growth for *Aspergillus niger* up to 47 hours. It was very difficult to measure linear growth for this fungus since the spores spread to different locations on the agar surface. Hence, growth was not seen at a central source outward but rather at different localities on the surface. No inhibition occurred at any level of treatment. At 168 hours all of the plates had growth and sporulation which appeared to be identical.

Table 8 shows the growth of *Fusarium oxysporum* at 31, 48, 60, 120, and 168 hours after inoculation. No inhibition occurred at any treatment level.

Table 9 summarizes the growth data for *Pythium debaryanum* at 23, 31, 47, and 55 hours from inoculation. Again, no inhibition occurred at any treatment level.

Table 6. x̄ Rate of mycelial growth (cm) of *Rhizopus arrhizus* at six different Dimilin° treatment levels (ppm) for three sampling periods (x ± S.D.).

Time (hours)	cm of growth					
	0	inert	1 ppm	10 ppm	50 ppm	100 ppm
24	1.7 ± 0.29	1.7 ± 0.11	1.6 ± 0.1	1.7 ± 0.15	1.6 ± 0.11	1.7 ± 0.11
33	2.9 ± 0.20	2.7 ± 0	2.8 ± 0.06	2.8 ± 0.15	2.7 ± 0.20	2.7 ± 0.10
48	4.0 ± 0	4.0 ± 0	4.0 ± 0	4.0 ± 0	4.0 ± 0	4.0 ± 0

P > 0.05 at all treatments.

Table 7. x̄ Rate of mycelial growth (cm) of *Aspergillus niger* at six different Dimilin° treatment levels (ppm) for three sampling periods (x̄ ± S.D.).

Time (hours)	cm of growth					
	0	inert	1 ppm	10 ppm	50 ppm	100 ppm
23	0.87 ± 0.06	0.77 ± 0.06	0.58 ± 0.13	0.63 ± 0.05	0.5 ± 0.09	0.8 ± 0.1
31	1.06 ± 0.11	1.03 ± 0.06	0.83 ± 0.11	0.73 ± 0.23	0.7 ± 0.1	1.0 ± 0.1
47	1.37 ± 0.25	1.33 ± 0.06	1.13 ± 0.15	1.07 ± 0.11	1.07 ± 0.11	1.4 ± 0.1

P > 0.05 at all treatments.

Table 8. x̄ Rate of mycelial growth (cm) of *Fusarium oxysporum* at six different Dimilin° treatment levels (ppm) for three sampling periods (x ± S.D.).

Time (hours)	cm of growth					
	0	inert	1 ppm	10 ppm	50 ppm	100 ppm
31	0.57 ± 0.12	0.57 ± 0.13	0.58 ± 0.08	0.55 ± 0.09	0.58 ± 0.11	0.67 ± 0.11
48	1.07 ± 0.12	1.1 ± 0.1	1.1 ± 0.1	1.23 ± 0.06	1.07 ± 0.15	1.07 ± 0.15
60	1.77 ± 0.6	1.77 ± 0.6	1.73 ± 0.06	1.77 ± 0.06	1.67 ± 0.15	1.7 ± 0.17
120	3.1 ± 0.2	3.1 ± 0.1	2.9 ± 1	2.9 ± 0.06	2.9 ± 0.12	2.7 ± 0
168	4.23 ± 0.06	4.2 ± 0.1	4.0 ± 0	4.17 ± 0.15	4.07 ± 0.06	3.9 ± 0.17

P > 0.05 at all treatments.

Table 9. X̄ Rate of mycelial growth (cm) of *Pythium debaryanum* at six different Dimilin® treatment levels (ppm) for three sampling periods (x̄ ± S.D.).

Time (hours)	cm of growth					
	0	inert	1 ppm	10 ppm	50 ppm	100 ppm
23	1.27 ± 0.12	1.3 ± 0.1	1.33 ± 0.12	1.33 ± 0.12	1.3 ± 0.1	1.3 ± 0
31	2.2 ± 0.1	2.1 ± 0.1	2.13 ± 0.06	2.2 ± 0.26	2.05 ± 0.09	1.93 ± 0.12
47	3.87 ± 0.15	4.0 ± 0	3.83 ± 0.12	4.0 ± 0	3.67 ± 0.06	3.53 ± 0.15
55	4.0 ± 0	4.0 ± 0	4.0 ± 0	4.0 ± 0	4.0 ± 0	4.0 ± 0

P > 0.05 at all treatments.

Table 10. X̄ Rate of mycelial growth (cm) of *Trichoderma viride* at six different Dimilin® treatment levels (ppm) for three sampling periods (x̄ ± S.D.).

Time (hours)	cm of growth					
	0	inert	1 ppm	10 ppm	50 ppm	100 ppm
31	1.37 ± 0.15	1.3 ± 0.26	1.23 ± 0.25	1.3 ± 0.1	1.13 ± 0.06	1.23 ± 0.12
79	4.27 ± 0.12	4.2 ± 0.1	4.17 ± 0.06	4.13 ± 0.06	4.13 ± 0.06	4.17 ± 0.06

P > 0.05 at all treatments.

Table 10 shows the growth of *Trichoderma viride* at 31 and 79 hours after inoculation. No inhibition occurred and growth was virtually linear.

Table 11 shows the growth of mycelia for the microrhizian, *Rhizopogan vinicolor* at 3, 4, 5, 7, 9, 12, 16, 21, and 29 days after inoculation. Considerable variation is seen in the growth of this organism, but no inhibition occurred at any level.

3.5.2.2 *Germination*. — Table 12 summarizes the germination data for all of the fungi except *Pythium debaryanum* and *Rhizopogan vinicolor* which did not sporulate under normal growth conditions. In general, there were no significant differences in germination rates. However, it appears that the inert ingredient actually increases growth in *Rhizopus* when used without DFB.

The size of the spores on *Trichoderma viride* were too small to determine accurate percentages. Thus, only the relative degree of germination, in comparison with the control, could be recorded. A remarkable confluency of germination occurred at 50 and 100 ppm for *Trichoderma*.

3.5.2.3 *Mycelial weight*. — Table 13 shows the weights of the mycelia for *Rhizopus arrhizus*. No inhibition occurred but significant increases in weight occurred at higher levels of DIMILIN®.

Table 13 summarizes the weights of *Aspergillus niger* data. No increases nor decreases in weight were observed for this oganism. The same pattern was true for *Fusarium oxysporum* (Table 13) and *Trichoderma viride* (Table 13). However, in *Pythium debaryanum* an increase in weight was observed for the inert ingredient (Table 13).

3.5.2.4 *CO_2 evolution from Rhizopus*. — Figure 15 depicts clearly that 100 ppm of DIMILIN® applied to an inoculation of *Rhizopus arrhizus* did not effect CO_2 respiration.

In summary, most fungi contain chitin and are important decomposers of leaf litter. Hence, potential toxic effects associated with DFB treatments are obvious. However, it is clear from these experiments that it does not negatively impact fungi. Additional experiments are needed to explain this lack of activity, although preliminary unpublished experiments from our laboratory have shown that *Rhizopus* eliminates ^{14}C-DIMILIN® rapidly. These data are suggestive that rapid excretion may account in part for the non-toxicity.

3.6 *Effect on bobwhite reproduction*

The bobwhite production data for the total 12-week period is summarized in Table 14.

Table 11. X̄ Rate of mycelial growth (cm) of *Rhizopogan vinicolor* (mycorrhizae) at six different Dimilin® treatment levels (ppm) for three sampling periods (x̄ ± S.D.).

Time (hours)	cm of growth					
	0	inert	1 ppm	10 ppm	50 ppm	100 ppm
3	0.5 ± 0	0.47 ± 0.06	0.4 ± 0	0.5 ± 0.1	0.4 ± 0	0.4 ± 0
4	0.6 ± 0	0.57 ± 0.06	0.5 ± 0	0.6 ± 0.1	0.5 ± 0.1	0.48 ± 0.19
5	0.7 ± 0	0.73 ± 0.12	0.6 ± 0	0.7 ± 0.1	0.6 ± 0.1	0.6 ± 0.1
7	0.92 ± 0.08	0.97 ± 0.15	0.8 ± 0	0.93 ± 0.12	0.8 ± 0.1	0.8 ± 0.1
9	1.28 ± 0.06	1.4 ± 0.10	1.02 ± 0.08	1.17 ± 0.06	0.92 ± 0.08	0.93 ± 0.15
12	1.82 ± 0.03	2.0 ± 0.10	1.57 ± 0.15	1.73 ± 0.23	1.40 ± 0.10	1.27 ± 0.21
16	2.47 ± 0.06	2.63 ± 0.15	2.17 ± 0.15	2.23 ± 0.12	1.93 ± 0.15	1.7 ± 0.30
21	2.9 ± 0.17	3.2 ± 0.17	2.77 ± 0.25	2.70 ± 0.10	2.47 ± 0.12	2.30 ± 0.30
29	3.17 ± 0.31	3.7 ± 0	3.23 ± 0.47	3.0 ± 0.10	2.93 ± 0.35	2.87 ± 0.29

P > 0.05 at all treatments.

Table 12. X̄ Percent of spores germinating per total spores observed in microscopic field at six different Dimilin® treatment levels (ppm) (x̄ ± S.D.).

Organism	0	inert	1 ppm	10 ppm	50 ppm	100 ppm
Rhizopus	48.6 ± 11.1	90.6 ± 8.14*	49.3 ± 13.01	42.0 ± 5.0	52.6 ± 17.9	65 ± 8.7
Aspergillus	94.3 ± 2.1	90.3 ± 16.7	98 ± 2.8	97.3 ± 4.6	100 ± 0	94.3 ± 9.8
Fusarium	26.1 ± 11.8	32.7 ± 3.8	41.5 ± 5.4	45.3 ± 5.1	49.0 ± 13.6	45 ± 8.2
Trichoderma[a]	+	++	+	+	+++	+++

P > 0.05 at all treatments except inert with *Rhizopus*.

*P < 0.05.

[a] The size of the *Trichoderma* spores were too small to determine accurate germination percentages. Hence, only relative percentages could be recorded by numbers of (+), i.e. +++ > ++ > + in terms of germination.

188

Table 13. X̄ Total mycelial weight (gms) of the respective fungi for six different Dimilin® treatment levels (ppm).

Organism	0	inert	1	10	50	100
Rhizopus	0.2864	0.3313	0.4459	0.5772	0.4423	0.3389
Aspergillus	0.4113	0.5527	0.3924	0.4530	0.4521	0.4953
Fusarium	0.4498	0.4145	0.4749	0.3969	0.4353	0.4671
Trichoderma	0.1677	0.2536	0.1423	0.1874	0.2272	0.2916
Pythium	0.0939	0.2177	0.1378	0.0693	0.0413	0.0694

$P > 0.05$ at all treatments.

A summary of statistical analysis of the data for eggs laid, eggs cracked, eggs set, eggs fertile, eggs hatched, and feed consumed is shown in Table 15.

There were clearly no differences on any of the response variables due to rooms or levels. Battery unit (rack) differences were insignificant on four of the six variables and the pooled interactions were insignificant on five of the six variables. This means that future bobwhite experiments could probably be conducted in one room and the treatments could be assigned to cages without concern for a rack or layer effect.

The treatment effects were significant at the 0.05 level for all six dependent variables. The five treatment means and the statistically significant differences among means are shown in Table 16 for each of the variables. There were 11 significant differences out of 60 possible differences which were examined. For eggs laid, set, and hatched, T2 (25 ppm) was significantly higher than both T1 (2.5 ppm) and C (control), while T2 was higher than C for fertile eggs.

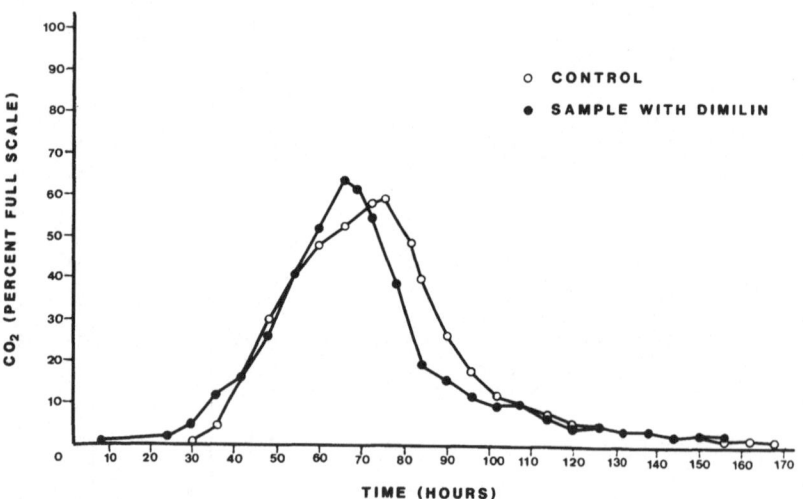

Figure 15. Plot of carbon dioxide (CO_2) evolution from *Rhizopus arrhizus* in treated (100 ppm) and control samples.

No other differences were statistically significant for these egg parameters.

The number of cracked eggs was significantly higher for Co (control + oil) than for C and T2, and the birds on C or Co consumed more feed than those on T1. While these differences were statistically significant, the actual differences appear to be small as to have no biological consequences and in most cases the best egg production or fertility was associated with T2.

Based on these results, DFB clearly has *no* effect on eggs laid, cracked, set, fertile, or hatched.

The differences in the egg parameters with respect to weeks was highly significant statistically but the actual observed differences were small and also appear to have no biological significance.

The egg shell thinning data showed that the treatment by weeks interaction was decidedly non-significant for the C6 (equator) and C8 (apex) measurement. A reduced model for those two measurements was used as:

$$Y_{ijk} = u + T_i + W_j + e_{ijk}$$

where:

u is the overall mean.
T_i is the effect of the ith treatment.
W_j is the effect of the jth week.
e_{ijk} is the effect of the kth egg in the ith treatment and jth week.

There are several statistically significant results in the data, but these data are not likely of any practical significance. It is well documented that very small differences can be found to be statistically significant if a large enough sample size is used. Five hundred eggs is a fairly large sample. The very small differences observed in the means suggest that weekly differences were just random variation and the statistical significance of weeks was mainly due to the large sample size and is of no practical significance.

The treatment comparisons were all non-significant with two exceptions. In C6, the three DFB treatments were associated with greater egg shell thickness than the two control treatments. No other comparisons were significant for C6. For C7, the 250 ppm level of DFB was associated with a greater egg shell thickness than the 2.5 ppm treatment. No other comparisons were significant for C7 and there were no significant treatment differences of any kind for C8.

It is questionable if the two significant treatment comparisons are of any practical or biological significance.

Since the experimental unit in all of these studies was each cage within a given level and battery unit, the data on hatched chicks could not be

Table 14. Summary of the data from the effect of diflubenzuron on bobwhite quail reproduction.

Variable	C (Control)	Co (Control + Oil)	T$_1$ (2.5 ppm)	T$_2$ (25 ppm)	T$_3$ (250 ppm)	TOT &/or Ave (High and Low)
# Birds						
Male/Female	89.5/89.5	86/88.13	79.04/77.96	87.42/88	85.11/84.91	427.1/428.5 (85.4/85.7)
# Eggs Laid	393.6	445.5	351.6	473.2	394.2	2058.7 (411.7)
# Cracked Eggs	3.8	6.4	4.5	4.1	4.3	23 (4.6)
# Eggs Set	389.8	439.1	347.1	469.1	389.9	2035.7 (407.14)
# Fertile Eggs	301	338	265	361	301	1566 (313)
# Hatched Eggs	251	277	216	297	241	1282 (256)
Shell Thickness Equator Ave (mm): (hi-low)	0.1932	0.1941	0.1941	0.1998	0.1965	0.1956

Cap Ave (mm): (hi-low)	0.2024	0.2070	0.2032	0.2006	0.2002	0.2003
Apex Ave (mm): (hi-low)	0.2322	0.2344	0.2404	0.2294	0.2273	0.2290
Feed Consumption Ave (g/bird/wk) (hi-low)	135.77	135.6	135.4	132.8	137.0	138.2
# Deaths (adults)	17	4	1	7	3	2
Percentage	0.02%	0.30%	0.08%	0.58%	0.25%	0.2%
# 14-Day Survival	219	204	259	189	236	209
# Eggs/Hen Day/Wk/Season	0.68/4.80/48.0	0.66/4.65/46.5	0.77/5.40/54.0	0.64/4.50/45.0	0.72/5.03/50.3	0.63/4.42/44.2
% Cracked	1.1	1.1	0.88	1.3	1.3	0.96
% Fertile	77.2	77.2	76.8	77.6	77.0	77.2
% Hatch	81.8	80.0	82.0	82.0	82.0	83.4
% 14-Day Survival	85.6	84.6	87.2	87.5	85.2	83.3
14-Day Survival/Hen	2.56	2.40	2.94	2.42	2.68	2.34

Table 15. Summary of significance levels for the 6 variables shown.

	Laid	Cracked	Set	Fertile	Hatched	Feed
Room	n.s.	n.s.	n.s.	n.s.	n.s.	n.s.
Rack	n.s.	n.s.	n.s.	0.05	0.05	n.s.
Level	n.s.	n.s.	n.s.	n.s.	n.s.	n.s.
Treatment	0.05	0.05	0.05	0.05	0.05	0.05
Pooled interactions	n.s.	0.05	n.s.	n.s.	n.s.	n.s.
Cages						
Weeks	0.01	0.01	0.01	0.01	0.01	0.01
Pooled int. × weeks	0.05	0.05	0.05	n.s.	n.s.	n.s.

analysed in the same manner as for the other parameters because the hatching chicks could not be traced precisely to a given egg in the incubator. Therefore, the % 14-day chick survival and the 14-day chick survival were analysed by Students "t" test by comparing each treatment mean with the control mean. Apparently, T2 produced a slightly larger chick/hen ratio. The other significant differences were meaningless since one of them involved differences between two controls and the other showed a small increase in chick survival in T1 compared to the control.

Based on these data, it is not likely that DIMILIN® applications will cause reproductive perturbations to bobwhite quail.

3.7 Teratogenic and biochemical effects on chick embryos

3.7.1 Teratology and membrane transfer

Table 17 shows the results of the teratology study conducted on 703 chick embryos. The control malformations were corrected using Abbot's formula. DFB did not show significant teratogenic activity over time when injected at 10 mg/egg. Table 18 shows the total % ^{14}C residues found in the

Table 16. Treatment means and significant differences among means for eggs laid, cracked, set, fertile, hatched, and feed consumption.

Treatment	Laid	Cracked	Set	Fertile	Hatched	Feed
C	8.79	0.081	8.71	7.45	6.60	139.02
Co	10.21	0.159	10.05	8.65	7.62	137.01
T_1	8.78	0.112	8.67	7.68	6.62	132.73
T_2	10.63	0.093	10.54	9.10	8.18	135.21
T_3	9.38	0.103	9.27	8.03	7.05	135.46
	T_2 T_1	Co C	T_2 T_1	T_2 C	T_2 C	C T_1
	T_2 C	Co T_2	T_2 C		T_2 T_1	Co T_1
C*	9.50	0.120	9.38	8.05	7.11	138.02
T**	9.60	0.103	9.49	8.27	7.28	134.47

C* = mean of C and Co for each parameter.

T** = mean of T_1, T_2 and T_3 for each parameter.

Table 17. Summary of teratological effects of TH-6040 on chick embryos.

Days After Treatment	Treatment	# in Sample	# Living Recovered	% Living	# Malformed	% Malformed	Corrected % Malformed[b]
1	P.O.	61	41	67	2	4.9	1.2
	10 mg	66	45	68	2	4.5	0.8
2	P.O.	53	32	60	2	6.3	2.7
	10 mg	65	44	68	3	6.8	3.2
3	P.O.	54	48	89	0	0	0
	10 mg	68	57	84	4	7	3.4
4	P.O.	61	50	82	2	4	0.3
	10 mg	73	64	88	3	4.7	1.0
6	P.O.	49	38	78	0	0	0
	10 mg	56	53	95	2	3.8	0.1
—	Control (no injection) Total Animals =	97 / 703	82	85	3	3.7	0

[a] Treatment included peanut oil (P.O.), 10 mg of diflubenzuron or no injection (control).

[b] Corrected % malformed = $\dfrac{\text{observed \% malformed} - \text{control \% malformed}}{100 - \text{control \% malformed}} \times 100$.

$P > 0.05$.

Table 18. Total % [^{14}C] residues found in wash, yolk, shell, membranes, and embryo at given incubation times with [^{14}C] diflubenzuron.

Hours Incub. After Treat.	Embryo Wash	Embryo Homog.	Yolk	Shell	Memb.	Total	% Recovery[a]
71	–	0.12	97.4	2.4	–	100	95
97	2.3	0.06	96.9	0.79	–	100	87
142	0.37	0.15	98.7	0.18	0.56	100	102
167	–	0.06	99.5	0.45	–	100	86
224	0.19	0.17	96.9	0.12	2.7	100	123

[a] These values represent the % [^{14}C] recovery of the injected dose.
$P > 0.05$.

embryo wash, embryo homogenate, yolk, shell, and extra-embryonic membranes. Less than 0.2% of the total amount injected partitioned into the embryo over the 224 hour period. Greater than 96% of the total counts remained in the yolk fraction, while only trace amounts remained in the embryo wash, shell and membrane. This lack of membrane transfer may explain why DFB is not a significant teratogenic agent to chick embryos (Table 17).

The yolk of several animals at the different sampling periods was extracted with diethyl ether (Table 19). The tlc analysis of the yolk extracts showed that > 90% of the residues was DFB at all sampling periods.

3.7.2 *Chrondroitin sulfate and hyaluronic acid analysis*

All of the data cited in Tables 20–27 under this section were analyzed by Students' "t" test, with at least 9 degrees of freedom for each value. Tables 20 and 21 summarize the [^{35}S] *in vitro* and *in ovo* incorporation into chondroitin over a 48-hour period after treatment with 10 mg of DFB. Sulfation of chondroitin clearly was not inhibited. The percent cpc precipitable cpm also confirms the normal incorporation of the [^{35}S] into macromolecules.

Tables 22 and 23 show that [^{3}H]-glucosamine was also incorporated into the glycosaminoglycan molecules *in vitro* and *in ovo* in a normal manner. No significant differences between the treatments and the control

Table 19. % Extractions of Chick Yolk Using Reagent Diethyl Ether.

Hours Incubated After Treatment	% Extraction
71	77
97	50
142	40
167	35
224	12

Table 20. In vitro [^{35}S] incorporation into chondroitin of chick tibias following *in ovo* treatment with 10 mg diflubenzuron/embryo.[a] Values are means ± S.E.

Parameter	Control	Diflubenzuron Treatment Period		
		2 Hour	24 Hour	48 Hour
X̄ total cpm	48,042 ± 1079	50,508 ± 807	48,498 ± 2463	45,263 ± 985
cpc prec.[b]	44,411 ± 1107	47,034 ± 988	44,405 ± 2180	41,215 ± 1220
% cpc prec.	92%	93%	92%	91%
% of control cpm	–	105%	101%	94.2%
% of control cpc prec.	–	106%	100%	92.8%

[a] Data expressed as total cpm/mg dry weight.
[b] Cpc prec. = cpm taken up by cartilage which are cpc precipitable (i.e. cpm bound to macromolecules).
P > 0.05.

Table 21. In ovo [^{35}S] incorporation into chondroitin of chick tibias following *in ovo* treatment with 10 mg diflubenzuron/embryo[a]. Values are means ± S.E.

Parameter	Control	DFB Treatment Period 48 Hours
X̄ total cpm	29,147 ± 7,784	30,235 ± 3,021
cpc prec.[b]	27,689 ± 4,104	29,358 ± 2,111
% cpc prec.	95%	97%
% of control cpm	–	104%
% of control cpc prec.	–	106%

[a] Data expressed as total cpm/mg dry weight.

[b] Cpc prec. = cpm taken up by cartilage which are cpc precipitable (i.e. cpm bound to macromolecules).

P > 0.05.

were observed. Table 24 shows that the total [^{3}H]-glucosamine incorporated into chondroitin of whole chick tibias was not affected.

Glucuronic acid total content estimate of chondroitin sulfate into macromolecules was also normal in the treated chick tibias (Table 25). Chondroitinase and hyaluronidase digestion of the [^{35}S] and [^{3}H]-glucosamine treated chick cartilage is summarized in Tables 26 and 27 respectively. Chondroitinase digested over 91% of the cpc precipitable counts as shown in Table 26. The enzyme digestion of chondroitin confirms the conclusion that DFB does not inhibit normal synthesis of chondroitin from metabolic precursors. The hyaluronidase treatment showed that about 0.6% of the control cpc precipitable counts were digested. The hyaluronidase recovered-cpm from the treatments ranged from 2.6% to 18.3%. Since the percent cpc precipitable cpm were normal, no disruption in hyaluronic acid synthesis could have occurred.

DFB has been shown to inhibit incorporation of ^{14}C labeled uridine diphosphate N-acetyl glucosamine (UDPAG) into chitin of insects (Deul et al., 1978) similar to the action of Polyoxin D.

Higher oranisms also make extensive use of UDPAG in the formation of glycosaminoglycans, ubiquitiously found in connective tissue matrix. For example, chondroitin sulfate is a high molecular weight constituent of cartilage matrix. The molecule is comprised of repeating disaccharide units of galactosamine and glucuronic acid which is covalently bonded at one end of the chain to a core protein via galactose and xylose. The amino sugar is sulfated at the 4 or 6 position. Hyaluronic acid, on the other hand, is of high molecular weight and consists of glucosamine and glucuronic acid, as repeating disaccharide units, and is an important component of cartilage and other connective tissue matrix.

These experiments have shown that DFB is not teratogenic to chick embryos, does not block the formation of hyaluronic acid and chondroitin sulfate in the chick embryo, nor does it cross membrane barriers of embryos.

Table 22. In vitro [³H]-glucosamine incorporation into chondroitin of chick tibias following in ovo treatment with 10 mg diflubenzuron/embryo.[a] Values are means ± S.E.

Parameter	Control	Diflubenzuron Treatment Period		
		2 Hour	24 Hour	48 Hour
X̄ total cpm	20,778 ± 1867	21,728 ± 1945	22,972 ± 3003	21,256 ± 2227
cpc prec.[b]	18,779 ± 1810	19,294 ± 2122	19,094 ± 2864	18,916 ± 2517
% cpc prec.	90%	89%	83%	89%
% of control cpm	—	105%	111%	102%
% of control cpc prec.	—	103%	102%	101%

[a] Data expressed as total cpm/mg dry weight.
[b] Cpc prec. = cpm taken up by cartilage which are cpc precipitable (i.e. cpm bound to macromolecules).
P > 0.05.

Table 23. In ovo [³H]-glucosamine incorporation into chondroitin of chick tibias following *in ovo* treatment with 10 mg diflubenzuron/embryo. Values are means ± S.E.

Parameter	Control	Diflubenzuron Treatment Period	
		24 Hours	48 Hours
X̄ total cpm[a]	1,083 ± 230	1,137 ± 318	1,385 ± 257
% of control cpm	–	105%	127%

[a] cpm/mg dry weight. The [³H]-glucosamine was incorporated during a two hour *in ovo* label with 10 uci.
$P > 0.05.$

Table 24. In vitro [³H]-glucosamine incorporation into chondroitin of whole chick tibias following *in ovo* treatment with 10 mg diflubenzuron/embryo[a]. Values are means ± S.E.

Parameter	Control	Diflubenzuron Treatment Period	
		24 Hour	48 Hour
X̄ total cpm	55,214 ± 3719	58,401 ± 1966	62,346 ± 2360
% of control cpm	–	106%	113%

[a] Values represent cpm in whole tibias dissolved in Protosol® and counted in toluene cocktail. Values given in cpm/mg tissue dry weight.
$P > 0.05.$

Table 25. Glucuronic acid assay data where values are expressed as ug glucuronic acid per mg dry weight. Values are means ± S.E.

Parameter	Control	Diflubenzuron Treatment Period		
		2 Hour	24 Hour	48 Hour
X̄ total cpm	38.9 ± 1.07	37.5 ± 0.92	36.1 ± 1.90	36.8 ± 0.95
% of control cpm	–	96.4%	92.8%	94.6%

$P > 0.05.$

Table 26. Chondroitinase treatment of labeled chick cartilage[a] following treatment with peanut oil (control) and diflubenzuron.

Parameter	[³⁵S] Label			[³H]-glucosamine		
		DFB Treatment			DFB Treatment	
	Control	24 Hour	48 Hour	Control	24 Hour	48 Hour
Total cpm	2906	1858	2435	1076	860	1048
cpc prec. cpm[b]	2571	1475	2119	920	699	912
% cpc prec.	88.5%	79.4%	87.0%	85.5%	81.3%	87.0%
cpc prec. cpm following chondroitinase	178	102	180	62	44	76
% cpm digested by chondroitinase	93.1%	93.1%	91.5%	93.3%	93.0%	91.7%

[a] Incubation media: 0.05 M Tris buffer, pH = 8.0 chondroitinase, 2 units/ml buffer 1:1 dilution of homogenate with enzyme solution
[b] cpc prec. = cpm taken up by cartilage which are cpc precipitable (i.e. cpm bound to macromolecules).
$P > 0.05.$

Table 27. Hyaluronidase treatment of labeled chick cartilage[a] following treatment with peanut oil (control) and diflubenzuron.

Parameter	[^{35}S] Label Control	DFB Treatment 24 Hour	48 Hour	[^{3}H]-glucosamine Control	DFB Treatment 24 Hour	48 Hour
Total cpm	954	723	798	473	361	547
cpc prec. cpm[b]	859	615	718	402	307	465
% cpc prec.	90.0%	85.1%	90.0%	85.0%	85.0%	85.0%
cpc prec. cpm following hyaluronidase	854	564	688	370	299	380
% cpm digested by hyaluronidase	0.6%	8.3%	4.2%	8.0%	2.6%	18.3%

[a] Incubation solution: 0.15 M NaCl buffer, pH = 5.7 Hyaluronidase, 0.1 mg 1 ml in 0.01 M sodium acetate 1:1 dilution of homogenate with enzyme solution
[b] cpc prec. = cpm taken up by cartilage which are cpc precipitable (i.e. cpm bound to macromolecules).
P > 0.05.

3.8 Investigations on rat serum testosterone, weight, and food consumption

3.8.1 Testosterone

Table 28 shows the mean testosterone levels in the rat serum by cage, day of sacrifice, and treatment; the overall mean for the replications within a given treatment and sacrifice day is also shown. It should be noted that the day 0 × treatment values were arbitrarily separated into 5 fractions for convenience in the analysis and for completeness of the matrix in Table 28. This procedure did not change the overall significance levels.

Significance levels for T and T × D did not approach the $\alpha = 0.05$ probability level, i.e. the overall F-test shows no significant differences in the testosterone levels in the experiment.

A log and square-root transformation of the testosterone data were also performed to adjust for heterogeneity of variance; however, this transformation did not change the above conclusion nor any of the significance levels.

A one-way ANOV was also conducted on each treatment compared to the control within a given sampling period. For example, on day 96 the testosterone levels at 75 ppm were compared to those from the control; those values at 150 ppm were compared to those from the control, and so forth. The F-tests were all non-significant therefore there were no significant differences between the control and treatment testosterone levels within each sampling period.

Table 28. Mean testosterone levels in rat serum (ng/100 ml).

Sampling Days	Treatments 0 ppm	75 ppm	150 ppm	300 ppm	3000 ppm
0	49.0	62.9	44.5	11.7	10.3
	53.8	64.9	57.4	50.6	84.6
	39.0	69.7	36.8	83.2	10.5
	48.6	126.1	62.3	48.5	94.2
	$\bar{x} = 47.6$	$\bar{x} = 80.9$	$\bar{x} = 50.3$	$\bar{x} = 48.5$	$\bar{x} = 49.9$
14	553.6	127.1	160.5	685.6	83.0
	118.1	91.0	86.6	209.9	195.9
	131.5	172.4	125.2	40.1	408.6
	9.3	398.1	68.8	150.3	81.2
	208.8	47.3	100.6	94.7	47.5
	$\bar{x} = 204.3$	$\bar{x} = 167.2$	$\bar{x} = 108.3$	$\bar{x} = 236.1$	$\bar{x} = 163.2$
28	1045.8	544.2	683.4	249.5	662.5
	306.4	495.9	600.9	985.9	923.1
	395.5	216.1	1168.5	600.8	704.8
	892.9	628.6	898.0	739.1	330.8
	774.5	974.0	716.1	771.4	276.9
	$\bar{x} = 683.0$	$\bar{x} = 571.8$	$\bar{x} = 813.4$	$\bar{x} = 669.3$	$\bar{x} = 579.6$
42	558.0	396.5	554.6	543.9	776.7
	415.0	250.9	255.9	593.5	324.7
	464.7	324.5	420.1	1400.6	338.4
	310.1	146.3	953.8	1011.0	250.8
	1026.3	955.6	1065.3	941.5	500.8
	$\bar{x} = 554.8$	$\bar{x} = 414.8$	$\bar{x} = 649.9$	$\bar{x} = 898.1$	$\bar{x} = 438.3$
96	347.0	432.0	653.0	154.6	143.2
	698.0	408.6	185.1	85.6	385.1
	292.2	255.1	178.0	552.3	379.4
	427.7	595.8	579.8	310.5	824.6
		443.5	484.8	102.4	
	$\bar{x} = 441.2$	$\bar{x} = 427.0$	$\bar{x} = 416.1$	$\bar{x} = 241.1$	$\bar{x} = 433.1$

$P > 0.05$.

3.8.2 *Weight*

Table 29 summarizes the mean weight of the rats by cage, day of sacrifice, and treatment; the overall mean for the replications within a given treatment and sacrifice day is also shown.

The F-ratios and significance levels of treatment, days, and the interaction of treatments with days were analyzed. The treatments were significant at $\alpha = 0.0272$, days at $\alpha = 0.0000$, but the T × D interaction was not significant. The average rat weight at 300 ppm was significantly lower than the weight at the 0 ppm level; but the decrease in weight was not dose-related since the weight at 3000 ppm was not significantly different from the control. The weights at the different days were significantly different from the controls because the rats were gaining weight over time. The only other linear combinations that were significant were the (75 ppm

Table 29. Mean weights of rats (grams).

Sampling Days	Treatments 0 ppm	75 ppm	150 ppm	300 ppm	3000 ppm
0	61.7	65.0	64.9	61.4	56.2
	65.3	65.5	63.1	56.1	64.3
	66.7	67.1	67.8	62.7	61.6
	55.4	60.6	61.4	66.4	60.2
	64.6	64.9		51.4	
	$\bar{x} = 62.7$	$\bar{x} = 64.6$	$\bar{x} = 64.3$	$\bar{x} = 59.6$	$\bar{x} = 60.6$
14	147.0	151.2	148.9	146.6	129.3
	142.2	142.4	139.1	158.8	150.6
	145.2	150.2	141.0	143.3	141.3
	153.5	140.8	145.6	141.6	139.1
	137.7	151.0	151.3	139.7	143.8
	$\bar{x} = 145.1$	$\bar{x} = 147.1$	$\bar{x} = 145.2$	$\bar{x} = 146.0$	$\bar{x} = 140.8$
28	218.9	236.2	238.4	223.0	233.5
	244.1	234.0	238.2	227.4	237.4
	233.6	234.5	245.1	231.2	227.7
	240.1	228.3	210.1	197.8	219.8
	219.1	225.8	239.0	229.9	227.2
	$\bar{x} = 231.2$	$\bar{x} = 231.8$	$\bar{x} = 234.2$	$\bar{x} = 221.9$	$\bar{x} = 229.1$
42	279.9	301.9	305.9	297.3	278.5
	298.2	279.7	282.6	287.4	294.9
	285.1	292.1	286.1	285.0	277.1
	296.5	307.8	273.6	302.4	277.4
	299.4	281.9	294.6	260.9	294.0
	$\bar{x} = 291.8$	$\bar{x} = 292.7$	$\bar{x} = 286.6$	$\bar{x} = 286.6$	$\bar{x} = 284.4$
96	393.5	334.1	406.8	313.9	362.5
	383.3	335.5	344.2	350.5	378.1
	395.5	379.9	376.8	366.4	384.6
	401.0	367.9	385.3	362.0	380.6
		393.2	403.0	383.9	391.7
	$\bar{x} = 393.8$	$\bar{x} = 362.3$	$\bar{x} = 383.2$	$\bar{x} = 355.3$	$\bar{x} = 379.5$

$P > 0.05.$

vs 0 ppm) vs (Day 96 vs Day 0) and (300 ppm vs 0 ppm) vs (Day 96 vs Day 0).

An additional square-root transformation of the rat weight data did not affect the final result.

The results of a one-way ANOV of the weight data within each sampling period showed no treatment effects.

3.8.3 *Food-consumption*

Table 30 shows the mean food-consumption of the rats by cage, day of sacrifice, and treatment; the overall mean for the replications within a given treatment and sacrifice day is also shown.

Table 30. Mean food consumed/rat/day (grams).

Sampling Days	Treatments 0 ppm	75 ppm	150 ppm	300 ppm	3000 ppm
0	–	–	–	–	–
14	17.01	14.34	10.91	10.03	13.47
	11.76	14.14	12.45	11.82	14.81
	19.00	15.65	15.57	13.29	16.76
	11.84	12.99	14.58	13.40	10.90
	15.12	11.37	11.57	9.68	15.26
	x̄ = 14.95	x̄ = 13.70	x̄ = 13.02	x̄ = 11.64	x̄ = 14.24
28	18.13	17.93	19.19	17.04	16.44
	19.94	19.07	17.87	16.69	15.98
	17.30	16.68	19.27	17.58	18.21
	17.49	17.58	18.45	17.82	17.47
	16.76	16.09	18.50	16.83	17.25
	x̄ = 17.92	x̄ = 17.47	x̄ = 18.66	x̄ = 17.19	x̄ = 17.07
42	20.85	19.96	22.39	17.64	19.21
	22.70	18.93	20.60	19.70	18.08
	20.39	18.42	22.19	20.03	18.55
	19.81	20.63	19.02	17.28	19.57
	18.74	18.83	19.46	19.16	18.80
	x̄ = 20.50	x̄ = 19.35	x̄ = 20.73	x̄ = 18.76	x̄ = 18.84
96	18.72	18.62	20.71	18.53	18.77
	19.92	20.45	21.64	18.76	18.86
	22.50	18.58	20.03	19.73	19.22
	19.66	20.11	20.54	20.11	19.06
		20.38	20.91	18.82	20.56
	x̄ = 20.20	x̄ = 19.63	x̄ = 20.77	x̄ = 19.19	x̄ = 19.29

$P > 0.05$ overall.

Data were summarized on the F-ratios and significance levels by treatment, days, and the interaction of treatments with days. The treatments were significant at $\alpha = 0.0017$ and of course the days were significant at $\alpha = 0.0000$ because as the rats grew, they obviously consumed more food. However, the T × D interaction was not significant. The estimated linear combinations showed the 300 ppm and 3000 ppm treatments to be significantly lower when compared to the controls.

Apparently, this depressed food consumption is not related to the amount of DFB actually consumed since the rats at day 14 and 3000 mg/kg diet, consumed the greatest amount of DFB and yet did not have depressed food consumption.

4.0 SUMMARY AND CONCLUSIONS

Applications of acetonic solutions of DFB to soil-coated glass slides causes remarkable crystal growth to greater than 10 microns wherein the

203

soil particles seem to act as centers for crystal growth. These large crystals decrease insect toxicity and tend to increase the persistency of the chemical. A high melting point and low water solubility results in a high energy of crystallization. Therefore, the rate of dissolution of these larger particles would be slow, causing the crystals to remain in a suspension for a longer period of time. However, particles of air-milled aqueous samples of DIMILIN® W-25 applied to soil remains 1–5 microns in size. All investigations of DIMILIN® should utilize only air-milled samples to preclude the possibility of increased crystal growth.

Environmental samples (water, forest litter, and sediment) generally contain less than 5 ppb shortly after application even though DFB adsorbs quickly to sediments. Apparently, the chemical degrades rapidly in soil via chemical and microbial processes. The major metabolite in an activated sludge system was CPU. This is also generally the major metabolite reported in most soil metabolism experiments.

The half life of DFB in the aqueous fraction of the 1 ppm sludge experiments was 4–15 hours. This rapid degradation is convincing evidence that DFB is dissipated quickly in wastewater-type environments.

Repeated applications of DIMILIN® W-25 and DIMILIN® 1-G to Utah Lake did not appear to disrupt population trends of oligochaetes and midges. Additionally, large doses of DIMILIN® did not affect (a) the growth of 6 species of fungi, (b) bobwhite quail reproduction, (c) incidence of chick embryonic malformations, (d) glycosaminoglycan formation in chicks, or (e) rat serum testosterone, weight, and food consumption.

These data provide additional information on the growing data base (Maas et al., 1980) on DFB. Accordingly, we conclude that DFB is an exceptionally environmentally safe molecule, largely because of its lack of persistence and unique mode of action.

5.0 REFERENCES

Bitter, T. and H. M. Muir. 1962. A modified uronic acid carbazole reaction. Analytical Biochem. 4: 330–334.
Booth, Gary. 1976. DIMILIN® and the environment. In: Gordon L. Berg (ed.), DIMILIN® Chitin Inhibitor: Breakthrough in Pest Control. Agri-Fieldman and Consultant, Willoughby, Ohio.
Booth, Cary M. 1977. An evaluation of DIMILIN® and diflubenzuron toxicity to avian species. pp. 28. Report submitted to Thompson-Hayward Chemical Co., Kansas City, KS
Booth, Gary M. and Duane Ferrell. 1977. Degradation of DIMILIN® by aquatic foodwebs, pp. 221–243. In: M. A. Q. Khan (ed.), Pesticides in Aquatic Environments. Plenum Press, Publisher, New York, New York.
Bryce, G. R. 1980. Data analysis in RUMMAGE — A user's guide. Department of Statistics, Brigham Young University, Provo, Utah.
Christman, Van. 1985. Personal communication.

Deul, D. H., B. J. De Jong and J. A. M. Kortenbach. 1972. Inhibition of chitin synthesis by two 1-(2,6-disubstituted benzoyl)-3-phenyl urea insecticides. II. Pest. Biochem. Physiol. 8: 98.

Heller, S. R., and G. W. A. Milne. "EPA/NIH Mass Spectral Data Base" Volume 1, NSRDS-NBS 63, U.S. Government Printing Office, Washington D.C., 1978, p. 234.

Longcore, Sampson and Whittendale. 1971. DDE thins eggshells and lowers reproductive success of captive black ducks. Bull. Environ. Contam. Tox. 6: 485–490.

Maas, W., R. van Hes, A. C. Grosscurt, and D. H. Deul. 1980. Benzopylphenyl Urea Insecticides, pp. 1–43. In: Wegler's Chemistry of Crop Protection Products. 6: 1–43.

Metcalf, R. L., P. Y. Lu, and S. Bowlus. 1975. Degradation and environmental fate of 1-(2,6-difluorobenzoyl)-3-(4-chlorophenyl) urea. J. Agric. Food. Chem. 23: 359.

Mulder, R. and M. J. Gijswijt. 1973. The laboratory evaluation of two promising new insecticides which interfere with cuticle deposition. Pest. Sci. 4: 737.

Rabenort, B., P. C. DeWilde, F. G. DeBoer, P. K. Korver, S. J. DiPrima and R. D. Cannizzaro. 1978. Diflubenzuron. pp. 57–72. In: Gunter Zweig and Joseph Sherma (eds.), Pesticides and Plant Growth Regulations, Academic Press, Inc., New York, New York.

Solursh, M. 1976. Glycosaminoglycan synthesis in the chick gastrula. Devel. Biol. 50: 525–530.

Verloop, A. and C. D. Ferrell. 1976. Benzoylphenyl ureas — A new group of larvicides interfering with chitin deposition. ACS Symp. Ser. 37: 237–270.

Verloop, A., W. B. Nimmo, and P. C. De Wilde. 1975. (Abstract). 8th Int. Plant Prot. Congress, Moscow.

Yagamata, T., S. H. Hidehiko, O. Habuchi, and S. Sazuki. 1968. Purification and properties of bacterial chondroitinases and chondrosulfatases. J. Biol. Chem. 243: 1523–1535.

9. Control of insect pests with benzoylphenyl ureas

A. Retnakaran and James. E. Wright

·8 INTRODUCTION

The introduction of chlorinated hydrocarbons, carbamates and organophosphates has revolutionized agricultural production by severely curbing the losses due to insects. Insect vectors that carry human and animal pathogens have been kept well under control with the use of these broad-spectrum, contact neurotixins. The long term effects of such wonder compounds on the environment became apparent when repeated large-scale applications resulted in visible effects and were dramatically brought into focus by the publication of Rachel Carson's "Silent Spring" (1962). There began an active search for environmentally safe methods of pest control centered mainly around biological control agents such as *Bacillus thuringiensis* preparations, nuclear polyhedrosis viruses, Entomopathogenic fungi, Entomogenous nematodes and parasites. Along with these methods emerged genetic approaches such as sterile male release and physiological approaches like pheromones and insect hormone analogs or insect growth regulators.

A serendipitous discovery by an observant individual at Duphar in Holland led to the discovery of an insect growth regulator, diflubenzuron in 1970. Although not a biorational in the true sense of the word (i.e. not a modified natural product), it shares many of the qualities of such materials. It acts on a predominantly unique biosynthetic system, chitin formation, in arthropods. Since it acts primarily on larval insects, it enjoys a narrow spectrum of activity, confined to immature stages but for a few exceptions. It has almost all the safety features of most microbial control agents yet has the efficacy and potency of broad-spectrum insecticides, a combination that is ideal for general use (Table 1).

Since benzoylphenyl ureas are fundamentally different in their structure and activity from that of the neurotoxic insecticides, they were subjected to rigorous scrutiny. Diflubenzuron is one of the few insecticides that has undergone thorough tests, including a repetition of the 2 year chronic

Wright, J. E. and Retnakaran, A. (Eds), Chitin and Benzoylphenyl ureas. ISBN 978-94-010-8638-7.
© *1987, Dr W. Junk Publishers, Dordrecht.*

Table 1. Relative merits of three classes of insecticides

Properties and use	Broad-spectrum contact insecticides (E.g. Carbamates, Organophosphates, Synthetic pyrethroids)	Biologicals (e.g. *Bacillus thuringinesis* preparations, Nuclear polyhedrosis virus)	Benzoylphenyl ureas (e.g. Diflubenzuron, Alsystin)
Activity on pests	Immediate effect and highly reproducible.	Slow effect and often not reproducible, varying from year to year one area to another	Effect slower than broad spectrum insecticide but highly reproducible
Effect of post-spray weather conditions	Since effect is immediate weather has minor influence	Post spray weather is of paramount importance. Poor rain fastness and labile to ultra violet radiation	Benzoylphenyl ureas adhere strongly to leaf or needle waxes and are not easily washed off. Sunlight has very little effect on breakdown.
Effect on parasites, predators and pollinators	Does not discriminate between pest and non-pest species. Often adverse effects.	Safe to most of the beneficials	Safe to most of the beneficials.
Effect on vertebrates	Since the action for the most part is neurotoxic, inhibiting cholinesterase activity, high exposure to these materials can be harmful. Non-allergenic to humans	No effect. Can be allergenic to humans because of the proteinaceous nature.	No effect. Non-allergenic

Effect on invertebrates	Varying degrees of effects.	No effect	Temporary decrease in population levels of certain juvenile crustacea.
Cost of production	Inexpensive because of large scale synthesis involving few steps.	Expensive because of labour intensive methods or fermentation process.	Moderately expensive because of several-step chemical synthesis.
Quality control	Easily monitored with chemical analyses	Not straightforward because of bioassays and batch to batch variations.	Easily monitored with chemical analyses.
Where most useful	Where immediate and total control is essential for saving a resource or preventing the spread of a pathogen.	Where total environmental protection is absolutely essential and a certain amount of resource loss can be tolerated.	For general purpose use in most areas
Use in integrated pest management (IPM)	Limited	Excellent	Excellent

toxicology studies. Safety to humans has been very well established (Keet and Kamp 1984). Some studies have even shown that diflubenzuron has anti-tumour activity in animal tests (Jenkins et al. 1984; Mayer et al. 1984).

Several new benzoylphenyl ureas have been introduced for commercial development since diflubenzuron made ts debut. Work on all these materials have been conducted in the laboratory and in the field around the world (Baronio and Pasqualini 1984; Busvine 1978; Maas et al. 1981; Retnakaran et al. 1985; Van Busschbach 1975). Some of the more important applications in the various areas of insect pest control together with the control potential of several benzoylphenyl ureas will be presented in this chapter.

2 MODE OF ACTION, ACTIVITY AND BIOASSAY

2.1 Mode of action

It has been well established that benzoylphenyl ureas inhibit chitin synthesis in larvae of insects (Grosscurt 1978b; Hajjar and Casida 1979; Maas et al. 1981; Retnakaran et al. 1985). The effects can be clearly seen in cross sections of cuticle of treated insects under the light microscope (Figure 1). Details of the chitin inhibition can be dramatically observed under the electron microscope (Figure 2). Whether or not the inhibition of chitin synthesis is the primary effect has been debated because of the difficulty in unequivocally establishing this mode of action at the biochemical level. Some of the theories that have been put forward are as follows:

1. Inhibits chitin synthase, the enzyme catalysing the polymerisation of UDP-N-acetyl glucosamine to chitin (Post et al. 1974; Van Eck 1979). This is the most widely held theory.
2. Inhibits the transport of UDP-N-acetylglucosamine across biomembranes (Mitsui et al. 1984).
3. Inhibits a protease that activates the chitin synthase zymogen (Leighton et al. 1981).
4. Inhibits ecdysone metabolizing enzymes resulting in accumulation of ecdysone. This stimulates chitinase production which results in moult disruption (Yu and Terriere 1977). O'Neill et al. (1977), however, found that diflubenzuron did not affect ecdysone titer.

Figure 1. Cross section of cuticle of diflubenzuron untreated (A) and untreated (B) Pieris brassicae (Cabbage butterfly) larvae. The endocuticle in the treated insect has very little chitin and has spherical bodies, probably proteinaceous. (From Duphar Technical Information report 7th ed. 1979).

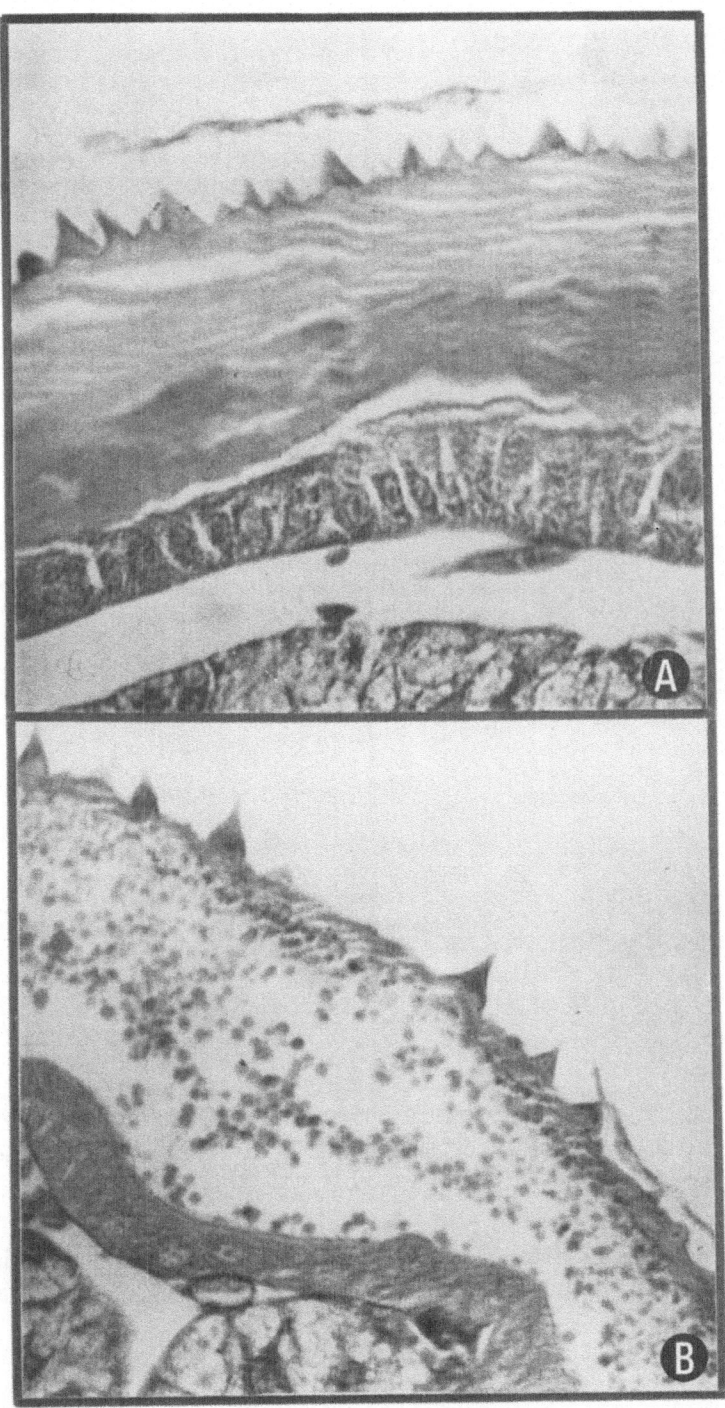

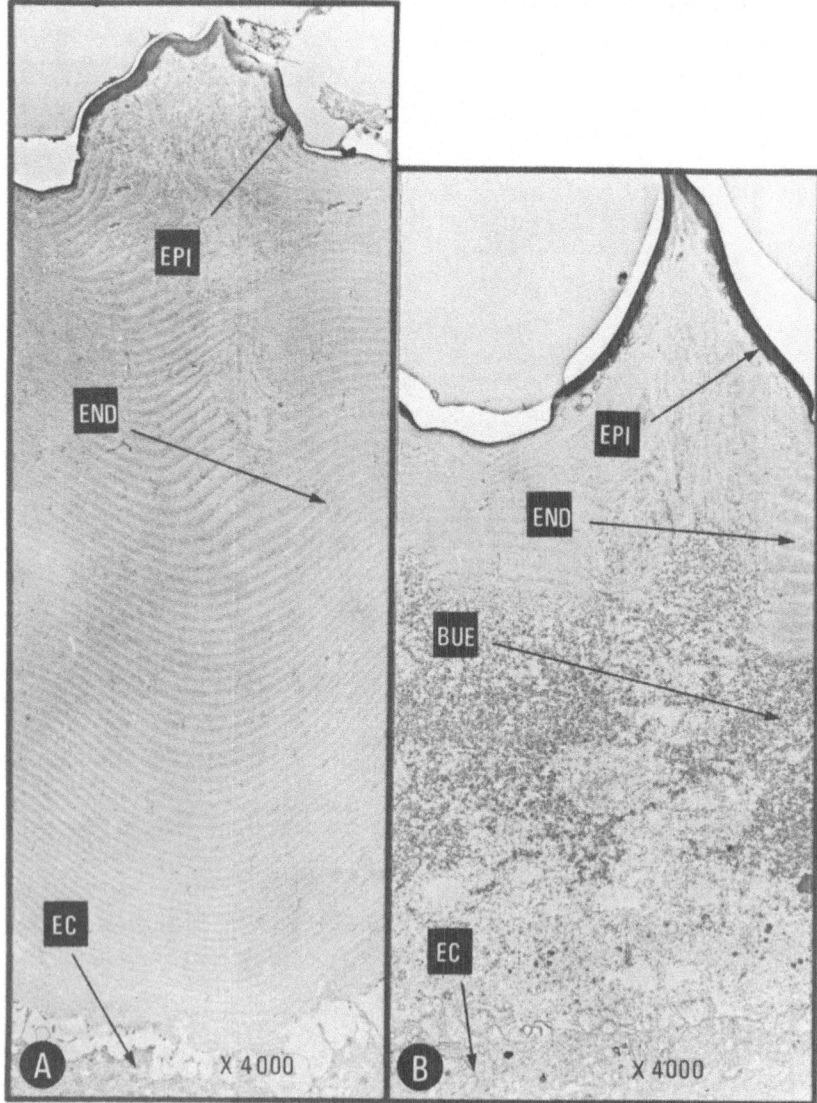

Figure 2. Ultrastructural effects of chlorfluazuron on the cuticle of the spruce budworm (*Choristoneura fumiferana*). (A) Untreated (B) Treated. (Percy et al. 1985).

5. Blocks conversion of glucose to fructose-6-phosphate resulting in inhibition of chitin synthesis; fructose level is depressed (Saxena and Kumar 1981).

6. An active metabolite of the benzoylphenyl urea is responsible for the inhibition of chitin synthesis (Cohen and Casida 1980).

7. *In vivo* regulatory system sensitive to benzoylphenyl urea is disrupted upon preparation of an *in vitro* system. Therefore, there is no longer any inhibition of chitin synthase by the benzoylphenyl urea (Cohen and Casida 1983).

8. Chitin synthesis in insects begins by the formation of oligosaccharides synthesized in the epidermal cell (Retnakaran and Hackman 1985) that are then transported to the cuticle where they are polymerized to form discrete chitin microfibrils and are covalently bound to proteins. Structural integrity is essential for the polymerization of the oligosaccharides. It is proposed that this second polymerization step is inhibited by benzoylphenyl ureas (Retnakaran 1985).

Once an optimal assay for chitin synthase is developed, the action of benzoylphenyl ureas on this enzyme can be tested. Any one of these theories or a combination of two or more will be ultimately demonstrated to be true.

2.2 *Activity*

The different activities can be grouped under two heads, Biochemical effects and Biological effects.

2.2.1 *Biochemical effects*

Several biochemical effects due to benzoylphenyl ureas have been reported and these may be all secondary effects. In several lepidopteran species, the pentose cycle was inhibited (Moreau et al. 1975). Again in certain lepidopterans, increase in cuticular lipids was observed (Salama et al. 1976). Yu and Terriere (1975) reported an increase in microsomal oxidase while Ishaaya and Ascher (1977) found that the levels of carbohydrate degrading enzymes such as trehalase, amylase and invertase were depressed. Also, there was an increase in phenoloxidase activity which probably accounts for the characteristic darkening (Deul et al. 1978; Ishaaya and Casida 1974). The level of glucose phosphate dehydrogenase is adversely affected by diflubenzuron (Wright unpublished). In the pine processionary caterpillar, *Thaumetopoea pityocampa*, it was shown that diflubenzuron increased the formation of cholesterol from ^{14}C-acetate but this may be microbial since it was inhibited by tetracycline (Denneulin and Lamy 1982). DNA synthesis was inhibited in at least 3 insect species (Deloach et al. 1981; Mitlin et al. 1977; Soltani et al. 1984).

2.2.2 *Biological effects*

The biological effects of benzoylphenyl ureas are used as end points in bioassays. In general the effects are slower than most neurotoxic insecti-

Figure 3. Larvicidal effects of diflubenzuron on the spruce budworm (*Choristoneura fumiferana*). Sixth instar larvae were allowed to feed on diet containing 1% diflubenzuron. (A) Most severe effect, the larva darkens and dies; apolysis complete. (B) Apolysis complete and abdomen almost pupal. (C) Poised for pupation, pharate. (D) Ecdysis has started. (E) Partial moult has occurred. (F) Moult is complete except for adhesion of exuviae.

cides. They can be grouped under 3 heads — larvicidal, ovicidal and other effects, the last being a catchall for peculiar effects.

2.2.2.a *Larvicidal effects.* — Benzoylphenyl ureas have several effects on the larvae and they all pertain to malformation of the cuticle (Figures 3 and 4).

1. Complete moult inhibition. — This is by far the most serious effect. The insect dies within the old cuticle and the new cuticle formed is extremely thin.
2. Partial moult inhibition. — This is the most common observation where the insect dies attempting to shed its old cuticle. Various degrees of the effect can be observed beginning with death at the earliest stage of dorsal splitting along the ecdysidal line to the stage where the old cuticle remains attached at one point or other.
3. Delayed effect: malformed pupa — The effect of larval treatment can be manifested at the pupal moult.
4. Failure to feed — Sometimes, no visible effects can be observed but the larvae fail to feed because of displaced mandibles (Retnakaran et al. 1985).

Larval effects result, for the most part, from ingestion of benzoylphenyl ureas. Egyptian leaf worm (*Spodoptera littoralis*) larva is unique in show-

Figure 4. Morphogenetic effects of diflubenzuron on the castor semilooper (*Achoea janata*). a) Normal larva, b) Larviform pupa with anterior portion completely larval, c–e) Larval-pupal mosaics, f) Forwardly bent pupa with larval head and two pairs of legs, g) Elongate pupa with short wing pads on one side only; sclerotized cuticle absent ventrally in the mid portion, h) Twisted larval-pupal intermediate, i) An almost normal pupa with larval head and legs, j) Curved pupa, k) Normal pupa (Rabindra and Balasubramaniam 1981).

ing topical effects (Ascher and Nemny 1976b) (Figure 5). Such topical effects on larvae are more the exception than the rule. Pupal effects again are extremely rare. Here again, *Spodoptera littoralis* is unique in showing topical effects on pupae (Abo-Elgar et al. 1978). Sometimes, the effects on pupae are delayed and expressed as adult deformities.

2.2.2.b *Ovicidal effects.* — Benzoylphenyl ureas inhibit chitin formation of the embryo which usually dies inside the egg shell as a fully formed larva. the term "ovicidal" may not exactly describe this effect and, instead, the term "embryocidal" has been suggested (Grosscurt 1977; 1978). Here again various degrees of effects can be observed:

1. Mortality of fully developed larvae inside the egg shell. This is the most common ovicidal effect that has been observed. The effect can occur by a) topical application to the egg, b) topical and/or feeding the material to the adult female, c) topical application to the male which is transferred to the female during mating (Grosscurt 1978; Wright and Villavaso 1983).

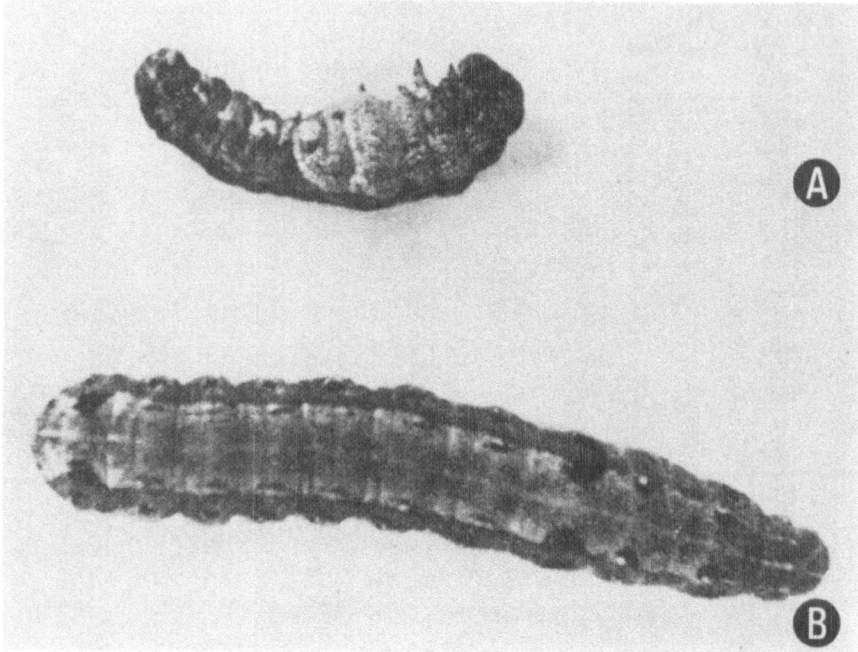

Figure 5. Effect of topical treatment of the larva of the Egyptian cotton leafworm, *Spodoptera littoralis* with an acetonic solution of diflubenzuron. This type of topical effect is unique to this species. (A) Treated (B) Untreated (Ascher and Nemny 1976).

2.   Delayed ovicidal effect where mortality occurs at hatch or soon after. This has been especially true in mosquitoes (Miura et al. 1976).

2.2.2.c *Other effects.* — Effects on adults is relatively rare, but has been shown in some coleopterans. Weakening of the elytra has been shown in the Colorado potato beetle, *Leptinotarsa decemlineata* (Grosscurt and Andersen 1980). Benzoylphenyl ureas also affect peritrophic membrane formation (Clark et al. 1977). Another effect reported is the absence of polysaccharides in the oenocytes in diflubenzuron treated insects (Denneulin and Lamy 1977).

2.3 *Bioassay*

The control potential of benzoylphenyl ureas is established in the laboratory by conducting suitable bioassays. These assays have to be long-term, unlike those for neurotoxic insecticides. Topical or ingestion effects are studied in the various instars and they fall under the categories of ovicidal or larvicidal effects (Retnakaran 1980; Wright and Villavaso 1983) (Figure 6).

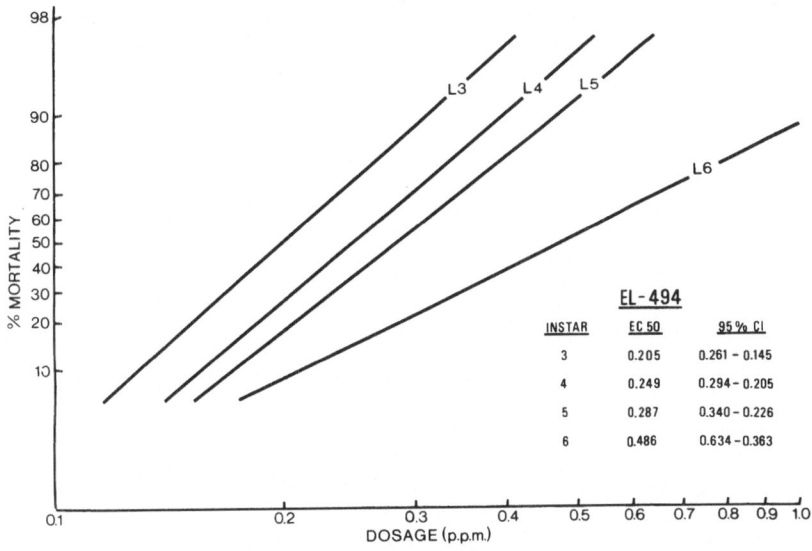

Figure 6. Effect of different concentrations of EL-494 in artificial diet on molt inhibition of 3rd, 4th, 5th and 6th instar larvae of the spruce budworm (*Choristoneura fumiferana*) (Retnakaran 1979).

The age of the instar often has an effect on the activity of the material. Younger instars are generally more sensitive than older instars but there are instances where the latter are more sensitive (Granett and Retnakaran 1977). This again depends on the analog used (Figures 7 and 8). Even within instars, the younger stages are generally more sensitive to benzoyl-phenyl ureas than the older stages. The differential sensitivity of several benzoyl ureas to selected lepidopterans is summarized in Table 2.

3 BENZOYLPHENYL UREAS AVAILABLE FOR COMMERCIALIZATION

In 1970, Philips Duphar Company in Holland made the first benzoyl-phenyl urea, DU-19111, by combining two herbicides, dichlobenil and diuron. During routine screening on insects, it was discovered that this compound was a slow acting stomach poison that induced moult deformities. This effect was later traced to the inability of the insect to synthesize chitin. Duphar then made their now famous diflubenzuron (Dimilin or PH 6040) and tested it on a world-wide basis. This compound is currently registered in 41 countries against an array of insects. Duphar has since made several other analogs, Penfluron (PH 6044) being the best known among these. However, this compound is currently not being developed for commercialization.

Table 2. Comparative toxicities of benzoylphenyl ureas to selected lepidopterans (Retnakaran et al. 1985).

Insect	Toxicity of insect growth regulator LC_{50} (95% CL) in ppm / slope					
	Diflubenzuron	Alsystin	EL 494	EL 1215	Chlorfluazuron	Penfluron
Eastern spruce budworm (Choristoneura fumiferana) (third instar)	36(20–67)[a] / 1.28	1.65(1.46–1.85)[b] / —	0.205(0.15–0.26)[c] / 1.16	0.08(0.07–0.10)[b] / 1.16	0.27(0.25–0.29)[d] / 7.26	0.31(0.28–0.39)[d] / —
Western spruce budworm (Choristoneura occidentalis) (third instar)	192(56–654)[x] / 0.61	6.95(3.48–11.2)[c] / 1.83	1.15(0.86–1.45)[c] / 2.97	0.76(0.68–0.84)[f] / 5.87	0.26(0.22–0.32)[f] / 5.47	1.94(0.93–3.11)[c] / 1.81
Omnivorous leafroller (Platynota stultana) (first instar)	69.38(3711–193)[g] / 0.95	173(139–219)[g] / 161		3.01(1.25–5.17)[h] / 2.84	1.36(1.19–1.57)[b] / 4.68	
Beet armyworm (Spodoptera exigua) (first instar)	1.36(1.13–1.51)[j] / 2.39	8.06(5.83–10.23)[j] / 2.39		1.07(0.84–1.26)[j] / 4.19	0.23(0.21–0.26)[j] / 5.47	0.27(0.24–0.29)[j] / 3.13
Douglas fir tussock moth (Orgyia pseudotsugata) (third instar)	0.058(0.032–0.083)[c] / 3.63	0.059(0.038–0.081)[c] / 2.03	2.56(1.05–7.84)[c] / 1.66			0.029(0.018–0.036)[c] / 4.88
Gypsy moth (Lymantria dispar) (third instar)	0.013(0.010–0.017)[j] / 4.90		0.30(0.27–0.32)[k] / 8.06	0.175(0.163–0.212)[k] / 5.66		

[a] Granett, J and Retnakaran, A., 1977.
[b] Retnakaran, A., 1980.
[c] Retnakaran, A., 1979.
[d] Brushwein, J., 1980.
[e] Rappaport, N. and Robertson, J., 1981
[f] Robertson, J., 1982.
[g] Granett, J. and Hejazi, M., 1983.
[h] Granett and Hejazi, unpublished data.
[i] Granett, J. et al., 1983.
[j] Granett, J. and Dunbar, D., 1975.
[k] Abdel-Monem, A. et al., 1980.

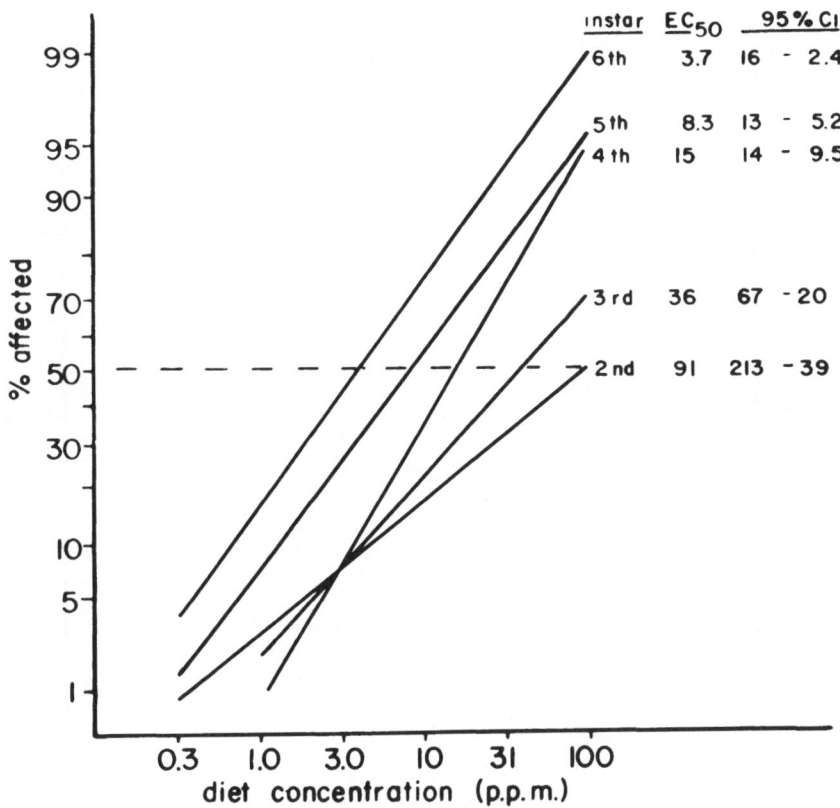

Figure 7. Effect of different concentrations of diflubenzuron in the diet of 2–6th larval instars of the spruce budworm (*Choristoneura fumiferana*). (Granett and Retnakaran 1977).

Eli Lilly and Company in the United States came up with a whole array of benzoylphenyl ureas, the most important of which is EL-494. Unfortunately, for various reasons, none of the compounds are being developed commercially at present.

Bayer Company in Germany introduced several analogs, the best known one being Alsystin (Triflumuron or BAY SIR 8514). Its activity is similar to that of diflubenzuron. The toxicology has been thoroughly investigated and the material is currently awaiting registration around the world.

Celamerck Company from Germany has introduced a new analog, CME-13401, which has been shown to be very active against several insects. Toxicology has been partially completed and the material will soon be commercialized.

Ishihara Sangyo Kaisha of Japan introduced a novel pyridoxyl benzoylphenyl urea, Chlorfluazuron (IKI 7899, CGA 112913, formerly UC

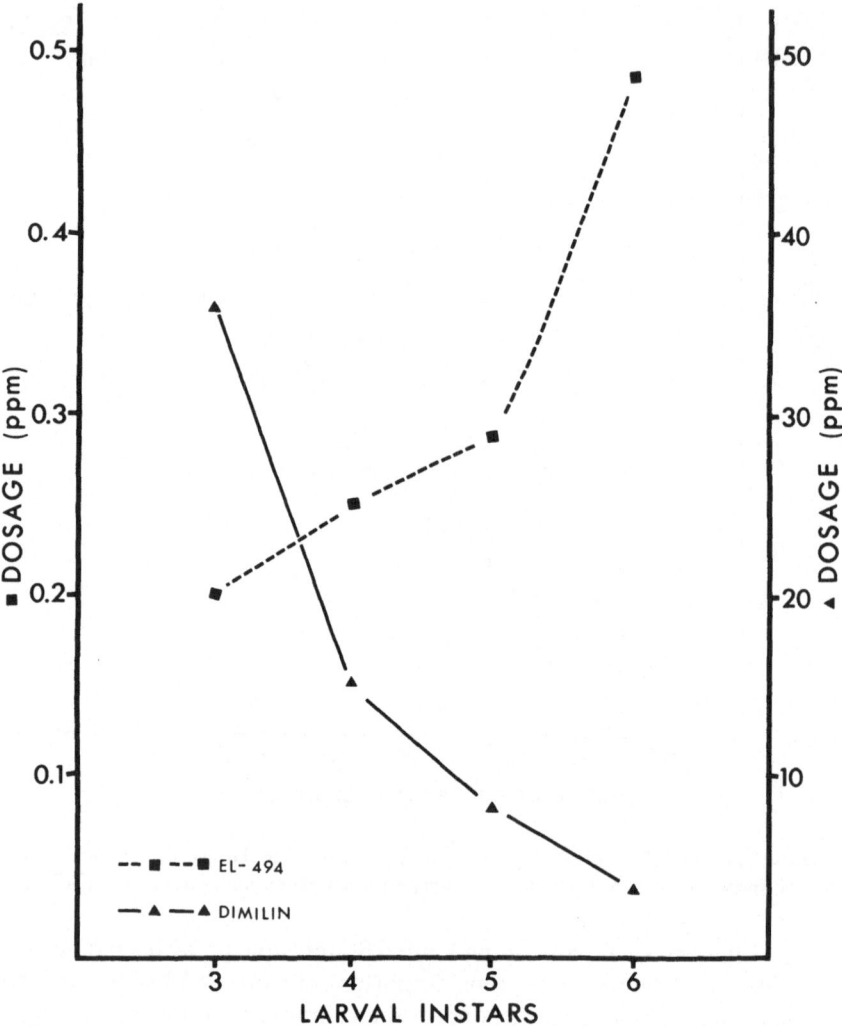

Figure 8. Response of the spruce budworm (*Choristoneura fumiferana*) larval instars to diflubenzuron and EL-494 expressed as EC_{50} (ppm in diet) for each instar tested (Retnakaran 1979).

62644). This compound was originally developed in N. America by Union Carbide but later was taken over by Ciba Geigy. This compound is very potent and is active on many species refractory to diflubenzuron. Toxicology is still in the early stages and the material has been shown to be non-mutagenic in the *Salmonella* assay (Retnakaran and Ennis 1985).

Dow Chemical Company in the United States has introduced a powerful new benzoylphenyl urea, DOW XRW 3136. It has shown excellent

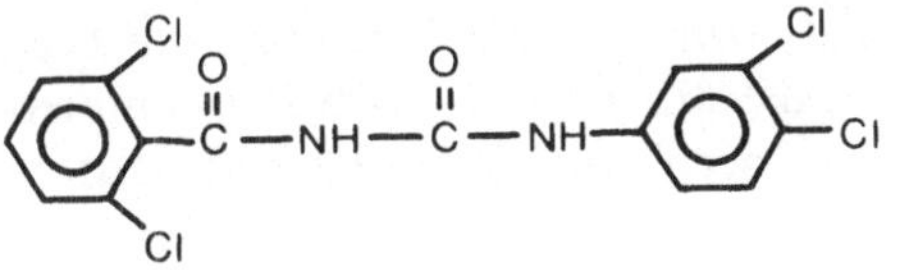

DU-19111

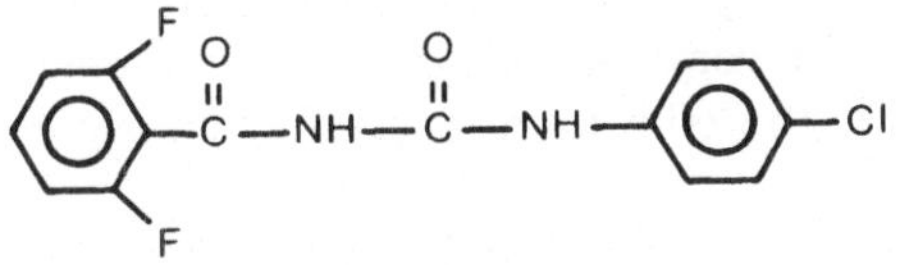

DIFLUBENZURON
Dimilin (PH 60-40)

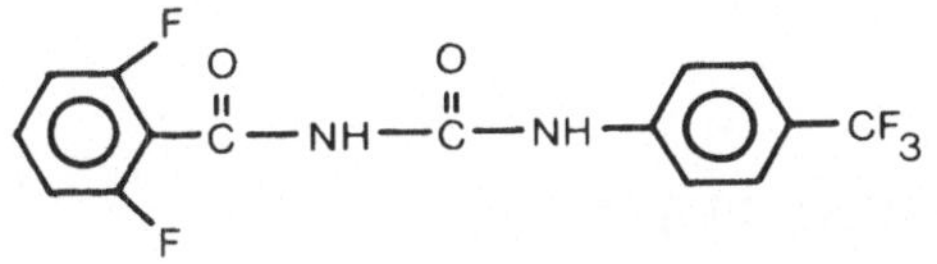

PENFLURON
PH 60-44

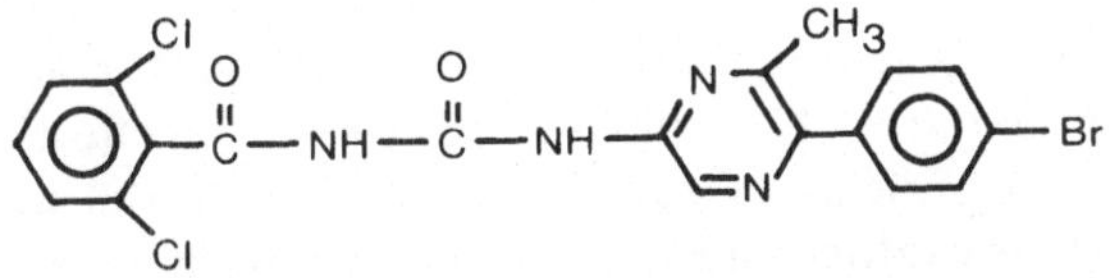

EL-494

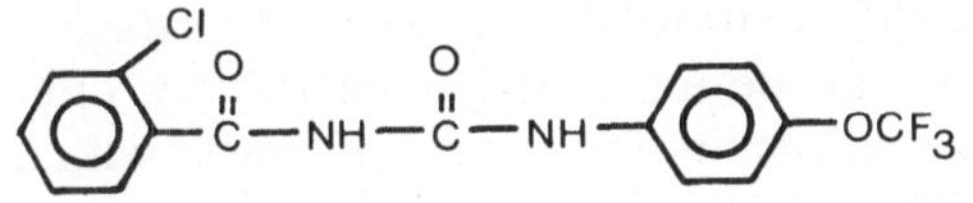

ALSYSTIN
Triflumuron
Bay Sir 8514

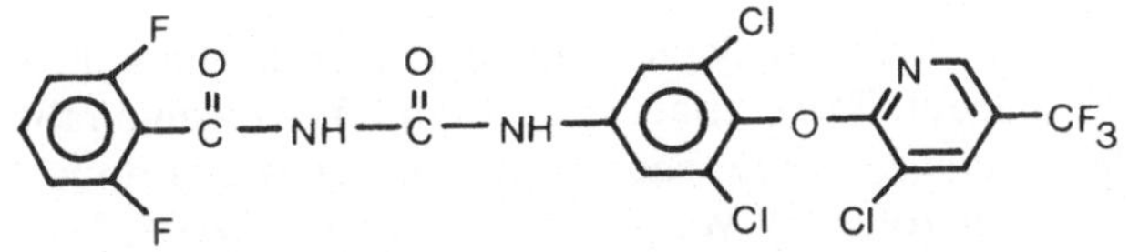

CHLORFLUAZURON
CGA-112913
IKI-7899 (UC62644)

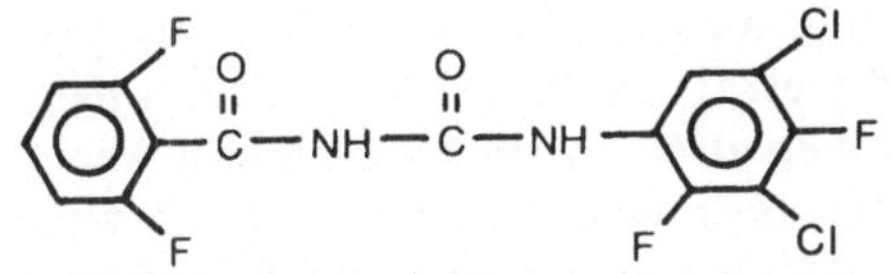

CME-13401

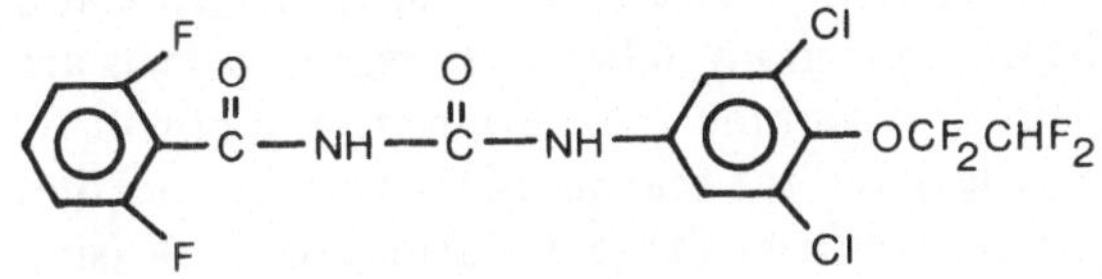

DOW XRW-3136

Figure 9. Chemical structures of some benzoylphenyl urea compounds with insecticidal activity.

activity against many pests and has a good chance of being commercialized. BASF in Germany has several new benzoylphenyl ureas in the early stage of development and some of them appear to be very active against certain insect pests. Depending on the range of activity and toxicology some of these analogs might be commercialized.

At present, diflubenzuron is the only benzoylphenyl urea that has been commercialized around the world. Alsystin is close to being commercialized, now that the toxicology has been completed. Chlorfluazuron, CME 13401 and DOW XRW 3136 are rapidly approaching the stage where a decision has to be made regarding their commercial development. BASF compounds are at present in the very early stages of development. The chemical structures of some of the compounds tested is shown in Figure 9.

The prognosis for commercialization of many benzoylphenyl urea appears promising.

4 CONTROL OF AGRICULTURAL PESTS

Insect control in agro-ecological situations is, for the most part, contained within well defined boundaries. Nevertheless, environmental effects are of paramount importance. Effects of runoff and residue as well as adverse effects on parasites, predators and pollinators have to be taken into consideration in the cost-benefit equation. Based on sound strategies, the application of control agents can be judiciously regulated by Integrated Pest Management to cause the maximum amount of benefit with a minimum of non-target effects. The control of several agricultural insect pests is summarized in Table 3.

The enormous returns on cotton make it an attractive area for any new control agent to make its debut. The contact effect of diflubenzuron on the female boll weevil, *Anthonomus grandis*, causing it to lay eggs that do not emerge, has been cleverly adapted by Wright and his co-workers to, not only control but also, in many situations eradicate this serious pest (Wright and Villavaso 1983; Wright et al. 1979; 1980a; 1980b; 1980c; 1983). The control potential of diflubenzuron in suppressing boll weevil populations by its ovicidal activity has been well demonstrated in several field trials (Table 4) (Ganyard et al. 1977; Mitchell et al. 1980; Taft and Hopkins 1975). In the eradication program, laboratory reared weevils are first sterilized by a combination of gamma irradiation from a cobalt-60 source, (which sterilizes the males) and the rearing of the weevils on a diet containing diflubenzuron, (which sterilizes the females) and then releasing them in the field (Figure 10). The weevils are marked by a red dye, Calco-red, in the diet to aid in the mark and recapture experiments (Haynes and Wright 1982; 1984; Wright and Robertson 1981).

The Egyptian cotton leafworm *Spodoptera littoralis*, a major pest of cotton and other crops in the middle east and elsewhere, can be effectively controlled with diflubenzuron, Alsystin or Penfluron (Ascher and Eliyahu 1981; Ascher and Nemny 1974; 1976a; 1976b; Ascher et al. 1982; El-Guindy et al. 1983a; 1983b; 1983c). The eggs are particularly susceptible

PERCENT
TOTAL EGG PRODUCTION

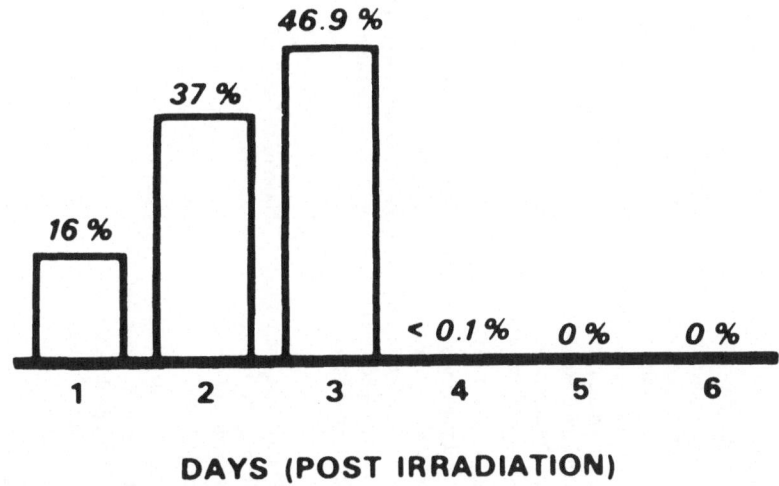

DAYS (POST IRRADIATION)

Figure 10. Pattern of egg production by boll weevils after the sterilization treatment of adults feeding on 100 ppm diflubenzuron for 5 days and then given 10 krad of radiation on day 6 after emergence (Wright and Roberson 1981).

to diflubenzuron (Table 5). It is one of the few lepidopterous species that shows larval and egg contact toxicity to benzoylphenyl ureas (Ascher et al. 1979; Radwan et al. 1978).

Spodoptera litura, the oriental leafworm, which is a pest of cotton and tobacco in Asia and Australia is susceptible to diflubenzuron (Sundaramurthy and Balasubramaniam 1978; Santharam and Balasubramaniam 1980) but to a lesser extent than the Egyptian cotton leafworm. *Spodoptera exigua*, the beet armyworm, is notoriously refractive to diflubenzuron but penfluron and chlorfluazuron are very effective (Table 6) (Granett et al. 1983). This species is able to metabolize or detoxify or eliminate diflubenzuron and escape its insecticidal effects. The fall armyworm, *Spodoptera frugiperda* might be similarly refractory to diflubenzuron (Jones 1975).

Pectinophora gossypiella, the pink bollworm, is a classic example where diflubenzuron has been shown to be very effective in the laboratory but not so in the field (Flint and Smith 1977. Flint et al. 1977). The larvae being internal feeders escape the effect of diflubenzuron. The same is true of *Earias insulana*, the Egyptian bollworm (Ascher et al. 1978). While most

Table 3. Control of some agricultural insect pests by benzoylphenyl ureas.

Insect species	Host	Analog	Activity	Reference
Anthonomus grandis (Cotton boll weevil) Fam. Curculionidae Ord. Coleoptera	Cotton	Diflubenzuron; Penfluron	Excellent control in the field by applying diflu-benzuron to cause female sterility upon ingestion. Also release of difluben-zuron-irradiation sterilized weevils in eradication programs gave excellent results. Penfluron also sterilizes the female weevils.	Earle *et al.* 1979; Ganyard *et al.* 1977; Haynes and Wright 1982; 1983; 1984; Haynes *et al.* 1981; McCoy and Wright 1979; Mitchell *et al.* 1980; Taft and Hopkins 1975; Wright and Roberson 1981; Wright and Villavaso 1983; Wright *et al.* 1979; 1980a; 1980b; 1980c; 1983.
Spodoptera littoralis (Egyptian cotton leafworm) Fam. Noctuidae Ord. Lepidoptera	Cotton; Also beet, tomato and alfalfa	Diflubenzuron	Excellent control in labor-atory and field trials. Has contact toxicity on eggs and larvae. Penfluron and Alsystin are slightly more effective than diflubenzuron.	Abbassy *et al.* 1980; Abo-Elgar *et al.* 1978; Ascher and Eliyahu 1981; Ascher *et al.* 1982; Ascher and Nemny 1974; 1976a; 1976b; Ascher *et al.* 1979; Assai *et al.* 1983; El-Guindy *et al.* 1983a; 1983b; 1983c; Radwan *et al.* 1978; Saad *et al.* 1981; Salama and El-Din 1977
Spodoptera litura (Oriental leafworm) Fam. Noctuidae Ord. Lepidoptera	Cotton and tobacco in Asia and Australia	Diflubenzuron	Good population control but some damage to tobacco leaves because of the slow mode of action.	Natesan and Balasubramamian 1980; Sundaramurthy 1977; Sundaramurthy and Balasubramian 1980; Santharam and Balasubramamian 1980.
Spodoptera exigua (Beet armyworm) Fam. Noctuidae Ord. Lepidoptera	Cotton, tobacco, Beet, Cabbage, Alfalfa, Tomato, Potato,	Diflubenzuron; Alsystin; Penfluron; Chlorfluazuron	Refractory to Alsystin and diflubenzuron but excellent control with penfluron and chlorfluazuron. Topical effect of larvae.	Granett *et al.* 1983.
Spodoptera mauritia (Rice swarming caterpillar) Fam. Noctuidae Ord. Lepidoptera	Rice	Diflubenzuron	Excellent laboratory results contact toxicity on larvae	Beevi and Dale 1980

Species	Crop	Compound	Remarks	References
Cydia pomonella (or *Laspeyresia pomonella*) (Codling moth) Fam. Olethreutidae Ord. Lepidoptera	Apple, pear, walnut, quince.	Diflubenzuron; Alsystin; Penfluron	Excellent control in the field by ovicidal action. No effect on adults; early instars less sensitive than later instars.	Audemard 1977; Audemard et al. 1975; 1976; Burts, 1983; Cranham 1978; Elliott and Anderson 1982; Glen et al. 1982; Hoying and Riedl 1980; Matalin and Kuldova 1982; Moffitt et al. 1983; 1984; Wilson 1979
Operophtera brumata (Winter moth) Fam. Geometridae Ord. Lepidoptera	Apple	Diflubenzuron	Pre-blossom spray provides excellent control	Wilson 1979
Adoxophyes orana (Summer fruit tortrix moth) Fam. Tortricidae Ord. Lepidoptera	Apple especially in Holland	Diflubenzuron	Refractory to diflubenzuron.	De Jong and Minks 1981
Orthosia hibisci (Green fruit worm) Fam. Noctuidae Ord. Lepidoptera	Apple	Diflubenzuron	Excellent control when sprayed at the pink bud stage	Paradis 1978
Grapholitha molesta (Oriental fruit moth) Fam. Olethreutidae Ord. Lepidoptera	Peach	Diflubenzuron; Alsystin	Excellent potential as control agent by ovicidal activity	Broadbent and Pree 1984a.
Choristoneura rosaceana (Oblique banded leafroller) Fam. Tortricidae Ord. Lepidoptera	Peach	Diflubenzuron; Alsystin	Refractory to larvicidal activity of both compounds	Broadbent and Pree 1984b
Psylla mali (Apple sucker) Fam. Psyllidae Ord. Homoptera	Apple	Diflubenzuron; Alsystin	Not effective	Glen et al. 1982.

Table 3. Control of some agricultural insect pests by benzoylphenyl ureas (continued)

Insect species	Host	Analog	Activity	Reference
Spodoptera frugiperda (Fall armyworm) Fam. Noctuidae Ord. Lepidoptera	Corn	Diflubenzuron	Poor control in field	Janes 1975
Heliothis virescens (Tobacco budworm) Fam. Noctuidae Ord. Lepidoptera	Cotton, tobacco, tomato	Diflubenzuron	Excellent laboratory results but poor field results because of the larvae being inside the bud.	Wolfenbarger *et al.* 1977.
Heliothis armigera (Old world bollworm) Fam. Noctuidae Ord. Lepidoptera	Cotton in Australia. Also corn, tomato and lucerne (alfalfa)	Diflubenzuron; Alsystin.	Both analogs very effective in laboratory trials.	Radwan and Rizk 1976
Pectinophora gossypiella (Pink bollworm) Fam. Gelechiidae Ord. Lepidoptera.	Cotton	Diflubenzuron	Excellent results in the laboratory but very little control in the field because the larvae feeding inside the boll.	Flint and Smith 1977; Flint *et al.* 1977; Rizk and Radwan 1976
Earias insulana (Egyptian bollworm) Fam. Noctuidae Ord. Lepidoptera	Cotton	Diflubenzuron	Because of the larval feeding inside the boll very little control in the field was achieved	Ascher *et al.* 1978; Rizk and Radwan 1975
Bucculatrix thurberiella (Cotton leaf perforator) Fam. Lyconetiidae Ord. Lepidoptera	Cotton	Diflubenzuron	Excellent control in the field.	Flint *et al.* 1977
Trichoplusia ni (Cabbage looper) Fam. Nocutidae Ord. Lepidoptera	Cotton, cabbage, potato, tomato, Chrysanthemum	Diflubenzuron	Refractory to diflubenzuron; requires high dosages.	Flint *et al.* 1977

Species	Host	Chemical	Effect	References
Anticarsia gemmatalis (Velvetbean caterpillar) Fam. Noctuidae Ord. Lepidoptera	Soybean, jointvetch	Diflubenzuron	Excellent control in the field.	Bullock and Kretschmer Jr., 1982. Degaspari and Gomez 1982; Greene 1975; Heinrichs and Dasilva 1978; Heinrichs et al. 1979; Lara et al. 1977; Turnipseed et al. 1974.
Plathypena scabra (Green cloverworm) Fam. Noctuidae Ord. Lepidoptera	Soybean	Diflubenzuron	Excellent control in the field	Pediglo and Hammond 1976; 1978; 1979; Smith 1976; Turnipseed et al. 1974.
Pseudoplusia includens (Soybean looper) Fam. Noctuidae Ord. Lepidoptera	Soybean	Diflubenzuron	Good control at higher doses (than velvetbean caterpillar)	Reed and Bass 1980; Turnipseed et al. 1974.
Epilachna varivestis (Mexican bean beetle) Fam. Coccinellidae Ord. Coleoptera	Soybean and other beans	Diflubenzuron	Significant reduction in numbers in field trials.	Smith 1976; Turnipseed et al. 1975; Zungoli et al. 1983.
Nephantis serinopa (Coconut black-headed caterpillar) Fam. Xyloryctidae Ord. Lepidoptera	Coconut	Diflubenzuron	Promising in laboratory tests	Sundaramurthy and Santhanakrishnan 1979.
Boarmia (Ascotis) selenaria (Avocado looper) Fam. Geometridae Ord. Coleoptera	Avocado Cotton, peanut, mulberry, tea, coffee	Diflubenzuron	Promising in laboratory tests	Ascher et al. 1978
Ostrinia nubilalis (European cornborer) Fam. Pyralidae Ord. Lepidoptera	Corn; also other plants	Diflubenzuron	Effective as ovicide and larvicide. Larvae have to be treated before they become internal feeders.	Berry et al. 1980; Büchi 1978; Faragalla et al. 1980.

Table 3. Control of some agricultural insect pests by benzoylphenyl ureas (continued)

Insect species	Host	Analog	Activity	Reference
Harrisina brillians (Western grape leaf skeletonizer) Fam. Zygaenidae Ord. Lepidoptera	Grapes	Diflubenzuron	Excellent in the field	Stern *et al.* 1983.
Lobesia (Polychrosis) botrana (Vine moth) Fam. Tortricidae Ord. Lepidoptera	Grapes	Diflubenzuron	Ovicidal activity when eggs are treated at 27° (much les at lower temperature)	Ascher *et al.* 1978.
Cryptoblabes gnidiella (Honeydew moth) Fam. Phycitidae Ord. Lepidoptera	Grapes, citrus, loquat, pomegranate	Diflubenzuron	Ovicidal acitivty at 27° and much less at 22°	Ascher *et al.* 1983.
Achoea janata (Castor semilooper) Fam. Noctuidae Ord. Lepidoptera	Castor, rose, pomegranate	Diflubenzuron	Excellent control in laboratory trials	Rabindra and Balasubramamian 1981.
Leptinotarsa decemlineata (Colorado potato beetle) Fam. Chrysomelidae Ord. Coleoptera	Potato	Diflubenzuron; Alsystin	Good control in the field by both larvicidal and ovicidal effects.	Ammar 1984; Cooper *et al.* 1983; Grosscurt 1978a; Tamaki *et al.* 1984.
Diaprepes abbreviatus (Citrus root weevil) Fam. Curculionidae Ord. Coleoptera	Citrus, Sugar cane	Diflubenzuron	Good control in field trials by ovicidal activity	Beavers *et al.* 1976; Schroeder *et al.* 1976; Schroeder and Sutton 1977.
Hypera postica (Alfalfa weevil) Fam. Curculiondae Ord. Coleoptera	Alfalfa	Diflubenzuron	Good control in the field	Chu and Brindley 1981; Neal Jr. 1974.

Species	Crop	Compound	Effect	Reference
Hypera brunneipennis (Egyptian alfalfa weevil) Fam. Curculionidae Ord. Coleoptera	Alfalfa	Diflubenzuron	Poor control	Summers 1975
Graphognathus leucoloma (White-fringed weevil) Fam. Curculionidae Ord. Coleoptera	Alfalfa	Diflubenzuron	Excellent control by ovicidal and larvicidal effects	Henzell *et al.* 1977; 1979
Conotrachelus nenuphar (Plum curculio) Fam. Curculionidae Ord. Coleoptera	Peaches, plums, nectarines apples	Diflubenzuron	Poor control in the field	Calkins *et al.* 1977.
Lissorhoptrus oryzophilus (Rice water weevil) Fam. Curculionidae Ord. Coleoptera	Rice	Diflubenzuron	Good control by ovicidal activity.	Tsuzuki and Asayama 1983
Oxya japonica (Rice grasshopper) Fam. Acrididae Ord. Orthroptera	Rice	Diflubenzuron	Good control potential in laboratory studies. Loss of hind legs restricts movement	Lim and Lee 1982a; 1982b
Schistocerca gregaria (Migratory locust) Fam. Acrididae Ord. Orthoptera	Variety of crops	Diflubenzuron	Limited control potential	Strebler 1979.
Melanoplus sanguinipes (Migratory grasshopper) Fam. Acrididae Ord. Orthoptera	Variety of crops	Diflubenzuron	Good control potential in laboratory trials	Elliott and Iyer 1982
Psila rosae (Carrot fly) Fam. Psilidae Ord. Diptera	Carrot	Diflubenzuron	Good laboratory results but poor control in field trials	Overbeck 1979

228

Table 3. Control of some agricultural insect pests by benzoylphenyl ureas (continued)

Insect species	Host	Analog	Activity	Reference
Tipula paludosa (European crane fly) Fam. Tipulidae Ord. Diptera	Turf	Diflubenzuron	Poor effect but may act as an antifeedant	Campbell 1975
Heteropteza pygmaea (Mushroom gnat) Fam. Cecidomyiidae Ord. Diptera	Mushroom	Diflubenzuron	Limited effect in laboratory studies	White 1977
Hylema platura (Seedcorn maggot) Fam. Anthomyiidae Ord. Diptera	Beans, sweet corn	Diflubenzuron	Not effective in field trials.	Vea *et al.* 1976; Veire *et al.* 1975
Ceratitis capitata (Mediterranean fruit fly) Fam. Trypetidae Ord. Diptera	Citrus and other fruits	Diflubenzuron; Alsystin	Insufficient activity for control.	Arambourg *et al.* 1977; Farghal *et al.* 1983; Sarasua and Santiago-Alvarez 1983

Dacus oleae (Olive fruit fly) Fam. Tryptidae Ord. Diptera	Olive	Diflubenzuron	Very high dosage required in laboratory tests	Fytizas 1976.
Lycoriella mali (Mushroom fly) Fam. Sciaridae Ord. Diptera	Mushroom	Diflubenzuron; Alsystin	Excellent control. EPA in U.S. has established 0.2 ppm as tolerance level.	Argauer *et al.* 1980; Cantelo 1979; 1981; 1983; Kalberer and Vogel 1978; White 1981; E.P.A. 1983.
Homopterans such as aphids, scale insects, white flies and mealy bugs Ord. Homoptera	Various plants	Diflubenzuron	No control potential with no contact effect.	Collmann and All 1982; Peleg and Gothilf 1981; Price 1979; Radwan *et al.* 1982.
Phyllocoptruta oleivora (Citrus rust mite) Fam. Eriophyidae Ord. Acarina	Citrus	Diflubenzuron	Excellent control	Bullock and McCoy 1978; McCoy 1978.

Table 4. Field evaluation of diflubenzuron for the control of boll weevil (*Anthonomus grandis*) reproduction (Ganyard *et al.* 1977).

Application rate (g AI/ha)	Number of replicates (fields)	Number of boll weevil larvae and pupae/ha on month/day				
		7/8	7/22	8/5	8/19	9/3
141	2	0	309	0	0	161
282	2	0	0	0	0	161
564	3	0	0	0	143	0
Control	4	3507	6049	22,833	92,022	55,787

noctuids are susceptible to diflubenzuron, *Trichoplusia ni*, the cabbage looper, is relatively refractory to this compound requiring high doses (Flint et al. 1977).

Diflubenzuron is an excellent control agent for the codling moth, *Cydia (Laspeyresia) pomonella* (Figure 11). (Audemard 1977; Audemard et al. 1975; 1976; Cranham 1978; Elliott and Andersib 1982; Moffitt et al. 1983). Other benzoylphenyl urea such as Alsystin are equally effective (Glen et al. 1982). Both larvicidal activity by ingestion and ovicidal activity by contact with eggs have been shown to be responsible for the good control obtained in the field (Matolin and Kuldova 1982; Moffitt et al. 1983). Because these compounds are non toxic to most adult parasites, predators and pollinators, they can be incorporated in an Integrated Pest Management Program or a "soft pesticide program" (Burts 1983). The winter moth,

Table 5. Effect of diflubenzuron on the eggs of the Egyptian cotton leafworm (*Spodoptera littoralis*) (Salama and El-Din, 1977).

Concentration (ppm)	Age of egg (days)	% hatching (delayed effect)
100	Fresh	0
100	1	2 (all died)
100	2	13.5
50	Fresh	0
50	1	10 (all died)
50	2	23
30	Fresh	0
30	1	40 (all died)
30	2	35.5
10	Fresh	0
10	1	36 (all died)
10	2	32
5	Fresh	0
5	1	60 (all died)
5	2	52
1	Fresh	35 (all died)
1	1	54 (all died)
1	2	69
control	—	97.9

Table 6. Toxicity of benzoylphenyl urea insecticides to 1st through 2nd instar beet armyworms (*Spodoptera exigua*) (Granett *et al.* 1983).

Chemical	N	Slope ($\pm$ SD)	LC$_{50}$ (ppm in diet)	95% CL
Chlorfluazuron	597	5.47 $\pm$ 0.62	0.23	0.21–0.26
Penfluron	648	3.13 $\pm$ 0.26	0.27	0.24–0.29
EL-1215	695	4.19 $\pm$ 0.50	1.1	0.84–1.3
Diflubenzuron	636	2.99 $\pm$ 0.30	1.4	1.1–1.6
Alsystin	565	2.32 $\pm$ 0.30	8.1	5.8–10

Operophtera brumata (Wilson 1979), *Orthosia hibisci*, the green fruitworm (Paradis 1978) and *Grapholitha molesta*, the oriental fruit moth (Broadbent and Pree 1984) can all be well controlled in apple and other fruit orchards with diflubenzuron. Two pests of apple that do not respond to benzoylphenyl ureas are *Choristoneura rosaceana*, the oblique banded leafoller (Broadbent and Pree 1984) and *Psylla mali*, the apple sucker (Glen et al. 1982).

Soybean is another important crop where the economic returns justify massive control efforts to improve production. Benzoylphenyl ureas are very effective in controlling a whole complex of pests and has been well demonstrated in Brazil and the United States. The velvet bean caterpillar,

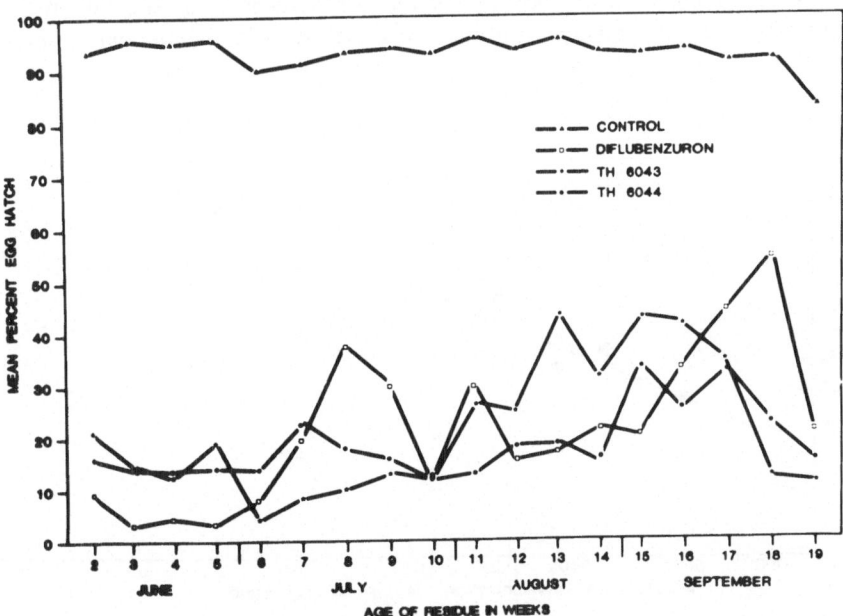

Figure 11. Ovicidal effect of diflubenzuron, penfluron (TH-6044) and another analog (TH 6043) on the codling moth, *Cydia* (*Laspereysia*) *pomonella* eggs. Pear trees were sprayed with the material and the residual activity was monitored (Moffitt et al. 1984).

Anticarsia gemmatalis is one of the most important pests and the larvae are very susceptible to diflubenzuron (Green 1975; Heinrichs and Da Silva 1978; Heinrichs et al. 1979; Lara et al. 1977; Turnipseed et al. 1974). Other important pests are *Plathypena scabra*, green cloverworm, *Pseudoplusia includens*, soybean looper and the Mexican bean beetle, *Epilachna varivestis* and they can all be controlled with diflubenzuron (Pediglio and Hammond, 1979; Reed and Bass 1980; Smith 1976; Turnipseed et al. 1975; Zungoli et al. 1983).

Pests of grapes are well controlled by diflubenzuron. *Harrisina brillians*, the western grapeleaf skeletonizer is very susceptible to this material (Stern et al. 1983). Diflubenzuron has better ovicidal activity at 27°C than at 22°C when eggs of *Lobesia (Polychrosis) botrana*, the vine moth and *Crytoblabes gnidiella*, the honeydew moth are treated with this material (Figure 12) (Ascher et al. 1978; 1983).

The castor semilooper, *Achoea janata* is a pest on castor, rose and pomegranate. Dramatic larvicidal effects showing various degrees of malformation are observed when treated with diflubenzuron (Rabindra and Balasubramaniam 1981).

Both diflubenzuron and Alsystin have larvicidal and ovicidal effects on the Colorado potato beetle, *Leptinotarsa decemlineate* (Ammare 1984; Cooper et al. 1983; Tamaki et al. 1984).

Weevils can be controlled by the ovicidal action of diflubenzuron but the effects are not always dramatic. The boll weevil, *Anthonomus grandis* is the real success story whereas the others are not nearly as good. The

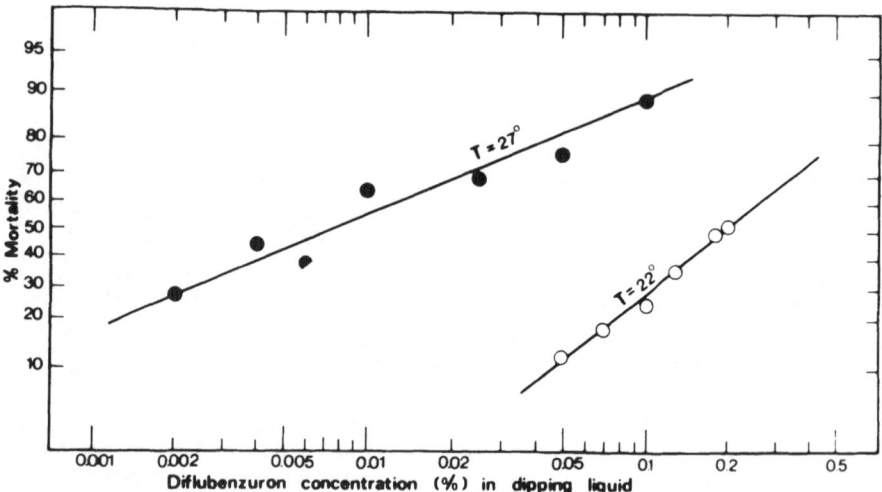

Figure 12. Effect of temperature on the ovicidal activity of diflubenzuron on the vine moth, *Lobesia botrana.* Eggs were dipped in aqueous solution of diflubenzuron and held at two different temperature regimens (Ascher et al. 1978).

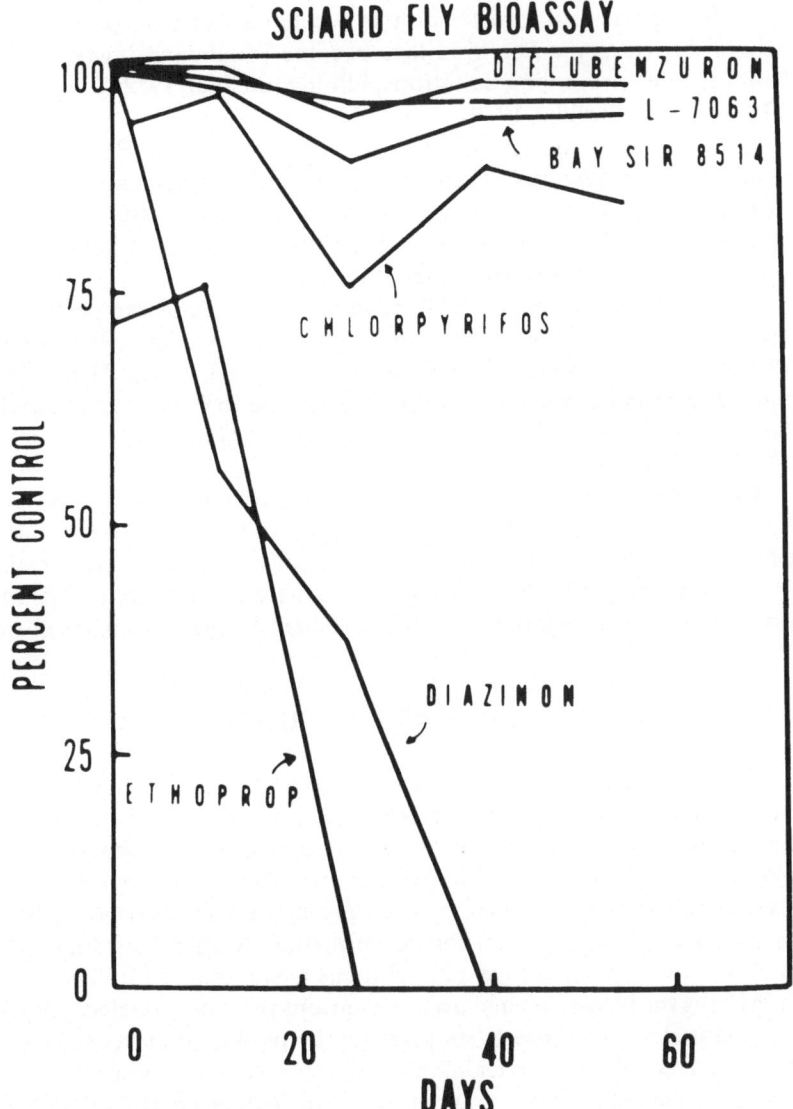

Figure 13. Effect of diflubenzuron on the mushroom fly, *Lycoriella mali*. Diflubenzuron, L-7063 and BAY SIR 8514 (3 IGRs) were incorporated in the compost used for growing mushroom (Argauer and Cantelo 1980).

white fringed weevil, *Graphognathus leucoloma* can be well controlled in alfalfa (lucerne) fields in New Zealand by the ovicidal property of diflu-benzuron (Henzell et al. 1977; 1979). The alfalfa weevil, *Hypera postica* can be reasonably well controlled with this material (Chu and Brindley 1981; Neal 1974).

Grasshoppers and locusts show a peculiar effect in response to diflu-benzuron treatment. The hind legs become fragile and break off at the femoral joint preventing saltation (Elliott and Iyer 1982; Lim and Lee 1982a; 1982b; Strebler 1979).

Diflubenzuron and Alsystin are very effective in controlling the mush-room fly, *Lycoriella mali* (Figure 13). The benzoylphenyl ureas are incor-porated in the compost which results in total suppression of the pest (Cantelo 1979; 1981; 1983). Most other dipteran pests in agriculture are poorly controlled by benzoylphenyl ureas.

Benzoylphenyl ureas are ineffective against homopteran pests such as aphids, scale insects, white flies and mealy bugs (Collman and All 1982; Peleg and Gothilf 1981; Price 1979; Radwan et al. 1982). The citris rust mite, *Phyllocoptruta oleivora* is very susceptible to diflubenzuron and can be well controlled in the field with this product (Figure 14) (Bullock and McCoy 1978; McCoy 1978).

Contact activity is an exception rather than the rule and the activity of benzoylphenyl ureas is either by disrupting moulting in larvae or hatching of eggs (larvicidal and ovicidal action). Because of their inactivity against most parasites, predators and pollinators these compounds can be incor-porated into an Integrated Pest Management Program in agriculture.

5 CONTROL OF STORED PRODUCT INSECTS

The control of stored product insects is one area where ecological interactions are not major concern. The primary concern is the safety factor. With the increasing incidence of resistance to products like mal-athion, there is a search for alternate control agents. Benzoylphenyl ureas have been tested against many stored product insects as a safe substitute and appear to have good control potential. Control of some stored product insects with benzoylphenyl ureas is summarized in Table 7.

Mixing the benzoylphenyl urea with grains provides excellent control of most stored product insects suppressing the population level to near zero by both larvicidal and ovicidal effects (Table 8). Various species of *Sito-philus* are very sensitive to diflubenzuron and can be controlled with a low dosage of about 1 ppm (Carter 1975; McGregor and Kramer 1976). *Tri-bolium* species are also quite sensitive and diflubenzuron appears to be slightly more active than Alsystin at least in the case of the red flour beetle, *Tribolium castaneum* (Carter 1975; Ishaaya et al. 1981; McGregor and Kramer 1976; Mian and Mulla 1982a).

The cigarette beetle, *Lasioderma serricorne* and the khapra beetle, *Trogoderma granarium* are quite refractory to benzoylphenyl ureas (Carter 1975; McGregor and Kramer 1976; Rajendran and Shivaramiah 1983; Sexena and Kumar 1982). Another dermestid, *Dermestes maculatus*, the

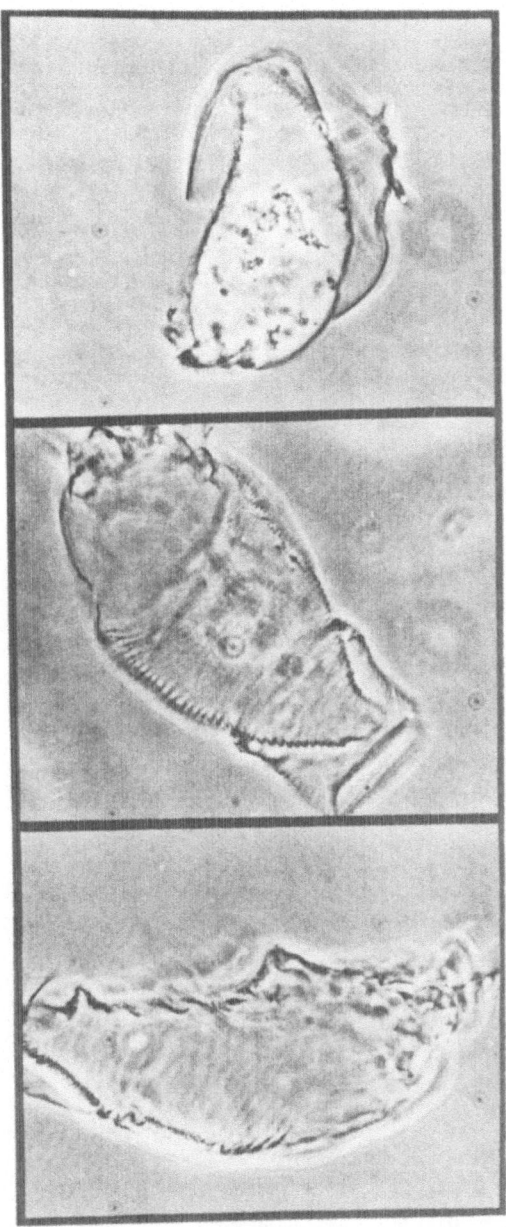

Figure 14. Effect of diflubenzuron on the citrus rust mite, *Phyllocoptruta oleivora*. The mites that were exposed as eggs are entrapped at the 2nd nymphal stage inside the exuviae (McCoy 1978).

Table 7. Control of some stored product insects with benzoylphenyl ureas.

Insect species	Host	Analog	Activity	Reference
Sitophilus oryzae (Rice weevil) Fam. Curculionidae	Rice, wheat, corn and other grain	Diflubenzuron; Alsystin	Eggs and younger larvae were affected. 5–1 ppm gave total control.	Carter 1975; McGregor and Kramer 1976; Mian and Mulla 1982a; 1982b.
Sitophilus granarius (Granary weevil) Fam. Curculionidae Ord. Coleoptera	Wheat, corn, and other grain.	Diflubenzuron	1–2 ppm gave total control.	Carter 1975; McGregor and Kramer 1976.
Sitophilus zeamais (Maize weevil) Fam. Curculionidae Ord. Coleoptera	Maize, wheat, corn and other grain	Diflubenzuron	1 ppm gave total control	McGregor and Kramer 1976.
Rhyzopertha dominica (Lesser grain borer) Fam. Bostrichidae Ord. Coleoptera	Wheat, corn and other grain	Diflubenzuron; Alsystin	1–5 ppm of both analogs gave total control and the effect lasted for 12 months	McGregor and Kramer 1976; Mian and Mulla 1982a; 1982b
Tribolium confusum (Confused flour beetle) Fam. Tenebrionidae Ord. Coleoptera	Wheat, corn and other grain as well as milled products	Diflubenzuron; Alsystin	0.4–1 ppm of either of the analogs gave total control	Ishaaya *et al.* 1981; McGregor and Kramer 1976
Tribolium castaneum (Red flour beetle) Fam. Tenebrionidae Ord. Coleoptera	Wheat and other grains, especially milled products	Diflubenzuron; Alsystin	Alsystin was more effect and at 5 ppm gave total control. Diflubenzuron was slightly less active.	Carter 1975; Mian and Mulla 1982a; Saxena and Mathur 1981a; 1981b.
Oryzaephilus surinamensis (Saw toothed grain beetle) Fam. Cucujidae Ord. Coleoptera	Wheat, corn and other grains	Diflubenzuron; Alsystin	1–5 ppm of either of the analogs gave total control	McGregor and Kramer 1976; Mian and Mulla 1982a

Species	Commodity	Insecticide	Effect	Reference
Lasioderma serricorne (Cigarette beetle) Fam. Anobiidae Ord. Coleoptera	Grain and tobacco	Diflubenzuron	Relatively refractory requiring higher dosages	Carter 1975; McGregor and Kramer 1976
Trogoderma granarium (Khapra beetle) Fam. Dermestidae Ord. Coleoptera	Wheat and other grains	Diflubenzuron; Penfluron	Relatively refractory and even at higher doses 100% mortality was never	Rajendran and Shivaramiah 1983; Saxena and Kumar 1982; Webley and Airey 1982.
Dermestes maculatus (Hide beetle) Fam. Dermestidae Ord. Coleoptera	Hides, fish and meat products	Diflubenzuron	Excellent control; eggs and younger larvae are more susceptible	Webley and Airey 1982.
Callosobruchus chinensis (Cowpea weevil) Fam. Bruchidae Ord. Coleoptera	Various pulses and beans.	Diflubenzuron	Excellent control of early stages	Webley and Airey 1982.
Acanthoscelides obtectus (Bean weevil) Fam. Acanthoscelidae Ord. Coleoptera	Beans and pulses	Diflubenzuron	Good control.	Webley and Airey 1982.
Stegobium paniceum (Drugstore beetle) Fam. Anobiidae Ord. Coleoptera	Wheat and other stored grains; Packaged goods and drugs.	Diflubenzuron	Ineffective on this species	Carter 1975.

Table 8. Activity of diflubenzuron against 7 species of stored grain beetles (McGregor and Kramer 1976).

Insects	Control progeny	% reduction at ppm of		
		0.1	1.0	10.0
Rice weevil (*Sitophilus oryzae*)	1437	86	99	99
Granary weevil (*S. granarius*)	746	68	99	100
Maize weevil (*S. zeamais*)	1745	68	100	100
Lesser grain borer (*Rhyzopertha dominica*)	414	88	98	98
Confused flour beetle (*Tribolium confusum*)	964	91	100	100
Sawtoothed grain beetle (*Oryzaephilus surinamensis*)	5683	91	99	99
Cigarette beetle (*Lasioderma serricorne*)	5074	44	72	71

hide beetle on the other hand is very susceptible to diflubenzuron (Webley and Airey 1982). So are the cowpea weevil, *Callosobruchus chinensis* and the bean weevil, *Acanthoscelides obtectus* (Webley and Airey 1982). The drug store beetle, *Stegobium paniceum* is very resistant to benzoylphenyl ureas and cannot be controlled with diflubenzuron (Carter 1975).

6 CONTROL OF FOREST INSECT PESTS

The forest ecosystem is complex and in many respects quite fragile. The intricate mesh of varied flora and fauna that make up the system are resilient only within limits. Repeated and widespread application of powerful pest control agents can result in severe damage to the forest ecology. The main reason for the environmental hazard of chlorinated hydrocarbons were their broad spectrum of activity, persistence in the environment and characteristics for biological concentration in the environment (Pimentel 1971). The undesirable side effects of DDT became apparent in the early sixties, 20 years after it was first introduced (Carson 1962). In Canada, for instance, it has been a practice to spray over $2\frac{1}{2}$ million acres every year to control the spruce budworm, *Choristoneura fumiferana* (Angus 1973). Repeated spraying of DDT in the province of New Brunswick during 1954-56 resulted in significant reduction in woodcock populations, and young salmon were almost eliminated in the Miramichi river. Similar spray operations in British Columbia resulted in total mortality of the Coho salmon (O'Brien 1967).

When benzoylphenyl ureas are sprayed on a forest they selectively act on insect larvae sparing the adult insects such as predators, parasites and pollinators that may fly in the area and come in contact with the material. They combine the best of two worlds, powerful activity on the larval pest like broad spectrum insecticides yet gentle to the environment like the biological insecticides. These compounds are uniquely suited for insect pest control in the forests. Control of some forest insect pests with benzoylphenyl ureas is summarized in Table 9.

Benzoylphenyl ureas have dramatic effects on some forest insects and yet have very little effect on others. Also certain analogs are more active than others and this again varies from insect to insect. In 1975, Granett and Dunbar showed that the gypsy moth, *Lymantria dispar* can be controlled in the field by applying diflubenzuron at the rate of 1 oz AI/acre

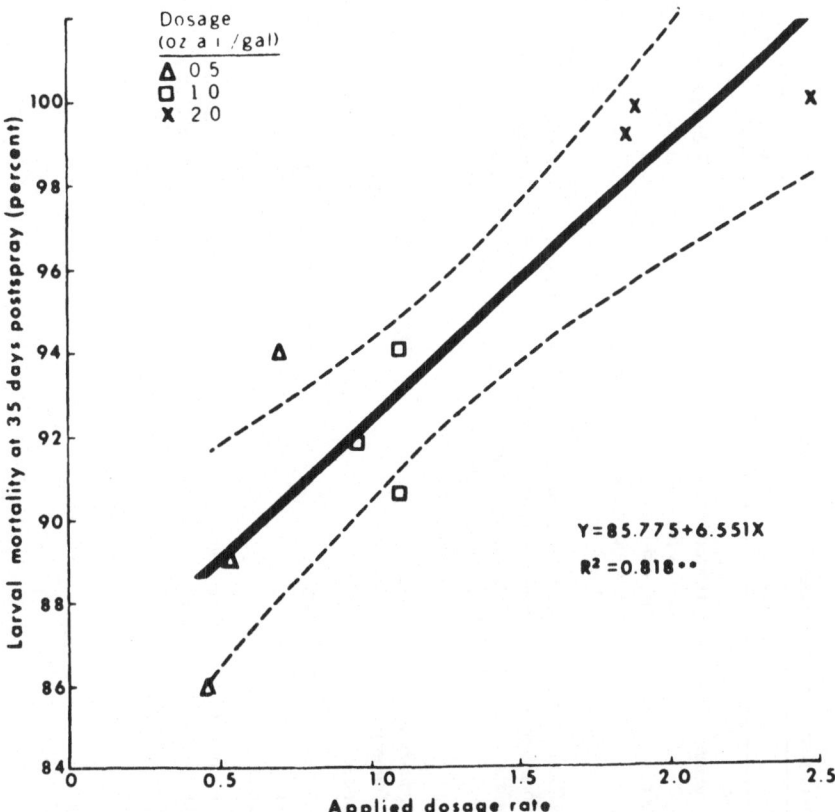

Figure 15. Effect of aerial application of diflubenzuron (indicated as ounces/0.5 U. S. gal) on Douglas-fir infested with the larvae of the Douglas-fir tussock moth, *Orgyia pseudotsugata.* At 2 oz AI/0.5 U. S. gal/acre (140 g AI/4.7 L/ha) close to 100% control was achieved (Hard et al. 1978).

Table 9. Control of some forest insect pests with benzoylphenyl ureas.

Insect species	Host	Analog	Activity	Reference
Order Lepidoptera				
Operophtera brumata (Winter moth) Fam. Geometridae	Hardwood trees such as red oak, elm, red maple, basswood, poplar and willows	Diflubenzuron; Alsystin	Excellent control in small scale trails	Albert 1984; Pree 1976; Tanks *et al.* 1978;
Lambdina fiscellaria lugubrosa (Western hemlock looper) Fam. Geometridae	Western hemlock, Sitka spruce, Engelmann spruce, white spruce, Douglas-fir, western red cedar and true firs.	Diflubenzuron	Very effective in laboratory tests	Sahota and Shepherd 1975
Lithocolletis hamadryadella (White oak leafminer) Fam. Gracillaridae	White oak	Diflubenzuron	Very effective in laboratory tests.	Appleby 1981
Gracillaria syringella (Lilac leafminer) Fam. Gracillaridae	Ash	Diflubenzuron	Excellent control in field trials.	Kiziroglu 1976
Thaumetopoea pityocampa (Pine processionary moth) Fam. Notodontiade	Pine and other conifers.	Diflubenzuron	Excellent control in field trials.	Badiali 1979; Demolin and Millet 1981a; 1981b; 1983; Ferrari and Tiber 1979; Gadais *et al.* 1978; Georgevitis 1979; 1982; Halperin 1980; Millo 1978; Ribrioux and Dolbeau 1975; Robredo 1980a; 1980b; 1980c; Spaic 1981

Species / Family	Host	Chemical	Efficacy	References
Lymantria dispar (Gypsy moth) Fam. Lymantriidae	Oak, birch, poplar and most other hardwoods; also conifers	Diflubenzuron; Alsystin	Excellent control in field trials.	Abel-Monem and Mumma 1981; Cameron and McManus 1979; Forgash *et al.* 1978; Granett and Dunbar 1975; Granett *et al.* 1976; Jobin and Caron 1982; Skatulla 1975; Ticehurst *et al.* 1982; Vasic 1977
Lymantria monacha (Nun moth) Fam. Lymantriidae	Oak and other hardwoods; sometimes pine.	Diflubenzuron	Excellent control in field trials	Krushev and Marchenko 1981; Skatulla 1975; Szmidt and Sliwan 1981.
Euproctis chrysorrhoea (Brown tail) Fam. Lymantriidae	Hardwoods and sometimes conifers.	Diflubenzuron with B.t.	Very effective	Canivet *et al.* 1978.
Orgyia pseudotsugata (Douglas fir tussock moth) Fam. Lymantriidae	Douglas fir	Diflubenzuron	Excellent control in field trials	Ciesla 1977; Gillette *et al.* 1978; Hard *et al.* 1978; Limnane 1979; Markin and Wilcox 1978; Niesses *et al.* 1976; Robertson and Boelter 1979a; 1979b
Leucoma (Stilpnotia) salicis (Satin moth) Fam. Lymantriidae	Poplar and willow	Diflubenzuron	Excellent control in field trials.	Vasic 1979
Leucoma wiltshirei (Iranian oak moth) Fam. Lymantriidae	Oak	Diflubenzuron	Excellent control in field trials.	Abai 1981
Rhyacionia frustrana (Nantucket pine tip moth) Fam. Olethreutidae	Pines except long leaf and Eastern white pine.	Diflubenzuron	Very effective	Ellis 1979; Richmond and Cunningham 1983

Table 9. Control of some forest insect pests with benzoylphenyl ureas (continued)

Insect species	Host	Analog	Activity	Reference
Rhyacionia buoliana (European pine shoot moth) Fam. Olethreutidae	Pines especially red and mugho	Diflubenzuron	Very effective	Fodor and Halmagyi 1978
Choristoneura occidentalis (Western spruce budworm) Fam. Torricidae	Douglas fir and true firs	Diflubenzuron, Alsystin and others	Not very effective	Gillette *et al.* 1978; Rappaport and Robertson 1981; Robertson and Haverty 1981; Robertson and Smith 1984; Robertson *et al.* 1984
Choristoneura fumiferana (Spruce budworm) Fam. Tortricidae	White spruce and balsam fir	Diflubenzuron, Alsystin, Chlorfluazuron, and others	Chlorfluazuron shows excellent control in the field. All others are not effective	Madore *et al.* 1976; Granett and Retnakaran 1977; Granett *et al.* 1980; Retnakaran 1979; 1980; 1981; 1982b; Retnakaran and Smith 1975; Retnakaran *et al.* 1980
Tortrix viridana (Green oak tortrix) Fam. Tortricidae	Oak	Diflubenzuron	Excellent control in field trials.	Horstman 1982; Malphettes and Martouret 1979.
Zeiraphera isertana (Oak budmoth) Fam. Tortricidae	Oak	Diflubenzuron	Excellent control in field trials.	Horstman 1982
Croesia semipurpurana (Oak-leaf tier) Fam. Tortricidae	Oak	Diflubenzuron	Excellent control in field trials with diflubenzuron	Retnakaran and Grant 1985; Retnakaran and Tomkins 1982
Malacosoma disstria (Forest tent caterpillar) Fam. Lasiocampidae	Hardwoods especially trembling aspen.	Diflubenzuron	Excellent control in the field	Harper and Abrahamson 1979; Retnakaran and Smith 1976; Retnakaran *et al.* 1979
Dichomeris ligulellus (Palmerworm) Fam. Gelechiidae	Hardwoods	Diflubenzuron	Good control in laboratory trials	Nielson and Balderston 1979.

Species	Host	Chemical	Effect	Reference
Coleophora laricella (Larch casebearer) Fam. Coleophoridae	Larch	Diflubenzuron	Good control in laboratory trials.	Page *et al.* 1982.
Hyphantria cunea (Fall webworm) Fam. Arctiidae	Over 100 species of forest and shade trees but mostly weed species.	Diflubenzuron	Excellent control potential.	Szanto 1978.
Order Hymenoptera				
Diprion similis (Introduced pine sawfly) Fam. Diprionidae	Pines especially Scots pine	Diflubenzuron	Excellent effects in the greenhouse	Fogal 1977; Valovage and Kulman 1978a
Neodiprion sertifer (European pine sawfly) Fam. Diprionidae	Pines	Alsystin	Laboratory effects.	Luber 1983
Pikonema alaskensis (Yellowheaded spruce sawfly) Fam. Tenthredinidae	Spruce	Diflubenzuron	Good control in laboratory trials	Valovage and Kulman 1978b
Pristiphora erichsonii (Larch sawfly) Fam. Tenthredinidae	Larch	Diflubenzuron	Good control in laboratory trials.	Valovage and Kulman 1978c
Order Coleoptera				
Hylobius abietis (Pine weevil) Fam. Curculionidae	Pine	Diflubenzuron; Alsystin.	Sterility effect in laboratory tests.	Kolbe and Hartwig 1982; Novak and Sehnal 1978; 1979.
Pissodes strobi (White pine weevil) Fam. Curculionidae	Pines especially white and Scots; also Sitka spruce	Diflubenzuron; Alsystin.	Sterility effects observed.	Retnakaran and Smith 1982; Sahot and McMullen 1979.
Dendroctonus frontalis (Southern pine beetle) Fam. Scolytidae	Pine	Diflubenzuron and Penfluron	Sterility effect in laboratory tests	Richmond *et al.* 1976.
Dendroctonus rufipennis (Pine bark beetle) Fam. Scolytidae	Pine	Diflubenzuron	Sterility effect in laboratory effects	Sahota and Ibaraki 1980

244

Figure 16. Effect of aerial spraying of diflubenzuron on forest tent caterpillar, *Malacosoma disstria* (Hübner). (A,C) Control plot showing massive defoliation of trembling aspen. (B,D) treatment plot showing foliage protection. (E) A dead forest tent caterpillar after feeding on sprayed foliage. (F,G,H) Various degrees of larval moult disruption induced by treatment. Second instar larvae showing incomplete head capsule slippage (F), partial eclosion of old cuticle (G), and enlargement of thoracic region (H). (Retnakaran et al. 1979).

(70 g AI/acre). The material also had no effect on the adult parasites of this insect (Granett et al. 1976). This spectacular effect has since been confirmed by others (Jobin and Caron 1982; Skatulla 1975). The nun moth, *Lymantria monacha* is also very sensitive to diflubenzuron and can be successfully controlled by using low dosages of this benzoylphenyl urea (Krushev and Marchenko 1981; Skatulla 1975; Szmidt and Sliwan 1981). The pine processionary caterpillar, *Thaumetopoea pityocampa* has been very well controlled in Spain and other parts by spraying diflubenzuron (Demolin and Millet 1983; Georgevitis 1982; Halperin 1980; Robredo 1980b). The Douglas-fir tussock moth is susceptible to diflubenzuron at a slightly higher rate (2 oz AI/acre or 140 g AI/ha) and this insect can be successfully controlled in the field (Figure 15) (Hard et al. 1978). The forest tent caterpillar, *Malocosoma disstria* is very sensitive to diflubenzuron and this insect can be controlled at 1 oz AI/acre (70 g AI/ha) or less (Figure 16) (Harper and Abrahamson 1979; Retnakaran and Smith 1976; Retnakaran et al. 1979).

There are many forest insects that can be well controlled, although not as dramatically as those described earlier. The green oak tortrix, *Tortrix viridana* (Horstman 1982; Malphettes and Martouret 1979) and the oak leaf shredder, *Croesia semipurpurana* (Figure 17) (Retnakaran and Grant 1975) belong to this category. There are many other species that appear susceptible to diflubenzuron in the laboratory.

Diflubenzuron and Alsystin are very effective on different sawflies and appear to have excellent control potential (Figure 18) (Fogal 1977; Luber 1983; Valovage and Kulman 1978a, b, c). The introduced pine sawfly, *Diprion similis* for instance, is extremely sensitive to diflubenzuron (Fogal 1977).

Coleopterous pests of forests especially curculionids and scolytids are difficult to control because they spend most of their life inside the host tree. One approach has been to use the sterility effect of diflubenzuron. Limited success has been achieved with a few pests.

Some insects such as the western spruce budworm *Choristoneura occidentalis* (Gillett et al. 1978; Rappaport and Robertson 1981; Robertson and Smith 1984) and the spruce budworm, *Choristoneura fumiferana* (Granett and Retnakaran 1977; Retnakaran 1979; 1981; 1982) are refractory to diflubenzuron (Figure 7). Younger instars are less susceptible to this material than older instars. This is not true with other benzoylphenyl ureas like EL-494 (Figures 6 and 8). The spruce budworm absorbs very little of the diflubenzuron ingested and metabolises the absorbed material very effficiently in contrast to susceptible species such as the forest tent caterpillar and the Douglas-fir tussock moth that absorb more and metabolize less (Granettt et al. 1980; Retnakaran et al. 1980). Chlorfluazuron on the other hand is very effective on the spruce budworm (Figure 19) and the insect can be effectively controlled at a dosage of 1 oz AI/acre or 70 g AI/ha (Retnakaran 1982b).

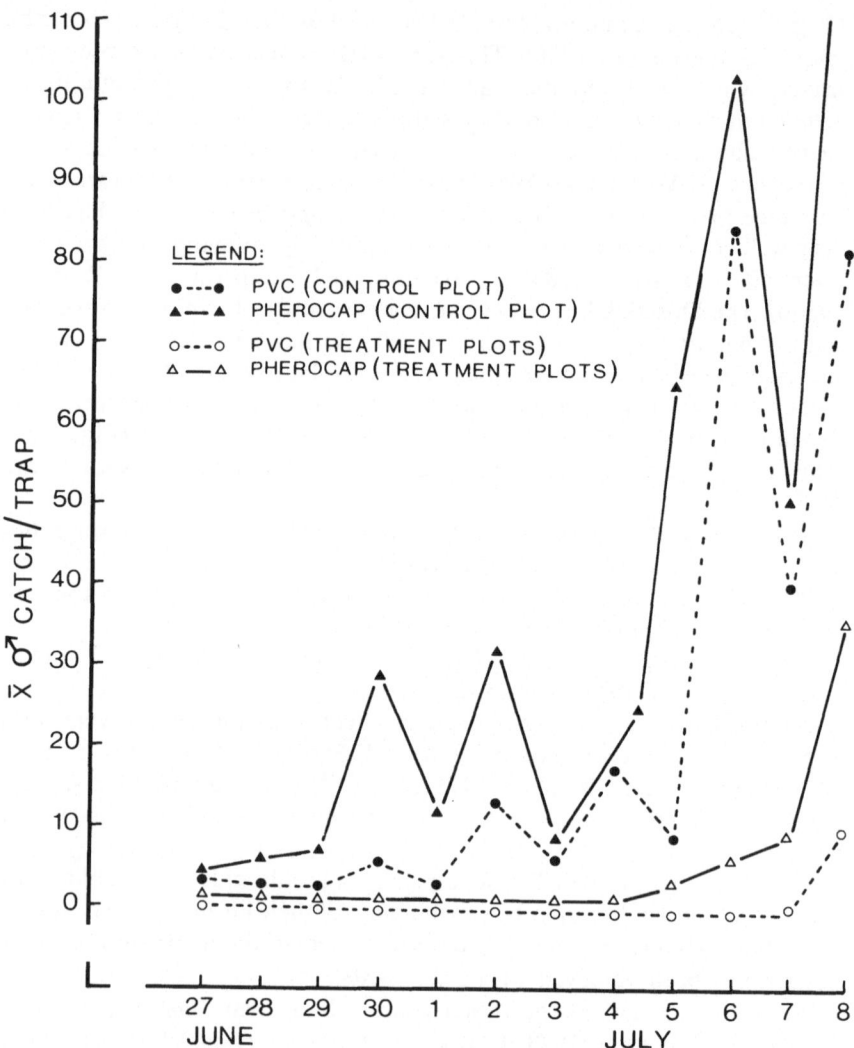

Figure 17. Pheromone-trap catches of the oak leaf shredder, *Croesia semipurpurana* (Kearfott) in the diflubenzuron treated plot and control plot for two different pheromone formulations, PVC and pherocap (Retnakaran and Grant 1985).

7 CONTROL OF INSECT PESTS OF PUBLIC HEALTH AND VETERINARY IMPORTANCE

Pests of public health and veterinary importance are intimately connected with our everyday life. Insects that carry the causative agents for some of the dreaded diseases such as malaria, filariasis, oncchocerosis,

Figure 18. Effect of diflubenzuron on the introduced pine saw fly, *Diprion similis* (Hartwig)
(1) Foliage protection (white pine) by treatment with 1% diflubenzuron; the introduced pine
saw fly larvae feeding on the untreated control shoot. (2) Pharate 3rd instar that ingested the
material as a 2nd instar; apolysis completed but no ecdysis. (3) Pharate last larval instar,
apolysis completed but failed to ecdyse and fluid accumulation beneath old cuticle, all effects
due to diflubenzuron. (4) 3rd instar at various stages of ecdysis. (5) Non-feeding 5th instar
larva dissected to show accumulation of fibrous material in the midgut (Fogal 1977).

Figure 19. Effect of chlorfluazuron on 5th instar spruce budworm (*Choristoneura fumiferana*). The larva on top has attempted to moult and shows severe malformation of the cuticle. The lower control larva has successfully moulted into the 6th instar (Retnakaran 1982).

trypanosomiasis, encephalitis and yellow fever have to be controlled. Insects that affect the health of cattle on which we are so dependent have to be controlled as well. Here again environment and safety factors are very important. Benzoylphenyl ureas are one of the few environmentally acceptable options available for the control of these insects. Control of some insect pests of public health and veterinary importance is summarized in Table 10.

The horn fly, *Haematobia irritans* is a serious nuisance insect on cattle. It can be well controlled by spraying the range cattle directly with diflubenzuron or making the animals rub their bodies on dust bags containing diflubenzuron hung on the sides of a gate, or incorporating the material in mineral blocks that they lick or spraying the fecal pats in which the fly breeds (Barker and Jones 1976; Hopkins and Chamberlain 1976; Kunz and Bay 1977; Kunz et al. 1976; 1977). The effect is primarily ovicidal and this can be accomplished either by treating the eggs or the adult males or females (Wright and Harris 1976). Larvae are susceptible to treatment as well (Schmidt and Kunz 1980).

The stable fly, *Stomoxys calcitrans* is a vicious biter and both sexes feed on the blood of many species of mammals. Besides causing severe annoyance to cattle it can also mechanically transmit several pathogenic organisms such as *Trypanosoma evansi* (parasite in horses and mules), *Borrelia recurrentis* (causative agent of relapsing fever) and *Bacillus anthracis* which causes anthrax in animals and man (Harwood and James 1979). Diflubenzuron provides excellent control by larvicidal and ovicidal effects (Table 11) (Schmidt and Kunz 1980; Spates and Wright 1980). Ingestion is the mode of entry in larvae; pupae show no topical effect unlike eggs. Wright and his co-workers (1975) did an interesting experiment at Lion Country Safari in Grand Prairie, Texas where house flies were a real problem. They fed white rhinoceros with $1 - 0.1$ mg/kg of diflubenzuron in the feed. The feces of these animals provided an unusually rich medium for the larvae of the house flies to thrive. The feces from diflubenzuron fed animals showed a total absence of house fly maggots.

Unlike horn flies where diflubenzuron treated adults of both sexes show congenital-ovicidal effects, stable flies show effects only through the females (Spates and Wright 1980).

The house fly, *Musca domestica* is a nuisance insect around the world and has been implicated as a mechanical carrier of many disease agents. It can be effectively controlled by feeding diflubenzuron to cattle or chicken along with feed or by spraying breeding sites or allowing cattle to lick mineral blocks containing the material (Barker and Newton 1976; Miller et al. 1975). Diflubenzuron has ovicidal effects when either fed to the female flies or when the females are topically treated with the material. There is no topical effect on eggs and males play no part (Grosscurt 1976). Penfluron and Alsystin are more effective than diflubenzuron (Chang 1979).

Musca autumnalis, the face fly, besides causing annoyance to man and animals has been shown to be capable of transmitting the bacterium *Moraxella bovis*, the pathogen that causes keratoconjunctivitis (pink eye) in cattle. In certain parts of the world it serves as an intermediate host to the eye worm, *Thelazia rhodesii* (Harwood and James 1979). Face fly can be well controlled with diflubenzuron by treating adults, spraying the breeding areas, dusting the cattle and "feed through" to cattle (Knapp and Herald 1982; Miller 1974; Pickens and De Milo 1977). The effect is primarily ovicidal and occurs when either the male or the female is treated with diflubenzuron (Pickens and De Milo 1977). Often the effect of treating the adults is delayed and the effect is manifested in the larvae in the form of molt inhibition (Knapp and Herald 1982).

The tsetse fly (= "fly destructive to cattle" in the Sechuana language of Botswana)., *Glossina morsitans* has been referred to as "Africa's bane". Both sexes feed on vertebrate blood and transmit the protozoan parasite, *Typamosoma* species that cause sleeping sickness. Each female fly carries a single larva in a "uterus" nourishing the larva with secretions from a "milk gland". The fully formed larva is deposited in a moist area and within an hour the larva forms a puparium (Harwood and James 1979).

Table 10. Control of some pests of public health and veterinary importance with benzoylphenyl ureas.

Insect species	Host/habitat	Analog	Activity	Reference
Haematobia irritans (Horn fly) Fam. Muscidae Ord. Diptera	Range cattle	Diflubenzuron	Excellent control though primarily ovicidal effect. This effect is congenital through both sexes.	Barker and Jones 1976; Hopkins and Chamberlain 1976; Kunz and Bay 1977; Kunz et al. 1976; Schmidt and Kunz 1980; Wright and Harris 1976.
Stomoxys calcitrans (Stable fly) Fam. Muscidae Ord. Diptera	Feed lot of livestock, poultry farms	Diflubenzuron	Excellent control by surface treatment of animals, feeding areas or incorporating the material in the feed ("feed through").	Campbell and Wright 1976; Hoyakawa 1976; Schmidt and Kunz 1980; Spates and Wright 1976; 1980; Wright 1974; Wright and Harris 1976; Wright and Spates 1976; Wright et al. 1975.
Musca domestica (House fly) Fam. Muscidae Ord. Diptera	Houses	Diflubenzuron, Alsystin, Penfluron	Excellent control by treating breeding areas or incorporating in the feed.	Barker and Newton 1976; Chang 1979; Grosscurt 1976; Keiding et al. 1976; Miller et al. 1975; Wright 1974; 1975; Wright and Spates 1976; Wright et al. 1975.
Musca autumnalis (Face fly) Fam. Muscidae Ord. Diptera	Livestock areas	Diflubenzuron	Excellent control by treating adults, breeding areas, dusting cattle or as feed through.	Knapp and Herald 1982; Miller 1974; Pickens and DeMilo 1977
Glossina morsitans morsitans (The tsetse fly) Fam. Glossinidae Ord. Diptera	Vertebrates (Humans and cattle)	Diflubenzuron; Penfluron	Good control potential. Treatment of adult flies or surfaces of areas frequented by the flies have good "larvicidal" effects.	Clarke 1982; Jordan and Trewern 1978; Jordan et al. 1979.
Simulium vittatum and other species (Black flies) Fam. Simuliidae Ord. Diptera	Vertebrates	Diflubenzuron; Alsystin	Excellent control by spraying streams.	McKague et al. 1978; Lacey and Mulla 1977; 1978a; 1978b.

Species	Habitat/Host	Chemical	Effect	Reference
Chironomid midges (Nuisance midges) Fam. Chironomidae Ord. Diptera	Ponds and lakes	Diflubenzuron; Alsystin	Generally not very effective requiring high dosages.	Ali and Mulla 1977; Johnson Mulla 1981; Mulla et al. 1976.
Lucilia cuprina (Sheep blowfly) Fam. Calliphoridae Ord. Diptera	Sheep in Australia	Diflubenzuron; Alsystin	Laboratory studies indicate that Alsystin might be effective.	Levot and Shipp 1983; Turnbull and Howelle 1980
Culicoides varipennis (Biting midge) Fam. Ceratopogonidae Ord. Diptera	Vertebrates	Diflubenzuron	Refractory to diflubenzuron requiring high doses.	Apperson and Yows 1976
Cochliomyia hominivorax (Screw worm) Fam. Calliphoridae Ord. Diptera	Cattle	Diflubenzuron	Requires high dosages and the effect is not long lasting; limited potential.	Crystal 1978.
Bovicola limbatus (Angora goat biting louse) Fam. Trichodectidae Ord. Mallophaga	Goat	Diflubenzuron	Limited effect; the louse is probably able to metabolize the material.	Hopkins and Chamberlain 1977.
MOSQUITOES (Fam. Culicidae; Ord. Diptera)				
Asdes aegypti (Yellow fever mosquito)	Vector for yellow fever, dengue and chikungunya viruses.	Diflubenzuron; Alsystin; Penfluron	Deformed pupae and failure of adult emergence were characteristic. Eggs in their early stage of development very sensitive. Excellent control (52 ppb)	Arias 1974; Bhakshi et al. 1982; Busvine et al. 1976; Gaaboub and Busvine 1976; Herald et al. 1980; Hsieh and Steelman 1974; Jakob 1973; Saleh et al. 1981; Sinegre et al. 1980.
A. caspius	—	Diflubenzuron	Excellent control	Sinegre et al. 1980

Table 10. Control of some pests of public health and veterinary importance with benzoylphenyl ureas (continued)

Insect species	Host/habitat	Analog	Activity	Reference
A. melanimon	—	Diflubenzuron	Excellent control at 28 g/ha applied aerially.	Schaefer *et al.* 1974; 1975.
A. nigromaculis	—	Diflubenzuron	Excellent control at 28 g/ha	Darwazeh *et al.* 1977; Schaefer *et al.* 1974; 1975.
A. sollicitans (Salt marsh mosquito)	—	Diflubenzuron	Extremely susceptible to diflubenzuron.	Hsieh and Steelman 1974
A. taeniorhynchus	—	Diflubenzuron	Eggs hatched by side split; Treated females showed congenital ovicidal effect. Excellent control at 11 g/ha	Axtell *et al.* 1979; Dame *et al.* 1976; Hsieh and Steelman 1974; Miura *et al.* 1976; Rathburn and Boike Jr. 1975; Rogers *et al.* 1976.
A. triseriatus	Vector for California encephalitis virus.	Diflubenzuron	Excellent control.	Hsieh and Steelman 1974.
Anopheles albimanus	Malaria vector in Central America.	Diflubenzuron	Excellent control at low dosages	Arias 1974; Jakob 1973; Mulla *et al.* 1974.
A. culicifacies	—	Diflubenzuron; Penfluron	Excellent control.	Bhakshi *et al.* 1982.
A. franciscanus	—	Diflubenzuron	Total control at 28 k/ha	Mulla and Darwazeh 1976.
A. gambiae	Malaria vector in Africa	Diflubenzuron	Excellent control.	Busvine *et al.* 1976.
A. nigromaculis	—	Diflubenzuron	Excellent control at 28 g/ha.	Schaefer *et al.* 1975.
A. quadrimaculatus (Common malaria mosquito)	—	Diflubenzuron	Total control at 28 g/ha	Busvine *et al.* 1976; Dame *et al.* 1976; Hsieh and Steelman 1974.

Species	Vector/Notes	Compound	Effectiveness	References
A. stephensi	—	Diflubenzuron; Penfluron	Excellent control at 5 ppm.	Bhakshi *et al.* 1982; Jakob 1973.
Culex nigripalpue	—	Diflubenzuron	Excellent control at 28 g/ha.	Dame *et al.* 1976; Rathburn and Boike Jr. 1975.
C. salinarius	—	Diflubenzuron	Excellent control.	Hsieh and Steelman 1974.
C. pipiens quinquefasciatus (= *C. pipiens fatigans*) (House mosquito)	Vector for *Wuchereria bancrofti*	Diflubenzuron; Penfluron.	Excellent control at levels as low as 0.25 ppm	Arias 1974; Bhakshi *et al.* 1982. Busvine *et al.* 1976; Hsieh and Steelman 1974; Jakob 1973; Miura *et al.* 1976; Mulla and Darwazeh 1975; 1976; Singere *et al.* 1980.
C. tarsalis	Vector for Equine encephalitis and St. Louis encephalitis viruses	Diflubenzuron	Excellent control at 28 g/ha.	Arias 1974; Hsieh and Steelman 1974; Jakob 1973; Miura *et al.* 1976; Schaefer *et al.* 1974; 1975.
C. triiaeniorhynchus	Vector for Japanese encephalitis virus	Diflubenzuron	Excellent control potential.	Takahashi and Ohtaki 1976.
Culicita inornata	Carrier of viral encephalitides	Diflubenzuron	Excellent control.	Hsieh and Steelman 1974.
Psorophora confinnis	—	Diflubenzuron	Excellent control at doses as low as 6 g/ha.	Darwazeh *et al.* 1977; Hsieh and Steelman 1974; Mulla and Darwazeh 1975; 1976.
P. ferox	—	Diflubenzuron	Excellent control.	Hsieh and Steelman 1974.
P. varipes	—	Diflubenzuron	Excellent control.	Hsieh and Steelman 1974.

Table 11. Effect of exposing male and female pupae of stable flies (*Stomoxys calcitrans*) and house flies (*Musca domestica*) to a layer of various concentrations of diflubenzuron on the egg production and hatch.

Concentration of diflubenzuron (% AI in WP)	Stable fly		House fly	
	No. eggs	% batch	No. eggs	% hatch
0	1800	83.5	500	92
0.1	842	0.05	3969	47.1
1.0	64	0	2155	38.5
10.0	1876	0.001	2652	18.3
25.0	754	0	3036	28.3
96.0 (Tech)	1622	0	—	—

*WP — Wettable powder formulation.

Benzoylphenyl ureas appear to have excellent control potential for this insect (Table 12). Topical treatment or tarsal contact with treated surfaces induce loss of *in utero* larva ("abortion"). Such larvicidal effects are less pronounced if males are treated (Clarke 1982; Jordan and Trewern 1978; Jordan et al. 1979).

The black flies, *Simulium vittatum* and other species are not only nuisance insects but also cause mortality in cattle due to a toxin contained in the fly's saliva. In Africa, *Simulium neavei*, is a vector of *Onchocera*

Table 12. Adult emergence rate from offspring of female tsetse fly (*Glossina morsitans morsitans*) following topical application of various doses of 3 IGRs. (63220 is the 4-bromophenyl analog of diflubenzuron). (Jordan *et al.* 1979).

Treatment day	Compound	Dose (g/fly)	Total offspring	% emergence	Significance (followed by same letter indicates no difference with each dose)
4	Diflubenzuron	5.0	173	3	9
4	63220	5.0	113	8	9
4	Penfluron	5.0	99	2	9
4	Diflubenzuron	0.5	148	17	9
4	63220	0.5	121	37	6
4	Penfluron	0.5	153	29	6
4	Diflubenzuron	0.05	156	85	9
4	63220	0.05	162	89	9
4	Penfluron	0.05	177	85	9
4	Control	—	159	97	—
18	Diflubenzuron	5.0	128	1	9
18	63220	5.0	137	2	9
18	Penfluron	5.0	108	3	9
18	Diflubenzuron	1.0	140	14	9
18	63220	1.0	125	15	9
18	Penfluron	1.0	165	6	6
18	Control	—	162	97	—

volvulus that causes blindness in humans, a condition referred to as "oncchocerosis" (Harwood and James 1979). Black flies can be effectively controlled by spraying streams (1 ppm for 15 min) with diflubenzuron (Figure 20) (McKague et al. 1978). Both ovicidal and larvicidal effects have been shown but there is no effect on pupae (Lacey and Mulla 1977; 1978a; 1978b).

Chironomid midges that occur in ponds and lakes are nuisance insects that have not been implicated as vectors of any pathogen. Excepting for a few, they are difficult to control with benzoylphenyl ureas and require high doses (Ali and Mulla 1977; Johnson and Mulla 1981; Mulla et al. 1976).

Lucilia cuprina, the Australian sheep blowfly is a major pest of sheep in Australia. Benzoylphenyl ureas appear to have control potential (Levot and Shipp 1983; Turnbull and Howells 1980).

Culicoides varipennis is an example of a biting midge that is particularly refractory to diflubenzuron requiring high doses (Apperson and Yows 1976). The screwworm, *Cochliomyia hominivorax* is another insect that requires a high dose of diflubenzuron and the affected females recover rapidly (Crystal 1978). The Angoragoat biting louse, *Bovicola limbatus* is

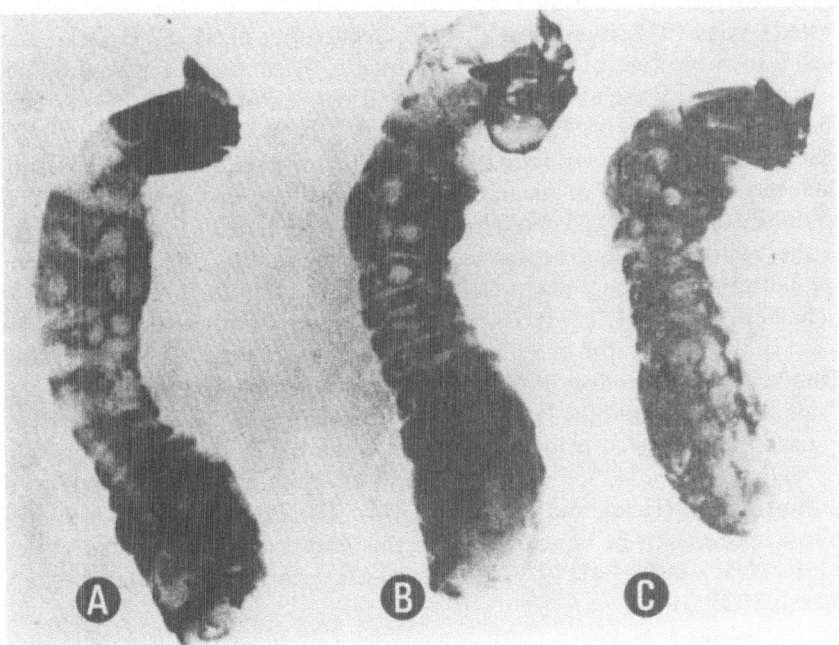

Figure 20. Moulting abnormality in penultimate instar larvae of the black fly, *Simulium vittatum* exposed to diflubenzuron. Larvae on the left (A) has a partially shed capsule, the one in the middle (B) has it almost completely shed and the one on the right (C) has the head capsule intact (Lacey and Mulla, 1978).

probably able to metabolize diflubenzuron and therefore this benzoyl-phenyl urea has limited effects (Hopkins and Chamberlain 1977).

Mosquitoes in general can be very well controlled with benzoylphenyl ureas (Miura et all. 1976; Schaefer et al. 1974; 1975). Special formulations are often necessary to increase residual effect in the water (Mulla and Darwazeh 1976; Saleh et al. 1981). These materials have larvicidal, congenital-ovicidal effects through female and topical ovicidal effects (Miura et al. 1976; Saleh et al. 1981).

Many of the Aedes species are carriers of arboviruses, the most notori-ous one being the yellow fever mosquito, *Aedes aegypti*. Diflubenzuron, Alsystin and Penfluron are all effective at low dosages (Bhakshi et al. 1982; Herald et al. 1980). The characteristic effect was "side hatching" from the egg instead of through the egg cap opening at one end. The embryos either died, fully formed inside the egg shell or emerged abnormally through the "side" and died at the time of moult especially at pupation. Also adults died attempting to come out of the pupal case (Arias 1974; Bhakshi et al. 1982; Busvine et al. 1976; Hsieh and Steelman 1974). *Aedes sollicitans*, the salt marsh mosquito, is one of the most sensitive species to diflubenzuron (Hsieh and Steelman 1974).

Benzoylphenyl ureas are very effective against all the *Anopheles* species tested (Arias 1974; Busvine et al. 1976; Schaefer et al. 1975). these are the well known vectors of the malarial parasite, *Plasmodium* species.

Culex species are also easily controlled with diflubenzuron (Hsieh and Steelman 1974; Jakob 1973; Sinegre et al. 1980). *Culex pipiens quinque-fasciatus* (= *C. pipiens fatigans*) is the vector for the filarial parasite, *Wuchereria bancrofti* in Asian countries and this mosquito is susceptible to diflubenzuron at 0.25 ppm (Arias 1974; Bhakshi et al. 1982). *C. tarsalis* is the vector for the arboviruses, equine encephalitis and St. Louis en-cephalitis and can be well controlled with this material (Miura et al. 1976; Schaefer et al. 1974; 1975). *C. tritaeniorhynchus*, another susceptible spe-cies, is the vector for the Japanese encephalitis virus (Takahashi and Ohtaki 1976). The effect of diflubenzuron on the eggs results in abnormal "side-hatching" amd the larvae usually die (Figure 21). Effect on the larva is usually manifested prior to pupation (Figure 22).

Culicita inornata, the carrier of viral encephalitides is susceptible to diflubenzuron (Hsieh and Steelman 1974). The large mosquitoes, *Psoro-phora* species can be controlled with this compound at doses as low as 6 g/ha (Darwazeh et al. 1977; Hsieh and Steelman 1974; Mulla and Dar-wazeh 1975; 1976).

8 CONTROL OF HOUSEHOLD AND URBAN INSECTS

While the human safety factor makes benzoylphenyl ureas attractive for household and urban insects, its slow mode of action does not provide the immediate protection often required.

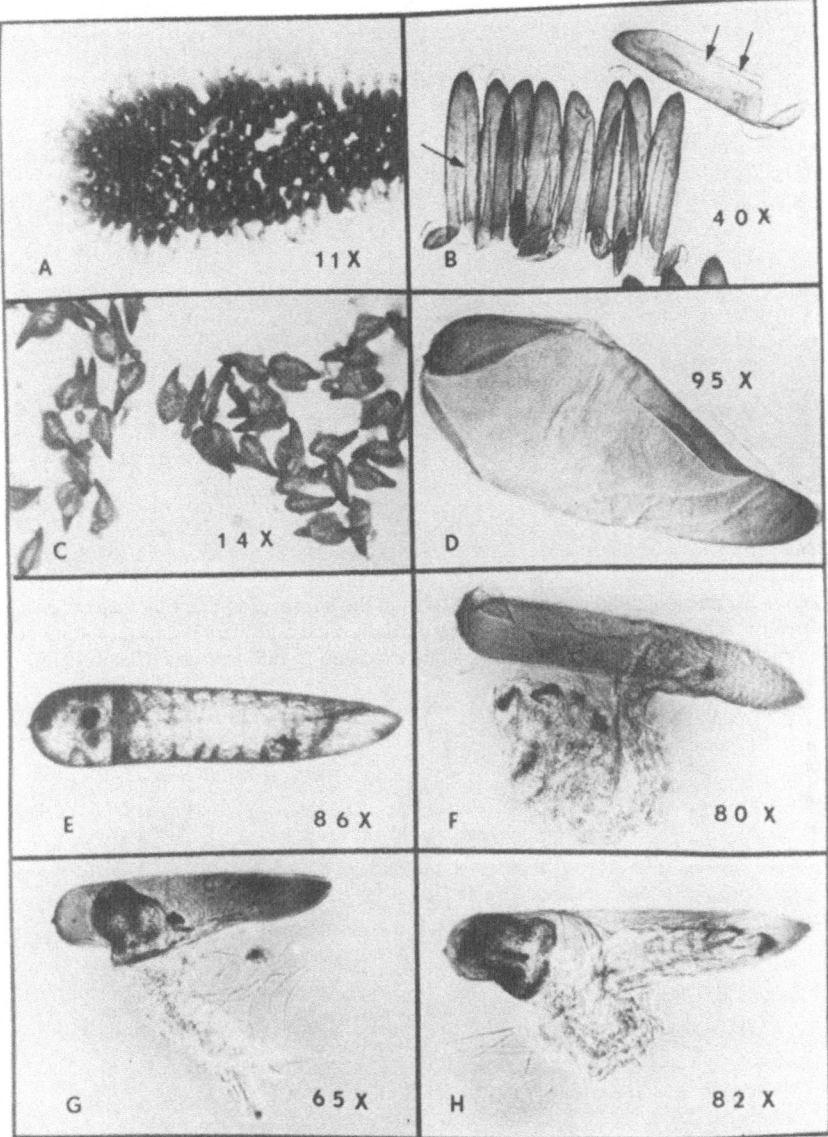

Figure 21. Effect of diflubenzuron on the hatching of the southern house mosquito (*Culex pipiens quinquefasciatus*) eggs. (A) Normal eggs with egg-caps. (B) Normally hatched egg shells showing opened caps and line of weakness on the mid-dorsal line. (C) Side-hatched egg shells caused by diflubenzuron; the larvae seldom survive. (D) Enlarged side-hatched egg shell with no egg-cap. (E) Unhatched, dead embryo within egg shell because of treatment. (F) Partial side-hatch with head capsule free. (G) Partial side-hatch with caudal end free. (H) Partial side-hatch with the thorax and abdomen free (Miura et al. 1976).

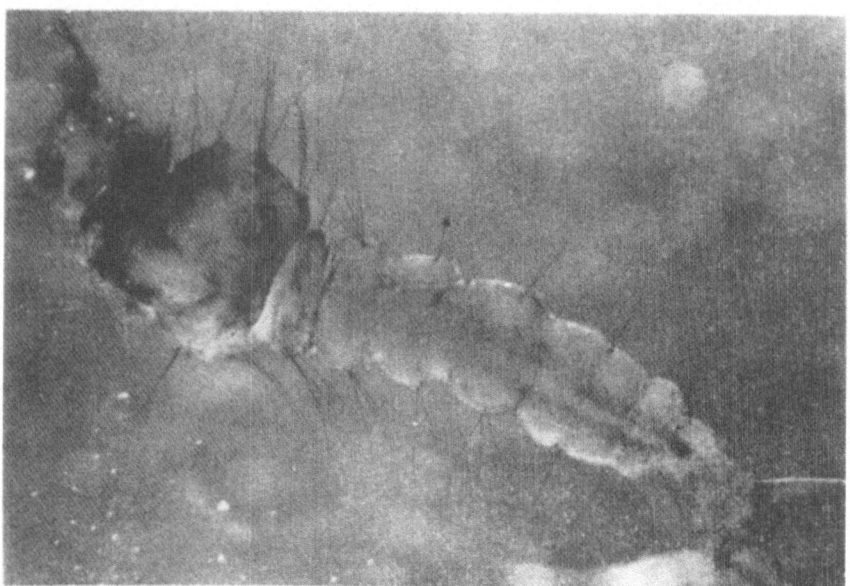

Figure 22. Effect of diflubenzuron on the larva of the house mosquito, *Culex pipiens quinque-fasciatus.* The abdominal cuticle has been partially shed and there is dorsal splitting of the thoracic cuticle, both characteristic of lethal exposure to diflubenzuron (Jakob 1973).

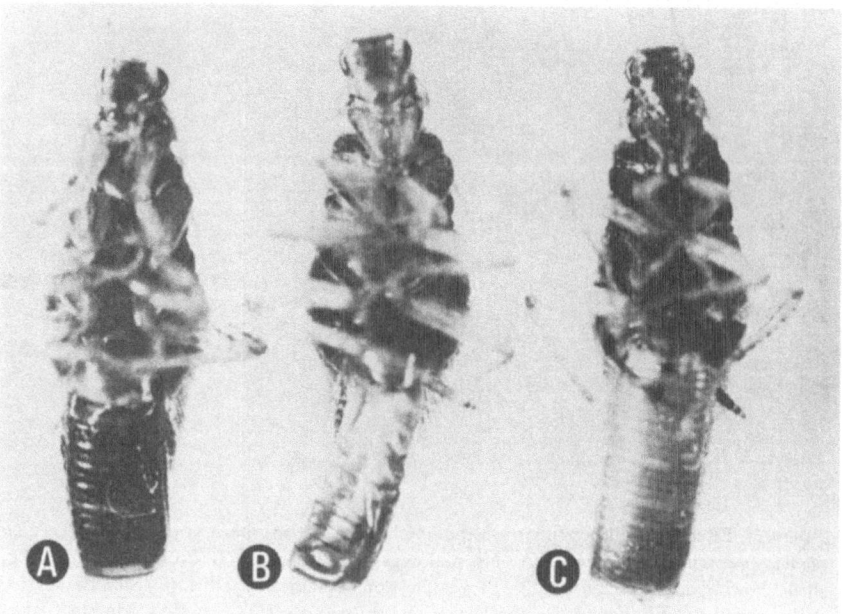

Figure 23. Effect of Alsystin on the German Cockroach, *Blattella germanica.* Adult cockroaches were fed Alsystin; the black oötheca on the left (A) was produced by the cockroach fed 0.5% AI of the material, the one in the middle (B) with the shrivelled oöthecum was fed 1.0% AI and the one on the right (C) was fed untreated diet (Weaver et al. 1984).

Diflubenzuron and Alsystin have been tested against several termites such as *Heterotermes indicola, Reticulitermes flavipes* and *Zootermopsis nevadensis* (Cymorek and Popischil 1982; Doppelreiter and Korioth 1981). At concentrations below 100 ppm fed to the termites via impregnated filter paper, there was suppression of caste differentiation. When it was increased to 1000 ppm there was complete control. These compounds appear promising as safe control agents for the termites.

Alsystin was tested on the German Cockroach, *Blattella germanica* (Figure 23). The compound was provided as a 1% bait which resulted in total mortality in 3 weeks. Adult females fed with this benzoylphenyl urea resulted in very poor hatch of the eggs they laid (Weaver et al. 1984).

9 EFFECT ON PARASITES, PREDATORS, POLLINATORS AND OTHER NON-TARGET SPECIES

Benzoylphenyl ureas manifest their effects on arthropods that are actively synthesizing chitin and this occurs for the most part during juvenile stages. Therefore it stands to reason that adult stages of the non-target species that occur alongside the target pest are seldom affected by diflubenzuron spray. This developmental compartmentalization that affords protection to the adults is by far the most important advantage over using broad-spectrum, contact-insecticides. However, it should be emphasized, that this protection is not universal and each pest in question has to be examined along with the co-existing non-target species be it beneficial or indifferent. Some of these effects on parasites, predators, pollinators and other non-target species are summarized in Table 13.

During diflubenzuron applications to control pests such as the house fly, *Musca domestica*, and the gypsy moth, *Lymantria dispar*, it was found that the parasites were unaffected (Ables et al. 1975; Shepherd and Kissam 1981; Skatulla 1975). Page and his co-workers (1982) found that the Eulophids, *Chrysocaris laricinellae* and *Dicladocerus nearcticus* were more tolerant than the host, the larch casebearer, *Coleophora laricella*, to diflubenzuron. Parasitic larvae inside treated host were affected (Broadbent and Pree 1984; Granett and Weseloh 1975). However adult parasites were unaffected (Granett and Weseloh 1975).

A classic example of differential tolerance for the benzoylphenyl urea insecticide is seen in the case of the predatory mites. *Typhlodromus occidentalis* and *Zetzellia mali* were unaffected during spray operations (Anderson and Elliott 1982). Adult predators such as the coccinellid, *Hippodamia convergens* (Jones et al. 1983), the lace wing, *Chrysopa oculata* (Broadbent and Pree 1984) and the assassin bug, *Acholla multispinosa* (Broadbent and Pree 1984) were unaffected when they fed on treated pest larvae. The only exception is the adult earwig, *Forficula auricularis* which

Table 13. Effect of benzoylphenyl ureas on parasites, predators, pollinators and other non-target species

Non-target species	Host-habitat	Analog	Activity	Reference
Muscidifurax raptor	Parasite on the house fly, *Musca domestica*	Diflubenzuron	No effect on parasite	Ables *et al.* 1975; Shepard and Kissam 1981.
Chrysocaris laricinellae (Eulophid)	Parasite on the larch casebearer, *Coleophora laricella*	Diflubenzuron	Parasite more tolerant than host.	Page *et al.* 1982.
Dicladocerus nearcticus (Eulophid)	Parasite on the larch casebearer *Coleophora laricella*	Diflubenzuron	Parasite more tolerant than host.	Page *et al.* 1982.
Apanteles melanoscelus (Braconid)	Parasite on the gypsy moth, *Lymantria dispar*	Diflubenzuron	Adult parasites unaffected but 2nd and 3rd moults inside treated host showed effects.	Granett and Weseloh 1975.
Aphanistes ruficornis (Ichneumonid)	Parasite on gypsy moth, *Lymantria dispar* and nun moth, *Lymantria monocha*	Diflubenzuron	No effect on parasite.	Skatulla 1975.
Barichneumon pachymerus (Ichneumonid)	Parasite on gypsy moth, *Lymantria dispar* and the nun moth, *Lymantria monocha*	Diflubenzuron Diflubenzuron	No effect on parasite. No effect on parasite.	Skatulla 1975. Skatulla 1975.
Macrocentrus ancylivorus (Braconid)	Parasite on the oriental fruit moth, *Grapholitha molesta*	Diflubenzuron; Alsystin	Reduced emergence of parasite from treated host.	Broadbent and Pree 1984b.
Typhlodromus occidentalis (Phytoseiid mite)	Predator of phytophagous mites.	Diflubenzuron	Non toxic to the predaceous mite.	Anderson and Elliott 1982.
Zetzellia mali (Stigmaeid mite)	Predator of phytophagous mites.	Diflubenzuron	Non toxic to the predaceous mite.	Anderson and Elliott 1982.

Organism	Role	Chemical	Effect	Reference
Hippodamia convergens (Coccinellid)	Predator of the cabbage looper, *Trichoplusia ni*.	Diflubenzuron; Penfluron; Alsystin	Adult predators suffered no toxic effects and the earlier instars approached equal tolerance with the pest.	Bones 1983. Jones *et al.* 1983.
Forficula auricularis (Earwig)	Predator of the woolly apple aphid, *Eriosoma lanigerum*	Diflubenzuron not detected.	Absence of Earwig in treated plot but exact sensitive stage was	Ravensberg 1981.
Chrysopa oculata (Lace wing)	Predator of the oriental fruit moth, *Grapholitha molesta*	Diflubenzuron; Alsystin	Topical treatment had adverse effect on 1st instar larvae. But when predator was fed treated larvae no effects were observed.	Broadbent and Pree 1984b.
Acholla multispinosa (Assassin bug)	Predator of the oriental fruit moth, *Grapholitha molesta*	Diflubenzuron; Alsystin	Topical treatment or feeding treated larvae had no effect on predator.	Broadbent and Pree 1984b.
Neoplectana carpocapsae (Entomogenous nematode)	Used as a biological control agent.	Diflubenzuron	No effect on the nematode	Hara and Kaya 1982.
Apis mellifera (Honey bee)	Pollinator	Diflubenzuron	Neither adult bees nor brood colonies were affected.	Emmett and Archer 1980; Schroeder *et al.* 1980.
Non-target insects such as Diving beetle larvae, Odonate naiads and May fly larvae.	Mosquito pool	Diflubenzuron	May fly larvae population was temporarily depressed.	Mulla *et al.* 1975.
Cladocerans such as *Daphnia* (crustacea)	Freshwater pool	Diflubenzuron	Affected temporarily; normal population in 2 to 11 weeks.	Apperson *et al.* 1978; Mulla *et al.* 1975.
Copepods (crustacea)	Freshwater pool and ocean pool.	Diflubenzuron	Not seriously affected and returned to normal levels soon. Dose dependent effects and recovery time.	Apperson *et al.* 1978; Mulla *et al.* 1975; Tester and Costlow 1981.
Ostracods (crustacea)	Freshwater pool	Diflubenzuron	No effect	Mulla *et al.* 1975.
Rotifers	Freshwater pool	Alsystin	No effect	Colwell and Schaefer 1981.

appears to be affected but the evidence is circumstantial (Ravensberg 1981).

Pollinators visiting treated flowers are not affected. Neither the adult nor the brood colonies of the honey bee, *Apis mellifera*, is affected by diflubenzuron at levels used for controlling pests (Emmett and Archer 1980; Schroeder et al. 1980).

The entomogenous nematode, *Neoplectana carpocapsae* which is used as a biological control agent is not affected by diflubenzuron. Other soil nematodes have been shown to be susceptible to this material. It has been hypothesized that diflubenzuron affects the egg shell of nematodes and since *Neoplectana* is viviparous it excapes the adverse effect (Hara and Kaya 1982).

One of the serious concerns of the environmental impact of benzoyl-phenyl ureas is their effects on crustacea. Here again there is varied tolerance. Cladocerans are very sensitive, copepods have some effects but ostracods are unaffected at levels of diflubenzuron used for controlling pests (Apperson et al. 1978; Mulla et al. 1975; Tester and Costlow 1981). Although the juveniles are severely affected, adults escape the brunt of the effect and repopulate the habitat within 2 to 11 weeks (Mulla et al. 1975). Other organisms such as rotifers are unaffected (Colwell and Schaefer 1981).

10 RESISTANCE, SYNERGISM AND ANTAGONISM

Resistance to benzoylphenyl ureas has been shown in many insects and this is not surprising. In 30 generations, resistance to diflubenzuron in *Spodoptera littoralis*, the Egyptian leafworm, increased by a factor of 290.7 (El-Guindy et al. 1983d). In many instances it is moderate (Amin and White 1984; Brown et al. 1978; Cerf and Georghiou 1974; Pimprikar and Georghiou 1979). Since the mode of action of benzoylphenyl ureas is basically different from that of broad-spectrum neurotoxins, it was felt that cross resistance was unlikely. However, studies with *Musca domestica*, the housefly, resistant to organophosphates, carbamates and organo-chlorines showed that there was cross resistance to diflubenzuron (Openoorth and Van Der Pas 1977).

A juvenile hormone analog, methoprene, was found to be synergistic with diflubenzuron in the boll weevil, *Anthonomus grandis* (Leopold and Marks 1980) but antagonistic in the western spruce budworm, *Choristo-neura occidentalis* (Robertson et al. 1984).

Using synergists as probes, attempts have been made to determine the detoxification mechanisms involved in the degradation of benzoylphenyl ureas. Mixed function oxidases are not involved in diflubenzuron degra-dation in the spruce budworm, *Choristoneura fumiferana* as indicated by the absence of synergism with piperonyl butoxide and SKF-525A (Retnakaran 1982a). The same was true in *Platynota sultana*, the omnivorous leafroller and *Spodoptera exigua*, the beet armyworm (Granett and Hejazi 1983). On the other hand in *Musca domestica*, the housefly, piperonyl butoxide and sesamex synergized diflubenzuron indicating that mixed function oxidases were involved in detoxification (Pimprikar and Georghiou 1979). S,S,S-tributyl phosphorotrithioate and diethyl maleate synergized diflubenzuron to a lesser extent in the housefly indicating that esterases and glutathione transferases played a minor role in diflubenzuron degradation (Pimprikar and Georghiou 1979). Contrary to these results, in *Platynota sultanta*, the omnivorous leafroller, S,S,S-tributyl phosphorotrithioate increased the activity of diflubenzuron 3,319-fold and diflubenzuron 45 fold indicating that esterases and glutathione transferases were involved in diflubenzuron degradation (Granett and Hejazi 1983).

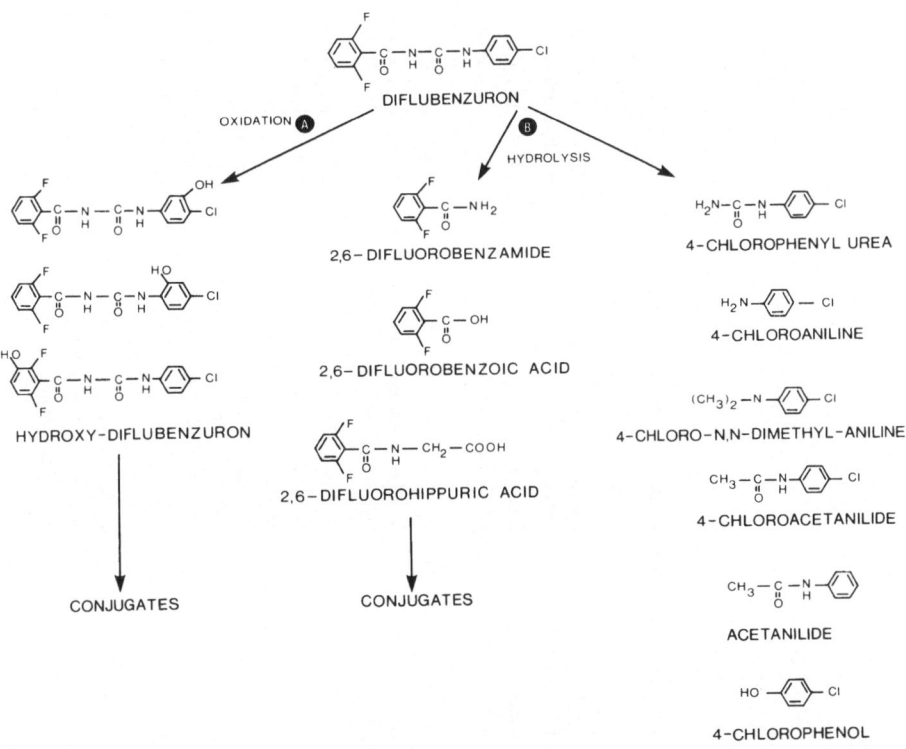

Figure 24. A generalized scheme of diflubenzuron metabolism in insects, mammals and micro-organisms (from Retnakaran et al. 1985).

Table 14. Resistance, synergism and antagonism to benzoylphenyl urea insecticides.

Insect species	Resistance/synergism	Mode of action or further details	Reference
Culex quinquefasciatus (House mosquito)	Resistance to diflubenzuron	Resistance developed slowly; 6.6 fold after 10 generations.	Amin and White 1984; Brown et al. 1978.
Spodoptera littoralis (Egyptian cotton leafworm)	Resistance to diflubenzuron	290.7 fold resistance in 30 generations. Piperonyl butoxide and S,S,S,Tributyl phosphoro-trithioate had marginal synergism.	El-Guindy et al. 1983d.
Tribolium confusum (Confused flour beetle)	Moderate resistance to diflubenzuron.	2 fold resistance in 8 generations.	Brown et al. 1978.
Musca domestica (House fly)	Resistance to diflubenzuron. Also cross resistance to flies resistant to organophosphate, carbamate and organochlorine insecticides	50 fold resistance in 10 generations. Since piperonyl butoxide and sesamex synergised the action, mixed function oxidases were involved. Esterases and glutathione transferase and glutathione transferase played a minor role.	Cerf and Georghiou 1974; Oppenoorth and Van Der Pas 1977; Pimpraker and Georghiou 1979.
Choristoneura fumiferana (spruce budworm)	No synergism with mixed function oxidase inhibitors and diflubenzuron	Piperonyl butoxide and SKF-525A did not increase diflubenzuron activity.	Retnakaran 1982a.

Species	Description	Reference
Choristoneura occidentalis (Western spruce budworm)	Juvenile hormone analog antagonistic to diflubenzuron and Alsystin.	Addition of methoprene lowers activity of diflubenzuron and Alsystin.	Robertson *et al.* 1984.
Anthonomus grandis (Cotton boll weevil)	Diflubenzuron synergistic to a juvenile hormone analog.	Diflubenzuron inhibits JH-esterase thereby increasing the activity of methoprene, a juvenile hormone analog.	Leopold and Marks 1980.
Brevicoryne brassicae (Cabbage aphid)	Synergism of an anionic surfactant to diflubenzuron	While non-ionic surfactants were antagonistic, Especrin, an anionic surfactant was synergistic.	Radwan *et al.* 1982.
Platynota sultana (Omnivorous leafroller)	S,S,S,tributyl phosphorotrithioate synergised diflubenzuron and Alsystin. No synergism with piperonyl butoxide.	This esterase inhibitor synergised the two benzoylphenyl ureas. 3,319 fold synergism with diflubenzuron and 45 fold with Alsystin.	Granett and Hejazi 1983.
Spodoptera exigua (Beet armyworm)	No synergism with S,S,S,tributyl phosphorotrithioate and piperonyl butoxide with diflubenzuron and Alsystin.	No esterase or mixed function oxidase inhibition.	Granett and Hejazi 1983.
Heliothis virescens (Tobacco budworm)	Diflubenzuron antagonistic to *Bacillus thuringiensis* and and has no effect on *Heliothis* NPV activity.	Mortality is halved to *Bacillus thuringiensis* alone.	Mohamed *et al.* 1983.

Table 15. Acute toxicity data for diflubenzuron (Wilcox and Coffey 1978).

Animal/ formulation	Oral LD_{50} (mg/kg)	Dermal LD_{50} (mg/kg)	Inhalational LD_{50}	Acute IP (mg/kg)
Mouse (tech)	4,640	—	—	2,150
Mouse (W-25)	10,000	—	—	—
Rat (tech)	4,640	—	no effect	—
Rat (W-25)	10,000	—	no effect	—
Rabbit (W-25)	—	4,640	no effect	—

Diflubenzuron was found to inhibit the activity of *Bacillus thuringiensis* in *Heliothis virescens*, the tobacco budworm but did not affect the activity of *Heliothis* Nuclear polyhedrosis virus (NPV) (Mohamed et al. 1983).

It becomes apparent that resistance to benzoylphenyl ureas does occur but cross-resistance is less likely. Species variation in regard to resistance, synergism and antagonism is widespread. Adjuvants such as anionic surfactants can somtimes increase the activity (Radwan et al. 1982). A summary of resistance, synergism and antagonism to benzoylphenyl ureas is shown in Table 14.

11 TOXICOLOGY AND FATE OF BENZOYLPHENYL UREAS

Diflubenzuron is one of a few insecticides that has been thoroughly investigated regarding its toxicology and environmental fate. The chronic toxicology of 2 benzoylphenyl ureas, diflubenzuron and Alsystin have been studied and they have been shown to be non-carcinogenic. Wilcox and Coffey (1978) have compiled a summary of information on the environmental impact of diflubenzuron (Table 15). In general, these compounds are extremely safe to vertebrates and most non-target species. Perhaps their only drawback is their action on juvenile crustaceans that actively synthesize chitin. The impact however, is temporary and the population recovers back to the original level within several weeks. Diflubenzuron is easily metabolized by many microorganisms (Seuferer et al. 1979). The metabolism of this material in plants and animals and the degradation and environmental fate have been well studied (Figure 24) and is the subject of another chapter in this book (Hammock and Quistad 1981; Ivie 1977; 1978; Mansager et al. 1979; Metcalf et al. 1975; Schaefer and Dupras 1976).

12 ACKNOWLEDGMENTS

The excellent art work and reproduction of illustrations by Mr. W. L. Tomkins and the untiring search for reference material by the librarian, Mrs. Sandy Burt and her assistant, Mrs. Nancy Jean Dukes are gratefully acknowledged.

13 REFERENCES

Abai, V. M. 1981. Studies on *Leucoma wiltshirei* Coll. (Lep., Lymantridae) a new pest in Iranian oak stands. 2. Biology, population dynamics and control measures. Z. Ang. Ent. 91: 86–99.

Abbassy, M. A., Ashry, M. and Salama, M. A. 1980. Selective effects of diflubenzuron and trifluron (SIR 8514) and diflubenzuron-resistance in *Spodoptera littoralis* Boisd. Med. Fac. Landbouww. Rijkuniv. Gent. 45/3: 721–726.

Abdel-Monem, A. H., Cameron, E. A. and Mumma, R. O. 1980. Toxicological studies on the molt inhibiting insecticide (EL-494) against the gypsy moth and effect on chitin biosynthesis. J. Econ. Entomol. 73: 22–25.

Abdel-Monem, A. H. and Mumma, R. O. 1981. Comparative toxicity of some molt inhibiting insecticides to the gypsy moth. J. Econ. Entomol. 74: 176–179.

Ables, J. R. West, R. P. and Shepard, M. 1975. Response of the house fly and its parasitoids to Dimilin (TH-6040). J. Econ. Entomol. 68: 622–624.

Abo-Elgar, M. R.., Radwan, H. S. A. and Ammar, I. M. A. 1978. Morphogenetic activity of an IGR disrupting chitin biosynthesis against pupae of the cotton leafworm, *Spodoptera littoralis* (Boisd.) Z. Ang. Ent. 86: 308–311.

Albert, R. 1984. Einsatz eines photoeklektors fuer junge baeume zur pruefung der wirkung der entwicklungshemmer Dimilin 25 WP und Bayer SIR 8514 auf *Operophtera brumata.* Anz. Schaedlingskde, Pflanzenschutz, Umweltschutz. 57: 51–54.

Ali, A. and Mulla, M. S. 1977. The IGR diflubenzuron and organophosphorus insecticides against nuisance midges in man-made residential-recreational lakes. J. Econ. Entomol. 70: 571–577.

Amin, A. M. and White, G. B. 1984. Resistance potential of *Culex quinquefasciatus* against the insect growth regulators methoprene and diflubenzuron. Ent. Exp. Appl. 36: 69–76.

Ammar, I. M. A. 1984. Ovicidal and latent effects following egg-treatment with sub-lethal concentrations of certain IGRs alone and in mixtures against the Colorado potato beetle, *Leptinotarsa decemlineata* Say, Z. Ang. Ent. 97: 464–470.

Anderson, D. W. and Elliott, R. H. 1982. Efficacy of diflubenzuron against the codling moth, *Laspeyresia pomonella* (Lepidoptera: Olethreutidae) and impact on orchard mites. Can. Ent. 114: 733–737.

Angus, T. A. 1973. Perspectives of biological insect control. Ann. N. Y. Acad. Sci. 217: 4–7.

Apperson, C. S., Schaefer, C. H., Colwell, A. E., Werner, G. H., Anderson, N. L., Dupras, E. F. Jr. and Longanecker, D. R. 1978. Effects of diflubenzuron on *Chaoborus astictopus* and nontarget organisms and persistence of diflubenzuron in lentic habitats. J. Econ. Entomol. 71: 521–527.

Apperson, C. S. and Yows, D. G. 1976. Laboratory evaluation of the activity of insect growth regulators against *Culicoides varipennis* (Diptera: Ceratopogonidae). Mosquito News. 36: 203–204.

Appleby, J. E. 1981. White oak: *Hamadryadella* control, 1977. Insecticide and Acaricide Tests. 3: 163.

Arambourg, Y., Pralavorio, R. and Dolbeau, C. 1977. Premières observations sur l'action du Diflubenzuron (PH 6040) sur la fécondité, la longévité et la viabilité des oeufs de *Ceratitis capitata* Wied. (Dipt. Trypetidae). Revue de Zoologie Agricole Pathol. Veg. 76: 118–126.

268

Argauer, R. J. and Cantelo, W. W. 1980. Stability of three ureide insect chitin synthesis inhibitors in mushroom compost determined by chemical and bioassay techniques. J. Econ. Entomol. 73: 671–674.

Arias, J. R. 1974. Biophysiological activity of insect growth regulators against mosquitoes. Diss. Abs. 35: 297B–298B.

Ascher, K. R. S. and Eliyahu, M. 1981. The residual contact toxicity of BAY SIR 8514 to *Spodoptera littoralis* larvae. Phytoparasitica. 9: 133–138.

Ascher, K. R. S., Eliyahu, M., Gurevitz, E. and Renneh, S. 1983. Rearing the honeydew moth, *Cryptoblabes gnidiella*, and the effect of diflubenzuron on its eggs. Phytoparasitica. 11: 195–198.

Ascher, K. R. S. Eliyahu, M., Nemny, N. E. and Ishaaya, I. 1982. The toxicity of some novel pesticides — synthetic pyrethroids and benzoyl phenylurea chitin synthesis inhibitors — for eggs of *Spodoptera littoralis* (Boisd.). Z. Ang. Ent. 94: 504–509.

Ascher, K. R. S., Gurevitz, E. and Eliyahu, M. 1978. The effect of diflubenzuron on eggs of the vine moth, *Lobesia (Polychrosis) botrana* Den. and Schiff. (Lepidoptera: Tortricidae). Phytoparasitica. 6: 25–27.

Ascher, K. R. S. and Nemny, N E. 1974. The ovicidal activity of PH 60–40 [1-(4-chlorophenyl)-3-(2,6-dirluorobenzoyl)-urea] in *Spodoptera littoralis* Boisd. Phytoparasitica. 2: 131–133.

Ascher, K. R. S., Nemny, N. E. 1976a. Toxicity of the chitin synthesis inhibitors, diflubenzuron and its dichloro-analogue, to *Spodoptera littoralis* larvae. Pestic. Sci. 7: 1–9.

Ascher, K. R. S., Nemny, N. E. 1976b. Contact activity of diflubenzuron against *Spodoptera littoralis* larvae. Pestic. Sci. 7: 447–452.

Ascher, K. R. S., Nemny, N. E., Eliyahu, M. and Ishaaya, I. 1979. The effect of BAY SIR 8514 on *Spodoptera littoralis* (Boisduval eggs and larvae. Phytoparasitica. 7: 177–184.

Ascher, K. R. S., Nemny, N. E., Kehat, M. and Gordon, D. 1978. The effect of diflubenzuron on eggs and larvae of *Earias insulana* (Boisd.). Phytoparasitica. 6: 29–33.

Ascher, K. R. S., Wyslki, M., Nemny, N. E. and Gur-Telzak, L. 1978. The effect of diflubenzuron upon larvae of the geometrid, *Boarmia (Ascotis) selenaria* on Avocado leaves. Pestic. Sci. 9: 219–224.

Assal, O. M., Radwan, H. S. A. and Samy, M. E. 1983. Egg hatch inhibition in the cotton leafworm with certain IGRs and synthetic pyrethroids. Z. Ang. Ent. 95: 259–263.

Audemard, H. 1977. Essai complémentaire sur le carpocapse (*Laspeyresia pomonella* L.) d'un nouvel insecticide le diflubenzuron. Defense Des Vegetaux. 31: 303–309.

Audemard, H., Causse, R. et Féron, M. 1976. Lutte intégrée en verger de pommiers orientations nouvelles dans la lutte contre la carpocapse (*Laspeyresia pomonella* L.) en 1974–1975. Academie d'agriculture de France, Comptes Rendus. 62: 228–235.

Audemard, H., Gunnelon, G., Baudouin, J. M., Dolbeau, C. et Schutz, D. 1975. Essais préliminaires sur le carpocapse (*Laspeyresia pomonella* L.) d'un insecticide présentant un nouveau mode d'action: Le diflubenzuron. Phytiatrie-Phytopharmacie. 24: 179–192.

Axtell, R. C., Dukes, J. C. and Edwards, T. D. 1979. Field tests of diflubenzuron, methoprene, Flit MLO and chlorpyrifos for the control of *Aedes taeniorhynchus* larvae in diked dredged spoil area. Mosquito News. 39: 520–527.

Badiali, G. 1979. Processionary caterpillar. Informatore Fitopatologico. 26: 21–27.

Barker, R. W. and Jones, R. L. 1976. Inhibition of larval horn fly development in the manure of bovines fed Dimilin mineral blocks. J. Econ. Entomol. 69: 441–443.

Barker, R. W. and Newton, G. L. 1976. Dimilin: Evaluation as a livestock dietary feed additive for control of *Musca domestica* larvae in cattle waste. J. Georgia Entomol. Soc. 11: 71–75.

Baronio, P. and Pasqualini, E. 1984. Diflubenzuron. Rivista Bimestrale di Fitoiatria, Fitofarmacia e Diserbo, Anno 7° (Supplemento al. n.2), Marzo–Aprile 1984, pp. 141.

Beavers, J. B., Schroeder, W. J. and Selhime, A. G. 1976. Effect of diflubenzuron on larvae of *Diaprepes abbreviatus* Florida Ent. 59: 434.

Beevi, S. P. and Dale, D. 1980. Effect of diflubenzuron on the larvae of rice swarming caterpillar, *Spodoptera mauritia* Boisd. (Lepidoptera: Noctuidae). J. Ent. Res. 4: 157–160.

Berry, E. C., Faragalla, A. A. and Guthrie, W. D. 1980. Field evaluation of diflubenzuron for control of first and second generation European corn borer. J. Econ. Entomol. 73: 634–636.

Bhakshi, N., Bhasin, V. K. and Pillai, M. K. K. 1982. Laboratory evaluation of insect growth regulating compounds against mosquitoes. Entomon. 7: 469–473.

Bones, V. M. 1983. Peropal, Alsystin and Cropotex: Evaluations of their effect on beneficial arthropods. Pflanzenschutz-Nachrichten. 36: 38–53.

Broadbent, A. B. and Pree, D. J. 1984a. Effects of diflubenzuron and BAY SIR 8514 on the oriental fruit moth (Lepidoptera: Olethreutidae) and the oblique banded leafroller (Lepidoptera: Tortricidae). J. Econ. Entomol. 77: 194–197.

Broadbent, A. B. and Pree, D. J. 1984b. Effects of diflubenzuron and BAY SIR 8514 on beneficial insects associated with peach. Environ. Entomol. 13: 133–136.

Brown, T. M., De Vries, T. H. and Brown, A. W. A. 1978. Induction of resistance to insect growth regulators. J. Econ. Entomol. 71: 223–229.

Brushwein, J. 1980. The effects of chitin-inhibitng insect growth regulators on the spruce budworm, Choristoneura fumiferana (Clemens), and two associated hymenopteran parasitoids, Apanteles fumiferana (Vireck) and Glypta fumiferanae (Vireck). M. S. thesis, University of Maine, Orono.

Büchi, Y. R. 1978. Ovizide and larvizide wirkung von Dimilin auf den maizünsler, Ostrinia nubilalis (Hbn.). Z. Ang. Ent. 86: 67–71.

Bullock, R. C. and Kretschmer, A. E. Jr.1982. Identification and control of the foliar pests of American jointvetch. Florida Ent. 65: 335–339.

Bullock, R. C. and McCoy, C. W. 1978. Activity of insect growth regulators versus citrus rust mite in Florida. Proc. Fla. State Hort. Soc. 91: 72–74.

Burts, E. C. 1983. Effectiveness of a soft-pesticide program on pear pests. J. Econ. Entomol. 76: 936–941.

Busvine, J. R. 1978. The prospects of pest control by disruption of arthropod development. Pestic. Sci. 9: 266–271.

Busvine, J. R., Rongsriyam, Y. and Bruno, D. 1976. Effects of some insect development inhibitors on mosquito larvae. Pestic. Sci. 7: 153–160.

Calkins, C. O., Hill, A. J., Huettel, M. D. and Mitchell, E. R. 1977. Effect of diflubenzuron on Plum curculio populations in laboratory and field tests. J. Econ. Entomol. 70: 463–466.

Cameron, E. A. and McManus, M. L. 1979. The gypsy moth (Lymantria dispar) in the United States: Current status of pest management. Proc. Int. Symp. IOBC/WPRS on Ineg. Control in Agriculture/Forestry. pp. 537–538.

Campbell, J. B. and Wright, J. E. 1976. Field evaluations of Insect Growth Regulators, insecticides, and a bacterial agent for stable fly control in feedlot breeding areas. J. Econ. Entomol. 69: 566–568.

Campbell, R. L. 1975. Insecticidal control of European crane fly in Washington. J. Econ. Entomol. 68: 386–388.

Canivet, J. P., Nef, L. and Lebrun, P. 1978. Utilisation combinee de Bacillus thuringiensis et d'insecticides chimiques a doses reduites contre Euproctis chrysorrhoea. Ang. Ent. 86: 85–97.

Cantelo, W. W. 1979. Lycoriella mali: Control in mushroom compost by incorporation of insecticides into compost. J. Econ. Entomol. 72: 703–705.

Cantelo, W. W. 1981. Advances in chemical control of the Sciarid fly, Lycoriella mali. Mushroom Science 11: 255–264.

Cantelo, W. W. 1983. Control of a mushroom-infesting fly (Diptera: Sciaridae) with insecticides applied to the casing layer. J. Econ. Entomol. 76: 1433–1436.

Carson, R. 1962. Silent Spring. Houghton Mifflin, Boston.

Carter, S. W. 1975. Laboratory evaluation of three novel insecticides inhibiting cuticle formation against some susceptible and resistant stored products beetles. J. Stored Prod. Res. 11: 187–193.

Cerf, D. C. and Georghiou, G. P. 1974. Cross-resistance to an inhibitor of chitin synthesis, TH 60-40, in insecticide-resistant strains of the housefly, J. Agr. Food Chem. 22: 1145–1146.

Chang, S. C. Laboratory evaluation of diflubenzuron, penfluron and BAY SIR 8514 as female sterilants against the house fly. J. Econ. Entomol. 72: 479–481.

Chu, C.-M. and Brindley, W. A. 1981. Effects of diflubenzuron on alfalfa weevil larvae and upon toxicity of Methidathion and Carbofuran. Iowa State J. Res. 55: 387–392.

Ciesla, W. M. Douglas fir tussock moth: Direct control with chemical and microbial insecticides. Bull. Entom. Soc. Amer. 23: 174–176.

Clarke, L. 1982. Factors affecting uptake and loss of diflubenzuron in the tsetse fly Glossina morsitans Westwood (Diptera: Glossinidae). Bull. Ent. Res. 72: 511–522.

270

Clark, L., Temple, G. H. R. and Vinent, J. F. V. 1977. The effects of a chitin inhibitor — Dimilin — on the production of peritrophic membrane in the locust, *Locusta migratoria*. J. Insect Physiol. 23: 241–246.

Cohen, E. and Casida, J. E. 1980. Inhibition of gut chitin synthetase. Pestic. Biochem. Physiol. 13: 129–136.

Cohen, E. and Casida, J. E. 1983. Insect chitin synthetase as a biochemical probe for insecticidal compounds. Int. Congr. Pesiticide Chemistry, 5th Proceedings, Ed. J. Miyamoto. Vol. 3: 25–32.

Collmann, G.L. and All, J. N. 1982. Biological impact of contact insecticides and insect growth regulators on isolated stages of the greenhouse whitefly (Homoptera: Aleyrodidae). J. Econ. Entomol. 75: 863–867.

Colwell, A. E. and Schaefer, C. H. 1981. Effects of the insect growth regulator BAY SIR 8514 on pest diptera and nontarget aquatic organisms. Can. Ent. 113: 185–191.

Cooper, R. M., Lindquist, R. K. and Simeonet, D. E. 1983. Timing applications for SIR 8514 for control of the Colorado potato beetle (Coleoptera: Chrysomelidae) on potatoes. J. Econ. Entomol. 76: 536–566.

Cranham, J. E. 1978. Control of codling moth with diflubenzuron. Mitteilungen aus der Biologischen Bundensanstalt für Land-und Forstwirtschaft Berlin-Dahlem. 180: 108–110.

Crystal, M. M. 1978. Diflubenzuron-induced decrease of egg hatch of screwworms (Diptera: Calliphoridae). J. Med. Entomol. 15: 52–56.

Cymorek, V. S. and Popischil, R. 1982. Über hormonmimetika und chitin synthesehemmer als biiologisch wirkende Schutzmittel für den Holzschutz. Holz-Zentralblatt, Stuttgard Nr. 34: 524–526.

Dame, D. A., Lowe, R. E., Wichterman, G. J., Cameron, A. L., Baldwin, K. F. and Miller, T. W. 1976. Laboratory and field assessment of insect growth regulators for mosquito control. Mosquito News. 36: 462–472.

Darwazeh, H. A., Mulla, M. S. and Dhillon, M. S. 1977. Granular Dimilin as prehatch treatment for the control of flood water mosquitoes. Proc. Ann. Conf. Calif. Mosquito Control Assoc. 45: 143–145.

Degaspari, N. and Gomez, S. A. 1982. Chemical control of the Soybean caterpillar in field conditions of Mato Grosso do Sul. Pequisas Agropecuaria Brasileira. 17: 513–517.

De Jong, D. J. and Mimks, A. K. 1981. Studies on *Adoxophyes orana*, the major leaf-roller pest in apple orchards in the Netherlands. Mitteilunger der Schweizerischen Entomologischen Gesellschaft. 54: 205–214.

DeLoach, J. R., Meola, S. M., Mayer, R. T. and Thompson, J. M. 1981. Inhibition of DNA synthesis by diflubenzuron in pupae of the stable fly *Stomoxys calcitrans* (L.). Pestic. Biochem. Physiol. 15: 172–180.

Demolin, G. and Millet, A. 1981a. Essais insecticides contre la processionnaire du pin (*Thaumetopoea pityocampa* Schiff.) Annales des Sciences Forestieres. 35: 107–110.

Demolin, G. and Millet, A. 1981b. Essais insecticides contre la Processionnaire du pin (*Thaumetopoea pityocampa* Schiff.). Action comparative de differentes spécialités commerciales Bactospéine, Dipel, Thuricide et Dimilin. Ann. Sci. Forest. 38: 389–404.

Demolin, G. and Millet, A. 1983. Le Dimilin utilise a trois doses sur la processionnaire du pin (*Thaumetopoea pityocampa*). Revue Francaise Forestieres. 35: 107–110.

Denneulin, J.-Cl. et Lamy, M. 1977. Oenocytes et mue la processionnaire du pin, *Thaumotopoea pityocampa* Schiff. (Lépidoptères): Effets du 1(4 chlorophényle)3 (2-6 difluro-benzoyle) urée. Annales d'Endocrinologie (Paris). 38: 405–406.

Denneulin, J.-Cl. and Lamy, M. 1982. Sterols of the pine processionary caterpillar; effects of diflubenzuron (Dimilin). Experientia. 38: 800–801.

Doppelreiter, V. H. and Korioth, M. 1981. Inhibition of development by the subterranean termites, *Heterotermes indicola* and *Reticulitermes flavipes* caused by diflubenzuron. Z. Ang. Ent. 91: 131–137.

Duel, D. H., De Jong, B. J. and Kortenbach, J. A. M. 1978. Inhibition of chitin synthesis by two 1-(2,6-disubstituted benzoyl)-3-phenyl- urea insecticides. Pestic. Biochem. Physiol. 8: 98–105.

Earle, N. W., Nilakhe, S. S. and Simmons, L. A. 1979. Mating ability of irradiated male boll weevils treated with diflubenzuron or Penfluron. J. Econ. Entomol. 72: 332–336.

271

El-Guindy, M. A., El-Rafai, A. R. M. and Abdel-Sattar, M. M. 1983a. The joint action of mixtures of insecticides, or of insect growth regulators and insecticides, on susceptible and diflubenzuron-resistant strains of *Spodoptera littoralis* Boisd. Pestic. Sci. 14: 246–252.

El-Guindy, M. A., El-Rafai, A. R. M. and Abder-Satter, M. M. 1983d. The pattern of cross-resistance to insecticides and juvenile hormone analogues in a diflubenzuron-resistant strain of the cotton leaf-worm *Spodoptera littoralis* Boisd. Pestic. Sci. 14: 235–245.

El-Guindy, M. A., Abdel-Sattar, M. M., Dogheim, S. M. A., Madi, S. M. and El-Assar, M. R. S. 1983c. Laboratory evaluation of the insect growth regulator dimilin (TH-6040) against susceptible and resistant strains of *Spodoptera littoralis* (Boisd.). International Pest Control. 25: 48–51.

El-Guindy, M. A., Abdel-Sattar, M. M. and El-Rafai, A. R. M. 1983b. The ovicidal action of insecticides and Insect Growth Regulator/Insecticide mixtures on the eggs of various ages of susceptible and diflubenzuron-resistant strains of *Spodoptera littoralis* Boisd. Pestic. Sci. 14: 253–260.

Elliott, R. H. and Anderson, D. W. 1982. Factors influencing the activity of diflubenzuron against the codling moth, *Laspeyresia pomonella* (Lepidoptera: Olethreutidae). Can. Ent. 114: 259–268.

Elliott, R. H. and Iyer, R. 1982. Toxicity of diflubenzuron to nymphs of the migratory grasshopper, *Melanoplus sanguinipes* (Orthoptera: Acrididae). Can. Ent. 144: 479–484.

Ellis, H. C. 1979. Pine tip moth control 1976. Insecticide and Acaricide Tests. 3: 147.

Emmett, B. J. and Archer, B. M. 1980. The toxicity of diflubenzuron to honey bee (*Apis mellifera* L.) colonies in apple orchards. Pl. Path. 29: 177–183.

EPA. 1983. Tolerances and exemptions from tolerance for pesticide chemicals in or raw agricultural commodities; diflubenzuron. Federal Register. 48: 37210.

Fargalla, A. A., Berry, E. C. and Guthrie, W. D. 1980. Ovicidal activity of diflubenzuron on European corn borer egg masses. J. Econ. Entomol. 78: 573–574.

Farghal, A. I., Morsy, M. A. A. and Ahmed, S. A. 1983. Influence of SIR 8514 and Altosid SR 10 on the immature stages of the Mediterranean fruit fly *Ceratitis capitata* (Weid.). International Pest Control (Nov/Dec 1983). pp. 178–180.

Ferrari, R. and Tiberi, R. 1979. Efficacy of a control test with diflubenzuron against *Thaumetopoea pityocampa* and first observations on the effect to hymenopterous chalcidoid egg parasites. Redia. 62: 315–323.

Flint, H. M. and Smith, R. L. 1977. Laboratory evaluation of TH 60–40 against the pine bollworm. J. Econ. Entomol. 70: 51–53.

Flint, H. M., Smith, R. L., Forey, D. and Horn, B. 1977. Diflubenzuron: Evaluation for control of the Pink bollworm, Cabbage looper, and Cotton leafperforator in a field cage test. J. Econ. Entomol. 70: 237–239.

Fodor, S. and Halmagyi, L. 1978. Die erprobung von Dimilin und zwei juvenoiden in der kiferntreibwicklerbekaemfung (*Rhyacionic buoliana*). Novenyvedlem. 14: 308–310.

Fogal, W. H. 1977. Effect of a phenyl-benzoylurea [1-(4-chlorophenyl)-3-(2,6-difluoro-benzoyl)-urea] on *Diprion similis* (Hymenoptera: Diprionidae) Can. Ent. 109: 981–986.

Forgash, A. J. Respicione, N. C. and Khoo, B. K. 1978. Contact action of diflubenzuron on eggs and larvae of gypsy moth, *Lymantria dispar*. J. N. Y. Entomol. Soc. 86: 287.

Fytizas, P. E. 1976. L'action du TH 6040 sur la métamorphose de *Dacus oleae* Gmel. (Diptera, Trypetidae). Z. Ang. Ent. 81: 440–444.

Gaaboub, J. A. and Busvine, J. R. 1976. Effects of larval treatment with the insect development inhibitor PH 60:40 on the vectorial capacity of *Aedes aegypti* (L.) for *Bruhia pahangi* (Buckley and Edeson). Ann. Trop. Med. Parasit. 70: 355–360.

Gadais, M., Mehieu, N. and Gilbert, R. 1978. La lutte contre la processionnaire du pin en loire-atlantique et en vendée. Phytoma-defense des cultures (Dec. 1978). pp. 19–21.

Gandhale, D. N., Patil, A. S. and Mali, A. R. 1983. Effect of diflubenzuron and SIR 8514 on *Heliothis armigera* Hub. Madras Agric. J. 70: 203–204.

Ganyard, M. C., Bradley, J. R. Jr., Boyd, F. J. and Brazzel, J. R. 1977. Field evaluation of diflubenzuron (Dimilin) for control of boll weevil reproduction. J. Econ. Entomol. 70: 347–350.

Georgevitis, R. P. 1979. Comparison of results in the control of *Thaumetopoea pityocampa* with Dimilin, Decis Bactospeine and Thuricide HP. Anakoinoseis Idrumaton Dasikon Ereunon. 7: 7–34.

272

Georgevitis, R. P. 1982. New perspectives for the control of the pine processionary moth (*Thaynetioiea pityocampa*) with Dimilin. Anakoinoseis Idrumaton Dasikon Ereunon. 8: 7–25.

Gillette, N. L., Robertson, J. L. and Lyon, R. L. 1978. Bioassays of TH 6038 and difluron applied to Western spruce budworm and Douglas-fir tussock moth. J. Econ. Entomol. 71: 319–322.

Glen, D. M., Milsom, M. F., Wiltshire, C. W. and Ruby, M. P. 1982. SIR 8514 compared with diflubenzuron for control of orchard moths and apple sucker. Ann. Appl. Biol. 100: 6–7.

Granett, J., Bisabri-Ershadi, B. and Hejazi, M. J. 1983. Some parameters of benzoylphenyl urea toxicity to Beet armyworms (Lepidoptera: Noctuidae). J. Econ. Entomol. 76: 399–402.

Granett, J., Bisabri-Ershadi, B. and Hejazi, M. J. 1983. Some parameters of benzoylphenyl urea toxicity to Beet armyworms. J. Econ. Entomol. 76: 403–406.

Granett, J. and Dunbar, D. M. 1975. TH 60–40: Laboratory and field trials for control of gypsy moths. J. Econ. Entomol. 68: 99–102.

Granett, J., Dunbar, D. M. and Weseloh, R. M. 1976. Gypsy moth control with Dimilin sprays timed to minimze effects on the parasite, *Apanteles melanoscelus*. J. Econ. Entomol. 69: 403–404.

Granett, J. and Hejazi, M. 1983. Synergism of two benzoylphenyl urea insect growth regulators. J. Econ. Entomol. 76: 399–402.

Granett, J. and Retnakaran, A. 1977. Stadial susceptibility of eastern spruce budworm, *Choristoneura fumiferana* (Lepidoptera: Tortricidae), to the insect growth regulator Dimilin. Can. Ent. 109: 893–894.

Granett, J., Robertson, J. L. and Retnakaran, A. 1980. Metabolic basis of differential susceptibility of two forest lepidopterans to diflubenzuron. Ent. Exp. Appl. 28: 295–300.

Granett, J. and Weseloh, R. M. 1975. Dimilin toxicity to the gypsy moth larval parasitoid, *Apanteles melanoscelus*. J. Econ. Entomol. 68: 577–580.

Greene, G. L. Pest management of the velvet-bean caterpillar in a soybean ecosystem. Proc. World Soybean Res. Conf. 602–610.

Grosscurt, A. C. 1976. Ovicidal effects of diflubenzuron on the housefly (*Musca domestica*). Med. Fac. Landbouww. Rijksuniv. Gent. 41/2: 949–963.

Grosscurt, A. C. 1977. Mode of action of diflubenzuron as an ovicide and some factors influencing its potency. Proc. 1977 British Crop Protection Conf. pp. 141–147.

Grosscurt, A. C. 1978a. Effects of diflubenzuron on mechanical penetrability, chitin formation, and strucuture of the elytra of *Lepitinotarsa decemlineata*. J. Insect Physio. 24: 827–831.

Grosscurt, A. C. 1978b. Diflubenzuron: Some aspects of its ovicidal and larvicidal mode of action and an evaluation of its practical possibilities. Pestic. Sci. 9: 373–386.

Grosscurt, A. C. and Andersen, S. O. 1980. Effects of diflubenzuron on some chemical and mechanical properties of the elytra of *Leptinotarsa decemlineata*. Entomology Proceedings C. 83: 143–150.

Hajjar, N. P. and Casida, J. E. 1979. Structure-activity relationships of benzoylphenyl ureas as toxicants and chitin synthesis inhibitors to *Oncopeltus fasciatus*. Pestic. Biochem. Physiol. 11: 33–45.

Halperin, J. 1980. Control of the pine processionary caterpillar (*Thaumetopoea wilkinsoni*) with diflubenzuron. Phytoparasitica. 8: 83–91.

Hammock, B. D. and Quistad, G. B. 1981. Metabolism and mode of action of juvenile hormone, juvenoids and other insect growth regulators. In: Progress in pesticide biochemistry. Ed. D. H. Hutson and T. R. Roberts. Vol. I. pp. 1–82. John Wiley and Sons, New York.

Hara, A. H. and Kaya, H. K. 1982. Effects of selected insecticides and nematicides on the *in vitro* development of the entomogenous nematode, *Neoplectana carpocapsae*. J. Nematol. 14: 486–491.

Hard, J. S., Ward, J. D. and Ilnytzky. 1978. Control of Douglas fir Tussock Moth by aerially applied Dimilin (TH 6040). Res. Paper PSW-130, 6 p. (Forest Serv., USDA).

Harper, J. D. and Abrahamson, L. P. 1979. Forest tent caterpillar control with aerially applied formulations of *Bacillus thuringiensis* and Dimilin. J. Econ. Entomol. 72: 74–77.

Harwood, R. F. and James, M. T. 1979. Entomology in human and animal health. MacMillan Publishing Co., Inc., New York. 548 pp.

Hayakawa, H. 1976. Evaluations of PH 60–40, an insect growth regulator, for control of the stable fly *Stomoxys calcitrans* (L.): (Diptera, Muscidae). Jap. J. Sanit. Zool. 27: 261–264.

Haynes, J. W. and Wright, J. E. 1982. Laboratory competitiveness of sterilized boll weevils. J. Georgia Entomol. Soc. 17: 382–388.

Haynes, J. W. and Wright, J. E. 1983. The effect of diflubenzuron on aged female boll weevils. Mississippi Agricultural and Forestry Experiment Station, Res. Rep. 8(14): 3 pp.

Haynes, J. W. and Wright, J. E. 1984. Competitiveness of diflubenzuron + irradiation treated vs. untreated boll weevils in alternate mating sequences. The Southern Ent. 9: 263–266.

Haynes, J. W., Wright, J. E. and Mattix, E. 1981. A diflubenzuron dust method of sterilizing female boll weevils (Coleoptera: curculionidae). Mississippi Agricultural and Forestry Experiment Station, Res. Rep. 6(4): 4 pp.

Heinrichs, E. A. and Da Silva, R. F. P. 1978. Control of *Anticarsia gemmatalis* with PH 6040 in low dosages. Agronomia Sulriograndense. 4: 261–267.

Heinrichs, E. A., Gastal, H. A. and Galileo, M. H. M. 1979. Incidence of natural control agents of the velvetbean caterpillar and response of its predators to insecticide treatments in Brazilian soybean fields. Pequisa Agropecuaria Brasileira. 14: 79–87.

Herald, F., Clark, J. L. III and Knapp, F. W. 1980. Susceptibility of *Aedes aegypti* to synthetic pyrethroids compared with a new Insect Growth Regulator. Mosquito News. 40: 380–382.

Henzell, R. F., Lauren, D. R. and East, R. 1979. Effect on the egg hatch of white-fringed weevil of feeding lucerne treated with the insect growth regulator diflubenzuron. N. Z. J. Agric. Res. 22: 197–200.

Henzell, R. F., Lauren, D. R. and Hall, W. T. 1977. Laboratory tests with the insect growth regulator, diflubenzuron, against white fringed weevil adults and army caterpillar larvae. Proc. 22nd Weed Pest Control Conf. (1976). pp. 143–147.

Hopkins, D E. and Chamberlain, W. F. 1976. Diflubenzuron: Relationship between age of exposed immature horn flies and inhibition of maturation. The Southwestern Ent. 1: 114–117.

Hopkins, D. E. and Chamberlain, W. F. 1977. Angora goat biting louse: Relationship between ingestion of diflubenzuron and ecdysis. J. Econ. Entomol. 71: 25–26.

Horstmann, V. K. 1982. Effects of an application of Dimilin on population of oak tortricids (Lepidoptera: Tortricidae) in Lower Franconia. Z. Ang. Ent. 94: 490–497.

Hoying, S. A. and Reidl, H. 1980. Susceptibility of the codling moth to diflubenzuron. J. Econ. Entomol. 73: 556–560.

Hsieh, M. Y. G. and Steelman, C. D. 1974. Susceptibility of selected mosquito species to five chemicals which inhibit insect development. Mosquito News. 34: 278–282.

Ishaaya, I. and Ascher, K. R. S. 1977. Effect of diflubenzuron on growth and carbohydrate hydrolases of *Tribolium castaneum*. Phytoparasitica. 5: 149–158.

Ishaaya, I., Ascher, K. R. S. and Yablonski, S. 1981. The effect of BAY SIR 8514, diflubenzuron and Hercules 24108 on growth and development of *Tribolium confusum*,. Phytoparasitica. 9: 207–209.

Ishaaya, I. and Casida, J. E. 1974. Dietary TH 6040 alters composition and enzyme activity of housefly larval cuticle. Pestic. Biochem. Physiol. 4: 484–490.

Ivie, G. W. 1977. Metabolism of insect growth regulators in animals. In: Fate of pesticides in large animals. Ed. G. W. Ivie and H. W. Dorough. pp. 111–126. Academic Press, New York.

Ivie, G. W. 1978. Fate of diflubenzuron in cattle and sheep. J. Agric. Food Chem. 26: 81–89.

Jakob, W. L. 1973. Developmental inhibition of mosquitoes and the house fly by urea analogues. J. Med. Ent. 10: 452:455.

Janes, M. J. 1975. Corn earworm and Fall armyworm: Comparative larval populations and insecticidal control on sweet corn in Florida. J. Econ. Entomol. 68: 657–658.

Jenkins, V. K., Mayer, R. T. and Perry, R. R. 1984. Effects of diflubenzuron on growth of malignant melanoma and skin carcinoma tumours in mice. Invest. New Drugs. 2: 19–27.

Jobin, L. and Caron, A. 1982. Results of aerial treatment with Dimilin and *Bacillus thuringiensis* for gypsy moth control in Quebec. C. F. S. Res. Notes. 2: 18–20.

274

Johnson, G. D. and Mulla, M. S. 1981. Chemical control of aquatic nuisance midges in residential-recreational lakes. Mosquito News. 41: 495–501.

Jones, D., Snyder, M. and Granett, J. 1983. Can insecticides be integrated with biological control agents of Trichoplusia ni in celery? Ent. Exp. Appl. 33: 290–296.

Jordan, A. M. and Trewern, M. A. 1978. Larvicidal activity of diflubenzuron in the tsetse fly. Nature. 272: 719–720.

Jordan, A. M., Trewern, M. A., Borkovec, A. B. and De Milo, A. B. 1979. Laboratory studies on the potential of three insect growth regulators for control of the tsetse fly, Glossina morsitans Westwood (Diptera: Glossinidae). Bull. Ent. Res. 69: 55–64.

Kalberer, P. and Vogel, E. 1978. Control of sciarids (Lycoriella auripila) in mushroom cultures (Agaricus bisporus) by diflubenzuron in the casing soil. Zeit. für Pflanzenkrankheiten und Pflanzenschutz. 85: 328–333.

Keet, C. and Kemp, A. 1984. Diflubenzuron, state of the art 1984. Paper presented at Dimilin Forestry Seminar, Duphar. Nov. 7, 8 (1984), Amsterdam, The Netherlands.

Keiding, J., Arevad, K. and Nielsen, H. 1976. Investigations on diflubenzuron as a housefly larvicide and ovicide. Danish Pest Infestation Laboratory Annual Report (1976). pp. 28–29.

Kiziroglu, V. I. 1976. On the knowledge of Gracilaria syringella (F.) (Lepid., Gracilariidae). 3. Ecology, gradology and control. Z. Ang. Ent. 81: 163–187.

Knapp, F. W. and Herald, F. 1982. Congenitally induced mortality in face flies (Diptera: Muscidae) following adult exposure to diflubenzuron-treated surfaces. J. Med. Entomol. 19: 191–194.

Kolbe, H. and Hartwig, I. 1982. Versuche mit den chitinsynthese-Hemmern BAYER SIR 8514 and Dimilin 25 WP zur vermiderung der schluepfrate von Hylobius abietis -larven. Zeit. Fuer Pflanzenkrankheiten und Pflanzenschutz. 89: 715–719.

Krushev, L. T. and Marchenko, Y. J. Control of the nun moth (Lymantria monacha). Zashchita Rastenii. 11: 35.

Kunz, S. E. and Bay, D. E. 1977. Diflubenzuron: Effects on the fecundity, production and longevity of the horn fly. The Southwestern Ent. 2: 27–31.

Kunz, S. E., Harris, R. L., Hogan, B. F. and Wright, J. E. 1977. Inhibition of development in a field population of horn flies treated with diflubenzuron. J. Econ. Entomol. 70: 298–300.

Kunz, S. E., Schmidt, C. D. and Harris, R. L. 1976. Effectiveness of diflubenzuron applied as dust to inhibit reproduction in horn flies. The Southwestern Ent. 1: 190–193.

Lacey, L. A. and Mulla, M. S. 1977. Larvicidal and ovicidal effect of Dimilin against Simulium vitatum. J. Econ. Entomol. 70: 369–373.

Lacey, L. A. and Mulla, M. S. 1978a. Factors affecting the activity of diflubenzuron against Simulium larvae (Diptera: Simuliidae). Mosquito News. 38: 264–268.

Lacey, L. A. and Mulla, M. S. 1978b. Biological activity of diflubenzuron and three new IGRs against Simulium vittatum (Diptera: Simuliidae). Mosquito News. 38: 377–381.

Lara, F. M., Bortoli, S. A. de and Nunes, D. 1977. Controle quimico de Anticarsia gemmatalis (H) 1818 na cultura da soya, Glycine max (L.). Anais da Sociedade Entomologica Brasileira. 6: 276–280.

Leighton, T., Marks, E. and Leighton, F. 1981. Pesticides: Insecticides and fungicides are chitin synthesis inhibitors. Science 213: 905–907.

Leopold, R. A. and Marks, E. P. 1980. Synergism of diflubenzuron (DFB) with juvenile hormone (JH) in the cotton boll weevil. Amer. Zool. 20: 859.

Levot, G. W,. and Shipp, E. 1983. Interference to egg and larval development of the Australian Sheep blowfly by three insect growth regulators. Ent. Exp. Appl. 34: 58–64.

Lim, S. -J. and Lee, S. -S. 1982a. The toxicity of diflubenzuron to Oxya japonica (Willemse) and its effect on moulting. Pestic. Sci. 13: 537–544.

Lim, S. -J. and Lee, S. -S. 1982b. Toxicity of diflubenzuron to the grasshopper Oxya japonica: Effects on reproduction. Ent. Exp. Appl. 31: 154–158.

Linnane, J. P. 1979. Ground tests with insecticides against the Douglas-fir tussock moth in New Mexico, 1977. Insecticide and Acanicide Tests. 3: 146.

Luber, V. B. 1983. Störungen bei der Kokonbildung von Neodiprion sertifer Geoffr. (Hym., Diprionidae) nach Aufnahme des Häutungshemmstoffs SIR BAY 8514 durch die Altlarven. Anz. Schädlingskde, Pflanzenschutz, Umweltschutz, 56: 9–10.

Maas, W., Van Hes, R., Grosscurt, A. C. and Duel, D. H. 1981. Benzoylphenyl urea insecticides. In: Chemie der Pflanzenschutz und Schadlingsbekampfungsmittel. Ed. R. Wegler. Vol. 6 pp. 423–470. Springer-Verlag, Heidelberg.

Madore, C. D., Boucias, D. G. and Dimond, J. B. 1976. Reduction of reproductive potential in spruce budworm (Lepidoptera: Tortricidae) by a chitin-inhibiting insect growth regulator. J. Econ. Entomol. 76: 708–710.

Malphettes, C. B. and Martouret, D. 1979. Essai pratique de lutte contre la tordeuse verte du chene (*Tortrix viridana*) contribution a la regeneration de lachenaie. Phytiatrie Phytopharmacie. 28: 263–274.

Mansager, E. R., Still, G. G. and Frear, D. S. 1979. Fate of [^{14}C] diflubenzuron on cotton and in soil. Pestic. Biochem. Physiol. 12: 172–182.

Markins, G. P. and Wilcox, H. 1978. Douglas-fir tussock moth ground spray tests, 1974. Insecticide and Acaricide tests 2: 113–114.

Matolin, S. and Kuldova, J. 1982. Effects of diflubenzuron and dimatif on eggs of codling moth, *Cydia pomonella* (Lepidoptera, Tortricidae). Acta Ent. Bohemoslov. 79: 267–273.

Meyer, R. T., Netter, K. J., Leising, H. B. amd Schactschable, D. O. 1984. Inhibition of the uptake of nucleosides in cultured Harding-Passey melanoma cells by diflubenzuron. Toxicology, 30: 1–6.

McCoy, C. W. 1978. Activity of Dimilin on the developmental stages of *Phyllocoptruta oleivora* and its performance in the field. J. Econ. Entomol. 71: 122–124.

McCoy, J R. and Wright, J. E. 1979. Evaluation of Bisazir and Penfluron as sterilants for the boll weevil. The Southern Ent. 4: 209–215.

McGregor, H. E. and Kramer, K. J. 1976. Activity of Dimilin (TH 60940) against coleoptera in stored wheat and corn. J. Econ. Entomol. 69: 479–480.

McKague, A. B., Pridmore, R. B. and Wood, P. M. 1978. Effects of Altosid and Dimilin on black flies (Diptera: Simuliidae): Laboratory and field tests. Can. Ent. 110: 1103–1110.

Metcalf, R. L., Lu, P. -Y. and Bowlus, S. 1975. Degradation and environmental fate of 1-(2,6-difluorobenzoyl)-3-)4-chlorophenyl) urea. J. Agric. Food Chem. 23: 359–364.

Mian, L. S. and Mulla, M. S. 1982a. Biological activity of IGRs against four stored-product coleopterans. J. Econ. Entomol. 75: 80–85.

Mian, L. S. and Mulla, M S. 1982b. Residual activity of insect growth regulators against stored product beetles in grain commodities. J. Econ. Entomol. 75: 599–603.

Miller, R. W. 1974. TH 6040 as a feed additive for control of the face fly and house fly. J. Econ. Entomol. 67: 697.

Miller, R. W., Corley, C. and Hill, K. R. 1975. Feeding TH 6040 to chickens: Effect on larval house flies in manure and determination of residues in eggs. J. Econ. Entomol. 68: 181–182.

Millo, B. 1978. Results of an application per helicopter of diflubenzuron against the pine processionary moth (*Thaumetopoea pityocampa*). Giornate Fitopatologico. pp. 555–559.

Mitchell, E. B., Merkl, M. E., Wright, J. E., Davich, T. B. and Heiser, R. F. 1980. Sterility of boll weevils in the field following treatment with diflubenzuron and gamma irradiation. J. Econ. Entomol. 73: 824–826.

Mitlin, N., Wiygul, G. and Haynes, J. W. 1977. Inhibition of DNA synthesis in boll weevils (*Anthonomus grandis* Boheman) sterilized by Dimilin. Pestic. Biochem. Physiol. 7: 599–563.

Mitsui, T., Nobusawa, C. and Fukuami, J. 1984. Mode of inhibition of chitin synthesis by diflubenzuron in the cabbage armyworm, *Mamestra brassicae* L. J. Pestic. Sci. 9: 19–26.

Miura, T., Schaefer, C. H., Takahashi, R. M. and Mulligan, F. S. III. 1976. Effects of the growth inhibitor, Dimilin, on hatching of mosquito eggs. J. Econ. Entomol. 69: 655–658.

Moffitt, H. R., Mantey, K. D. and Tamaki, G. 1983. Effects of chitin-synthesis inhibitors on oviposition by treated adults and on subsequent egg hatch of the codling moth, *Cydia pomonella* (Lepidoptera: Olethreutidae). Can. Ent. 115: 1659–1662.

Moffitt, H. R., Mantey, K. D. and Tamaki, G. 1984. Effects of residues of chitin-synthesis inhibitors on egg hatch and subsequent larval entry of the codling moth, *Cydia pomonella* (Lepidoptera: Olethreutidae). Can. Ent. 116: 1057–1062.

Mohamed, A. I., Young, S. Y. and Yearian, W. C. 1983. Susceptibility of *Heliothis virescens* (F.) (Lepidoptera: Noctuidae) larvae to microbial agent — chemical pesticide mixtures on cotton foliage. Environ. Entomol. 12: 1403–1405.

Moreau, R., Castex, C. and Lamy, M. 1975. First observations of the metabolic effects of a new insecticide among two mischievious insects, *Pierris brassicae* L. and *Thaumetopoea pityocampa* S. (Lepidoptera). Ann. Zool. — Ecol. Anim. 7: 161–169.

Mulla, M. S. and Darwazeh, H. A. 1975. Activity and Longevity of insect growth regulators against mosquitoes. J. Econ. Entomol. 68: 791–794.

Mulla, M. S., Darwazeh, H. A. and Norland, R. L. 1974. Insect growth regulators: Evaluation procedures and activity against mosquitoes. J. Econ. Entomol. 67: 329–332.

Mulla, M. S. and Darwazeh, H. A. 1976. The IGR Dimilin and its formulations against mosquitoes. J. Econ. Entomol. 69: 309–312.

Mulla, M. S., Kramer, W. L. and Barnard, D. R. 1976. Insect growth regulators for control of chironomid midges in residential-recreational lakes. J. Econ. Entomol. 69: 285–291.

Mulla, M. S., Majori, G. and Darwazeh, H. A. 1975. Effects of the insect growth regulator Dimilin or TH-6040 on mosquitoes and some nontarget organisms. Mosquito News. 35: 211–216.

Natesan, R. and Balasubramaniam, M. 1980. Effect of diflubenzuron on pupae of tobacco caterpillar, *Spodoptera litura* F. Entomon. 5; 211–213.

Neal, J. W. Jr. 1974. Alfalfa weevil control with the unique growth disruptor TH 6040 in small plot tests. J. Econ. Entomol. 67: 300–301.

Neisess, J., Markin, G. P. and Schaefer, R. 1976. Field evaluation of acephate and Dimilin against the Douglas-fir moth Tussock moth. J. Econ. Entomol. 69: 783–786.

Neilson, D. G. and Balderston, C. P. 1979. Oak, foliar sprays evaluated for control of palmerworm, Ohio 1977. Insecticide and Acaricide Tests. 3: 155.

Novak, V. and Sehnal, F. 1978. Sterilization of the pine weevil, *Hylobius abietis* with diflubenzuron. Acta Ent. Bohemslov. 75: 349–351.

Novak, V. and Sehnal, F. 1979. Sterilization of the pine weevil (*Hylobius abietis* L.) with diflubenzuron. Lesnictvi. 25: 811–828.

O'Brien, R. D. 1967. Insecticides — action and metabolism. 332 pp. Academic Press, New York.

O'Neill, M. P., Holman, G. M. and Wright, J. E. 1977. β-ecdysone levels in pharate pupae of the stable fly, *Stomoxys calcitrans* and interaction with the chitin inhibitor diflubenzuron. J. Insect Physiol. 23: 1243–1244.

Oppenoorth , F. J. and Van Der Pas, L. J. T. 1977. Cross-resistance to diflubenzuron in resident strains of housefly *Musca domestica*. Ent. Exp. Appl. 21: 217–228.

Overbeck, V. H. 1979. Zur wirkung von Dimilin auf das eistadium der Möhrenfliege *Psila rosae* F. (Diptera: Psilidae) Nachrichtenbl. Deut. Pflanzenschutzd. (Braunschweig) 31: 99–102.

Page, M., Ryan, R. B., Rappoport, N. and Schmidt, F. 1982. Comparative toxicity of Acephate, Diflubenzuron and Malathion to larvae of the larch casebearer, *Coleophora laricella* (Lepidoptera: Coleophoridae) and adults of its parasites, *Chrysocharis laricinellae* and *Dicladocerus nearcticus*. Environ. Entomol. 11: 730–732.

Paradis, R. O. 1978. *Orthosia hibisci* (Guenée) (Lépidoptères: Noctuidae) dans les pommeraies du sud-ouest du Québec II — Essais de lutte chimique. Phytoprotection. 59: 101–107.

Pediglo, L. P. and Hammond, R. 1976. Green clover-worm efficacy trials. Insecticide and Acaricide Tests. 1: 97.

Pediglo, L. P. and Hammond, R. 1978. Green clover-worm efficacy trials, 1976. Insecticide and Acaricide Tests. 2: 105.

Pediglo, L. P. and Hammond, R. 1979. Green clover-worm control, 1977. Insecticide and Acaricide Tests. 3: 135.

Peleg, B. -A. and Gothilf, S. 1981. Effect of the insect growth regulators Diflubenzuron and Methoprene on scale insects. J. Econ. Entomol. 74: 124–126.

Percy, J., Nicholson, D. and Retnakaran, A. 1985. The effect of ingested benzoylphenyl urea on the ultrastructure of cuticle deposited during the last larval instar of *Choristoneura fumiferana* (Lepidoptera: Tortricidae). Can. J. Zool. (submitted).

Pickens, L. G. and De Milo. A. B. 1977. Face fly: Inhibition of hatch by diflubenzuron and related analogues. J. Econ. Entomol. 70: 595–597.

Pimentel, D. 1971. Ecological effects of pesticides on non-target species. Executive office of the President, Office of Science and Technology, U. S. Govt. Printing Office, Washington, D. C.

277

Pimprikar, G. D. and Georghiou, G. P. 1979. Mechanisms of resistance to diflubenzuron in the house fly, *Musca domestica* (L.) Pestic Biochem. Physiol. 12: 10–22.

Post, L. C., De Jong, B. J. and Vincent, W. R. 1974. 1-(2,6-disubstituted benzoyl)-3-phenylurea insecticides: Inhibitors of chitin synthesis. Pestic. Biochem. Physiol. 4: 473–483.

Pree, D. J. 1976. Effects of two insect growth disruptors, PH 6038 and PH 6040 on the winter moth, *Operophtera brumata* (Lepidoptera: Geometridae). Can. Ent. 108: 49–52.

Price, J. F. 1979. Control of mealybugs on caladiums. Proc. Fla. State Hort. Soc. 92: 358–360.

Rabindra, R. J. and Balasubramaniam, M. 1981. The effect of Diflubenzuron on the castor semilooper, *Achoea janata* Linn. (Lepidoptera: Noctuidae). Entomon. 6: 15–18.

Radwan, H. S. A., Abo-Elghar, M. R. and Ammar, I. M. A. 1978. Reproductive performance of *Spodoptera littoralis* (Boisd.) treated topically with sublethal doses of an antimoulting IGR (Dimilin). Z. Ang. Ent. 86: 414–419.

Radwan, H. S. A., Mesbah, H. A., Abdel-Fattah, M. S., El-Mohymen, M. R. A. and Hassan, N. A. 1982. The effect of various adjuvants on the insecticidal activity of diflubenzuron against the cabbage aphid, *Brevicoyne brassicae* (L.). Z. Ang. Ent. 94: 420 423.

Radwan, H. S. A. and Rizk, G. A. M. 1976. Initial and chronic effects of the insect growth regulator, PH 60–40 on the American bollworm, *Heliothis armigera*. Rev. Appl. Entomol. 64: 293–297.

Rajendran, S. and Shivaramaiah, H. M. 1983. Effect of diflubenzuron on the productivity of the khapra beetle, *Trogoderma granarium*. Ent. Exp. Appl. 33: 15–19.

Rappaport, N. G. and Robertson, J. L. 1981. Lethal effects of five molt inhibitors fed to the western spruce budworm (*Choristoneura occidentalis* Freeman) (Lepidoptera: Tortricidae) and the Douglas-fir tussock moth (*Orgyia pseudotsugata* [McDonnough]) (Lepidoptera: Lymantriidae). Z. Ang. Ent. 91: 459–463.

Rathburn, C. B. and Boike, A. H. Jr. 1975. Laboratory and small plot field tests of Altosid and Dimilin for the control of *Aedes taeniorhynchus* and *Culex nigripalpus* larvae. Mosquito News. 35: 540–546.

Ravensberg, W. J. 1981. The natural enemies of the woolly apple aphid, *Eriosoma lanigerum* (Hausm.) (Homoptera: Aphidae) and their susceptibility to diflubenzuron. Med. Fac. Landbouww. Rijksuniv. Gent. 46: 437–441.

Reed. T. and Bass, M. H. 1980. Larval and postlarval effects of diflubenzuron on the soybean looper. J. Econ. Entomol. 73: 332–338.

Retnakaran, A. 1979. Effect of a new moult inhibitor (El-494) on the spruce budworm, *Choristoneura fumiferana* (Lepidoptera: Tortricidae). Can Ent. 111: 847–850.

Retnakaran, A. Effect of 3 new moult-inhibiting insect growth regulators on the spruce budworm. J. Econ. Entomol. 73: 520–524.

Retnakaran, A. 1981. Toxicology and efficacy of insect growth regulators aerially applied against the spruce budworm at Hearst (1978), Wawa (1979) and the French River area (1980). Inf. Rep. FPM-X-45. 60 pp.

Retnakaran, A. 1982a. Noninvolvement of mixed function oxidases in Dimilin metabolism. C. F. S. Res. Notes, 2: 1.

Retnakaran, A. 1982b. Laboratory and field evaluation of a fast-acting insect growth regulator against the spruce budworm, *Choristoneura fumiferana* (Lepidoptera: Tortricidae). Can. Ent. 114: 523–530.

Retnakaran, 1985. Chitin biosynthesis in insects and its disruption as a means of pest control. Abst. 3rd International Conference on Chitin/chitosan (April 1985), Senigalli (Ancona) Italy.

Retnakaran, A. and Ennis, T, J. 1985. *In vitro* mutagenicity testing of a potent, new, benzoyl urea insect growth regulator. Experientia (in press).

Retnakaran, A., Granett, J. and Ennis, T. 1985. Insect Growth Regulators. In: Comprehensive insect physiology, biochemistry and pharmacology. Ed. G. A. Kerkut and L. I. Gilbert. Vol. 12. Ch. 15. pp. 529–601. Pergamon Press, Oxford.

Retnakaran, A., Granett, J. and Robertson, J. 1980. Possible physiological mechanisms for the differential susceptibility of two forest lepidoptera to diflubenzuron. J. Insect Physiol. 26: 385–390.

278

Retnakaran, A. and Grant, G. G. 1985. Control of the oak-leaf shredder, *Croesia semipurpurana* (Kearfott) (Lepidoptera: Tortricidae), by aerial application of diflubenzuron. Can. Ent. 117: 363–369.

Retnakaran, A. and Hackman, R. H. 1985. Synthesis and deposition of chitin in larvae of the Australian sheep blowfly, *Lucilia cuprina*. Arch. Insect Biochem. Physiol. (In press).

Retnakaran, A. and Smith, L. 1975. Morphogenetic effects of an inhibitor of cuticle development on the spruce budworm, *Choristoneura fumiferana* (Lepidoptera: Torttricidae). Can. Ent. 107: 883–886.

Retnakaran, A. and Smith, L. 1976. Greenhouse evaluation of PH 60–40 activity on the forest tent caterpillar, C. F. S. Bi-mon. Res. Notes. 32: 2.

Retnakaran, A. and Smith, L. 1982. Reproductive effect of insect growth regulators on the white pine weevil, *Pissodes strobi* (Coleoptera: Curcolionidae). Can. Ent. 114: 381–383.

Retnakaran, A., Smith, L., Tomkins, B. and Granett, J. 1979. Control of forest tent caterpillar, *Malacosoma disstria* (Lepidoptera: Lasiocampidae), with Dimilin. Can. Ent. 111: 841–846.

Retnakaran, A. and Tomkins, W. 1982. Effectiveness of moult-inhibiting insect growth regulators in controlling the oak leaf shredder. C. F. S. Res. Notes. 2: 5–6.

Ribrioux, Y. and Dolbeau, C. 1978. Essai du lutte contre la processionnaire du pin (*Thaumetopoea pityocampa* Schiff.) a l'aide du diflubenzuron. Phytiatrie Phytopharmacie. 24: 193–204.

Richmond, J A. and Cunningham., P. A. 1983. Evaluation of diflubenzuron against egg and larval stages of the Nantucket pine tip moth. J. Georgia Entomol. Soc. 18: 280–284.

Richmond, J. A., De Milo, A. B., Thomas, H. A., and Borkovec, A. B. 1978. Mortality and sterility of Southern pine beetles treated with chemosterilants and insect growth regulators. J. Georgia Entomol. Soc. 13: 227–236.

Rizk, G. A. M. and Radwan, H. S. A. 1975. Potency and residuality of two antimoulting compounds against cotton leafworm and bollworms. Z. Ang. Ent. 79: 136–140.

Rizk, G. A. M. and Radwan, H. S. A. 1976. Response of pink bollworm to soil application of two unique growth disruptors. Proc. 8th British Insecticide and Fungicide Conference 1975: British Crop Protection Council, London.

Robertson, J. L. 1982. Toxicity of experimental molt inhibitors to Western spruce budworm. J. Georgia Entomol. Soc. 17: 413–416.

Robertson, J. L. and Boelter, L. M. 1979a. Toxicity of insecticides to Douglas-fir tussock moth, *Orgyia pseudotsugata*, residual toxicity and rainfastness. Can. Ent. 111: 1161–1175.

Robertson, J. L. and Boelter, L. M. 1979b. Toxicity of insecticides to Douglas-fir tussock moth, *Orgyia pseudotsugata*, contact and feeding toxicity. Can. Ent. 111: 1145–1159.

Robertson, J. L. and Haverty, M. I. 1981. Multiphase laboratory bioassays to select chemicals for field-testing on the western spruce budworm. J. Econ. Entomol. 74: 148–153.

Robertson, J. L. and Smith, K. C. 1984. Western spruce budworm: Joint action of pyrethroids and insect growth regulators by contact or ingestion. J. Georgia Entomol. Soc. 19: 454–462.

Robertson, J. L., Smith, K. C., Granett, J. and Retnakaran, A. 1984. Joint action of a juvenile hormone analogue with benzoylphenylureas ingested by western spruce budworm, *Choristoneura occidentalis* (Lepidoptera: Tortricidae). Can. Ent. 116: 1063–1068.

Robredo, F. J. 1980a. Tratamientos masivos con diflubenzuron contra la processionaria del pino en espana. Bol. Del Sevicio de defensa contra plagas E inspeccion Fitopatologica. 6: 141–154.

Robredo, F. J. 1980b. Extensive aerial treatments with diflubenzuron against the pine processionary caterpillar in Spain. Proc. VI International Agricultural Aviation Congress, Turin. pp. 37–55.

Robredo, F. J. 1980c. Control of the pine processionary moth with Dimilin, a new biological, non-toxic and non-contaminating insecticide. Agricultura. 49: 269–299.

Rogers, A. J., Rathburn, C. B. Jr., Beidler, E. J., Dodd, G., and Lafferty, A. 1976. Tests of two insect growth regulators formulated on sand against larvae of salt-marsh mosquitoes. Mosquito News. 36: 273–277.

Saad, A. S. A., Elewa, M. A., Aly, N. M., Auda, M. and El-Sebae, A. H. 1981. Toxicological studies on the Egyptian cotton leafworm, *Spodoptera littoralis* I. Potentiation and antagonism of synthetic pyrethroid, organophosphorus and urea derivative insecticides. Med. Fac. Landbouww. Rijksuniv. Gent. 46/2: 559–571.

279

Sahota, T. S. and Ibaraki, A. 1980. Prolonged inhibition of brood production in *Dendroctonus rufipennis* (Coleoptera: Scolytidae) by Dimilin. Can. Ent. 112: 85–88.

Sahota, T. S. and McMullen, L. H. 1979. Reduction in progeny production in the spruce weevil, *Pissodes strobi.* C. F. S. Bi-mon. Res. Notes. 35: 32–33.

Sahota, T. S. and Shepherd, R. F. 1975. Laboratory tests of 1-(4-chlorophenyl)-3-(2,6 difluorobenzoyl) urea on survival of western hemlock looper. C. F. S. Bi-Monthly Res. Notes. 31: 39–40.

Salama, H. S. and El-Din, M. M. 1977. Effect of the moulting inhibitor Dimilin on the cotton leafworm *Spodoptera littoralis* Boisd. in Egypt. Z. Ang. Ent. 83: 415–419.

Salama, H. S., Motagally, Z. A. and Skatulla, U. 1976. On the mode of action of Dimilin as a moulting inhibitor in some lepidopterous insects. Z. Ang. Ent. 80: 396–407.

Saleh, M. S., Gaaboub, I. A. and Kassem, SH. M. I. 1981. Larvicidal effectiveness of three controlled-release formulations of Dursban and Dimilin on *Culex pipiens* L. and *Aedes aegypti* (L.) J. Agric. Sci., Camb. 97: 87–96.

Santharam, G. and Balasubramaniam, M. 1980. Note on the control of *Spodoptera litura* Fabricius (Lepidoptera: Noctuidae) on tobacco with a nuclear poyhedrosis virus and diflubenzuron. Indian J. Agric. Sci. 50: 726–727.

Sarasua, M. J. and Santiago-Alvarez, C. 1983. Effect of diflubenzuron on the fecundity of *Ceratitis capitata.* Ent. Exp. Appl. 33: 223–225.

Saxena, S. C. and Kumar, V. 1981. Blockage in chitin biosynthetic chain in the grasshopper *Chrotogonous trachypterus* treated with diflubenzuron and penfluron. Ind. J. Exp. Biol. 19: 1199–1200.

Saxena, S. C. and Kumar, V. 1982. Effect of two chitin inhibitors on reproduction of *Trogoderma granarium.* Entomon. 7: 141–144.

Saxena, S. C. and Mathur, G. 1981a. Suppression of adult emergence on treating eggs of *Tribolim castaneum* Herbst by new, synthesized, disubstituted benzoylphenyl urea compounds. Current Science. 50: 336.

Saxena, S. C. and Mathur, G. 1981b. Suppression of adult emergence by administering new synthesized chitin synthesis inhibitor compounds in *Tribolium castaneum* Herbst. J. Environ. Biol. 2: 7–10.

Schaefer, C. H. and Dupras, E. F. Jr. 1976. Factors affecting the stability of Dimilin in water and the persistence of Dimilin in field waters. J. Agric. Food Chem. 24: 733–739.

Schaefer, C. H., Wilder, W. H. and Mulligan, F. S. III. 1975. A practical evaluation of TH 6040 as a mosquito control agent in California. J. Econ. Entomol. 68: 183–185.

Schaefer, C. H., Wilder, W. H., Mulligan, F. S. III and Dupras, E. F. Jr. 1974. Insect development inhibitors: Effects of Altosid, TH 6040 and H24108 against mosquitoes (Diptera: Culicidae). Proc. Ann. Conf. Calif. Mosquito Control Assoc. 42: 137–139.

Schmidt, C. D. and Kunz, S. E. 1980. Testing immature laboratory-reared stable flies and horn flies for susceptibility to insecticides. J. Econ. Entomol. 78: 702–703.

Schroeder, W. J., Beavers, J. B., Sutton, R. A. and Selhime, A. G. 1976. Ovicidal effect of Thompson-Hayward TH 6040 in *Diaprepes abbreviatus* on citrus in Florida. J. Econ. Entomol. 69: 780–782.

Schroeder, W. J. and Sutton, R. A. 1977. *Diaprepes abbreviatus*: Suppression of reproductive potential on citrus with an insect growth regulator plus spray oil. J. Econ. Entomol. 71: 69–70.

Schroeder, W. J., Sutton, R. A. and Beavers, J. B. (1980). *Diaprepes abbreviatus*: fate of diflubenzuron and effect on nontarget pests and beneficial species after application to citrus for weevil control. J. Econ. Entomol. 73: 637–638.

Seuferer, S. L., Braymer, H. D. and Quinn, N. J. 1979. Metabolism of diflubenzuron by soil microorganisms and mutagenicity of the metabolities. Pestic. Biochem. Physiol. 10: 174–180.

Shepard, M. and Kissam, J. B. 1981. Integrated control of house flies on poultry farms: Treatment of house fly resting surfaces with diflubenzuron plus releases of the parasitoid, *Muscidifurax raptor.* J. Georgia Entomol. Soc. 16: 222–227.

Sinègre, G., Julien, L. L., Gaven, B. et Crespo, O. 1980. Action larvicide et ovicide du diflubenzuron sur trois especes de culicides. Parasitologia. 22: 187–198.

Skatulla, V. U. 1975. Über die wirkung des entwicklungshemmers Dimilin auf forstinsekten. Anzeiger für Schädlingskunde Pflanzenschutz Umweltschutz. 48: 145–147.

280

Skatulla, V. U. 1975. Successful trials for controlling *Lymantria dispar* and *L. monacha* by help of PH 60–40, prohibiting the moulting of insect larvae. Anzeiger für schädlingskunde Pflanzenschutz Umweltschutz. 48: 17–18.

Smith, J. C. 1976. Insecticide effects on Mexican bean beetle, green cloverworm and predators in Soybean, Suffolk, Va.- 1975. Insecticide and Acaricide Tests. 1: 98.

Soltani, N., Besson, M. T. and Delachambre, J. 1984. Effects of diflubenzuron on the pupal-adult development of *Tenebrio molitor* L. (Coleoptera, Tenebrionidae): Growth and development, cuticle secretion, epidermal cell density, and DNA synthesis. Pestic. Biochem. Physiol. 21: 256–264.

Spaic, J. 1981. Results of the efficacy of new formulation of Dimilin against *Cnethocampa (Thaumetopoea) pityocampa*. Jugoslovensko Savetovanje O Primeno Pesticida Zbornik Radova. 2: 365–368.

Spates, G. E. and Wright, J. E. 1980. Residues of diflubenzuron applied topically to adult stable flies. J. Econ. Entomol. 73: 595–598.

Strebler, P. G. 1979. Utilisation du diflubenzuron en lutte antiacridienne. Z. Ang. Ent. 88: 124–131.

Stern, V. M., Flaherty, D. L. and Peacock, W. L. 1983. Control of the western grapeleaf skeletonizer (Lepidoptera: Zyganidae), a new grape pest in the San Joaquin Valley, California. J. Econ. Entomol. 76: 192–195.

Summers, C. G. 1975. Efficacy of insecticides and dosage rates applied for control of the Egyptian alfalfa weevil and Pea aphid. J. Econ. Entomol. 68: 846–866.

Sundaramurthy, V. T. 1977. Effect of an inhibitor of chitin deposition on the growth and differentiation of tobacco caterpillar *Spodoptera litura* Fb. (Noctuidae: Lepidoptera). Zeit. fur Pflanzenkrankheiten und Pflanzen schutz. 84: 597–601.

Sundaramurthy, V.T. and Balasubramaniam, M. 1978. Effect of an inhibitor of chitin deposition in tobacco caterpillar (*Spodoptera litura* Fb.) under induced hyper hormone condition. Z. Ang. Ent. 85: 317–321.

Sundaramurthy, V. T. and Santhanakrishnan, K. 1979. Morphogenetic effects of diflubenzuron, an inhibitor of chitin deposition, on the Coconut black-headed caterpillar (*Nephantis serinopa* Myer). Pestic. Sci. 10: 147–150.

Szanto, J. 1978. Dimilin 25 WP as a new option against the American white webbing moth (*Hyphantria cunea* Drury). Noveni Jvedelem. 14: 509–512.

Szmidt, A. and Sliwan, W. 1981. Susceptibility of various instars of caterpillars of *Lymantria monacha* to Dimilin WP. Roczniki Akademii Rolnicezej W Poznaniu. 82: 165–172.

Taft, H. M. and Hopkins, A. R. 1975. Boll weevils: Field populations controlled by sterilizing emerging overwintered females with TH-6040 sprayable bait. J. Econ. Entomol. 68: 551–554.

Takahashi, M. and Ohtaki, T. 1976. A laboratory evaluation of the IGR, PH 60–40, against *Culex pipiens pallens* and *Culex tritaeniorhynchus*. Eisei-Dobutsu. 27: 361–365. (English summary).

Tamaki, G., Chauvin, R. L., Moffitt, H. R. and Mantey, K. D. 1984. Diflubenzuron: Differential toxicity to larvae of the Colorado potato beetle (Coleoptera: Chrysomelidae) and its internal parasite, *Doryphorophaga doryphoae* (Diptera: Tachinidae). Can. Ent. 116: 197–202.

Tester, P. A. and Costlow, J. D. Jr. 1981. Effect of insect growth regulator Dimilin (TH 6040) on fecundity and egg viability of the marine copepod *Acartia tonsa*. Mar. Ecol. Prog. Ser. 5: 297–302.

Ticehurst, M., Fusco, R. A. and Blumenthal, E. M. 1982. Effect of reduced rates of Dipel 4L, Dylox, 5-oil and Dimilin W-25 on *Lymantria dispar* parasitism and defoliation. Environ. Entomol. 11: 1058–1062.

Tonks, N. V., Everson, P. R. and Theaker, T. L. 1978. Efficacy of insecticides against Geometrid larvae, *Operophtera* spp. J. Entomol. Soc. British Columbia, 75: 6–9.

Turnbull, I. F. and Howells, A. J. 1980. Larvicidal activity of chitin synthesis inhibitors in *Lucilia cuprina*. Am. Zool. 20: 1045.

Turnipseed, S. G., Heinrichs, E. A., Da Silva, R. F. P. and Todd, J. W. 1974. Response of soybean insects to foliar applications of a chitin synthesis inhibitor TH 6040. J. Econ. Entomol. 67: 760–762.

Tsuzuki, H. and Asayama, T. 1983. Influence of diflubenzuron on oviposition and hatching in the rice water weevil, *Lissorhoptrus oryzophilus* Kuschel (Coleoptera: Curculionidae). Jap. J. Appl. Ent. Zool. 27: 229–231.

Valovage, W. D. and Kulman, H. M. 1978a. *Diprion similis* control with Dimilin on white pine, 1976. Insecticide and Acaricide Tests. 2: 117.

Valovage, W. D. and Kulman, H. M. 1978b. *Pikonema alaskensis* control with Dimilin on white spruce, 1976. Insecticide and Acaricide Tests, 2: 117.

Valovage, W. D. and Kulman, H. M. 1978c. *Pristiphora erichsonii* control with Dimilin on larch, 1976. Insecticide and Acaricide Tests. 2: 116.

Van Busschbach, E. J. 1975. Apercu général des possibilités d'application du "Dimilin". Phytiatrie-phytopharmacie. 24: 159–178.

Van Eck, W. H. 1979. Mode of action of two benzoylphenyl ureas as inhibitors of chitin synthesis in insects. Insect Biochem. 9: 295–300.

Vasic, M. 1977. Gypsy moth and fall webworm control by Dimilin. Zastita Bilja. 28: 191–197.

Vasic, M. 1979. Control of *Stilpnotia salicis* L. by Dimilin. Proc. Inmt. Symp. IOBC/WRPS on Integrated Control. Oct. 8–12, 1979. p. 451.

Vea, E. V., Yu, C. -C., Webb, D. R., Eckenrode, C. J. and Kuhr, R. J. 1976. Laboratory and field evaluation of insecticides and insect growth regulators for control of the seedcorn maggot. J. Econ. Entomol. 69: 178–180.

Veire, M. V. D., Steene, F. V. D. and Hertveldt, L. 1975. Developmental inhibition of the seedcorn maggot, *Hylemya cilicrura* by the urea-analogue, Diflubenzuron. Med. Fac. Landbouww. Rijksuniv. Gent. 40: 403–408.

Weaver, J. E., Begley, J. W. and Kondo, V. A. 1984. Laboratory evaluation of Alsystin against the German Cockroach (Orthoptera: Blattellidae): Effects on immature stages and adult sterility. J. Econ. Entomol. 77: 313–317.

Webley, D. J. and Airey, W. A. 1982. A laboratory evaluation of the effectiveness of diflubenzuron against *Dermestes maculatus* De Geer and other storage insect pests. Pestic. Sci. 13: 595–601.

White, P. F. 1977. Sublethal effects of four pesticides on paedogenetic larvae of the cecidomyiidae. Ent. Exp. Appl. 22: 43–52.

White, P. F. 1981. Chemical control of the mushroom sciarid, *Lycoriella auripila* (Winn.) Mushroom Science. 11: 265–273.

Wilcox, H. III and Coffey, T. Jr. (Compiled by). 1978. Environmental impacts of diflubenzuron (Dimilin) insecticide. Forest insect and disease management, Forest Service, U. S. D. A., Broomall, Pa. 18 pp.

Wilson, D. 1979. The development of diflubenzuron in the UK for the control of lepidopterous pest in apple and pear. Proceedings, 1979 British Crop Protection Conference — Pests and Diseases. pp 69–75.

Wright, J. E. 1974. Insect growth regulators: Laboratory and field evaluations of Thompson-Hayward TH-6040 against the house fly and the stable fly. J. Econ. Entomol. 67: 746–747.

Wright, J. E, 1975. Insect growth regulators: Development of house flies in feces of bovines fed TH 6040 in mineral blocks and reduction in field populations by surface treatment with TH 6040 or a mixture of Stirofos and Dichlorvos at larval breeding areas. J. Econ. Entomol. 68: 322–324.

Wright, J.E. and Harris, R. L. 1976. Ovicidal activity of Thompson-Hayward TH 6040 in the stable fly and Horn fly after surface contact by adults. J. Econ. Entomol. 69: 728–730.

Wright, J. E., Haynes, J. W., McCoy, J. R. and Dawson, J. R, 1979. Boll weevil: Mating ability, sterility and survival of irradiated and fumigated adults of different ages. The Southern Ent. 4: 53–58.

Wright, J. E., Haynes, J., McCoy, J., Lindig, O., Wiygul, G. and Lloyd, E. P. 1983. Laboratory evaluation of sterile boll weevils (Coleoptera: Curculionidae) in the eradication trial in North Carolina. Mississippi Agricultural and Forestry Experiment Station, Technical Bull. 115.

Wright, J. E., McCoy, J. R., Dawson, J. R., Roberson, J. and Sikorowski, P. P. 1980b. Boll weevil sterility: Effects of different combinations of diflubenzuron, antibiotics, fumigation, and irradiation. The Southern Ent. 5: 84–89.

Wright, J. E., Moore, R., McCoy, J., Wiygul, G. and Haynes, J. 1980a. Comparison of three sterilization procedures on the quality of the male boll weevil. J. Econ. Entomol. 73: 493–496.

282

Wright, J. E., Ochler, D. D. and Johnson, J. H. 1975. Control of house fly and stable fly breeding in rhinocereos dung with an insect growth regulator used as a feed additive. J. Wildlife Diseases, 11: 522–524.

Wright, J. E. and Roberson, J. 1981. Laboratory evaluation of a method of sterilizing the boll weevil. J. Econ. Entomol. 74: 696–697.

Wright, J. E., Roberson, J. and Dawson, J. R. 1980c. Boll weevil: Effects of diflubenzuron on sperm transfer, mortality and sterility. J. Econ. Entomol. 73: 803–805.

Wright, J. E. and Spates, G. E. 1976. Reproductive inhibition activity of the Insect Growth Regulator TH 6040 against the stable fly and the house fly: Effects on hatchability. J. Econ. Entomol. 69: 365–368.

Wright, J. E. and Villavaso, E. J. 1983. Boll weevil sterility. In: Cotton insect management with special reference to the boll weevil. Ed. E. P. Lloyd, R. L. Ridgeway and W,. C. Cross. pp. 153–177, Plenum Press, New York.

Wolfenbarger, D. A., Gurerra, A. A. and Garcia, R. D. 1977. Diflubenzuron: Effect on the tobacco budworm and the boll weevil. J. Econ. Entomol. 70: 126–128.

Yu, S. J. and Terriere, L. C. 1975. Activities of hormone metabolizing enzymes in house flies treated with some substituted urea growth regulators. Life Sci. 17: 619–625.

Yu, S. J. and Terriere, L. C. 1977. Ecdysone metabolism by soluble enzymes from three species of diptera and its inhibition by the insect growth regulator TH-6040. Pestic. Biochem. Physiol. 7: 48–55.

Zungoli, P. A., Steinhaer, A. L. and Linduska, J. J. 1983. Evaluation of diflubenzuron for Mexican bean beetle control and impact on *Pediobius foveolatus*. J. Econ. Entomol. 76: 188–191.

10. Potential of benzoylphenyl ureas in integrated pest management

J. Granett

1 INTRODUCTION

Insect toxicologists, even those dealing with the most esoteric subjects, frequently are in a position to make discoveries which may have positive implications for society. In order for the more basic research (for example work concerning chitin and the benzoylphenyl ureas) to be useful, however, the science has to be incorporated into the constraints of existing agricultural technology and industrial economics. The technological links between the scientist, the farmer and the chemical industry are frequently the ones missing or underdeveloped. The people who deal with these technological links must not only understand the science, but must also understand the agricultural practices, the industry, and the economics involved.

As a matter of principle, most academically oriented scientists restrict their studies to the science involved. If, as scientists, we hope to influence technology we must try to understand the constraints on the utilization of our discoveries within the agricultural and industrial sectors. The purpose of this paper is to evaluate the potential use of benzoylphenyl ureas in IPM programs. This discussion will initially focus on the constraints to insecticide use imposed by IPM programs, the constraints imposed by the nature of modern agricultural technology, and the constraints imposed by economics and governmental regulations relating to insecticide registrations. The second focus of the paper concerns the relevance of these constraints to the chemical and biological characteristics of the benzoylphenyl ureas. From such an analysis the needs for additional research and for suggestions concerning the future use of the benzoylphenyl ureas will become apparent.

2 CONSTRAINTS

In discussing constraints on insecticide use imposed by IPM, agriculture, or the chemical industry, we must realize that there are two problems:

Wright, J. E. and Retnakaran, A. (Eds), Chitin and Benzoylphenyl ureas. ISBN 978-94-010-8638-7.
© *1987, Dr W. Junk Publishers, Dordrecht.*

first, reliable documentation and quantification of the problems are unavailable, and second, generalizations based on our current knowledge, are extremely risky.

For example, of the constraints on insecticide use imposed by IPM, selectivity of the chemical is paramount. There is a paucity of exacting field studies concerning selectivity and mechanistic studies concerning comparative insecticide pharmacodynamics in species within agricultural systems. Regarding agricultural constraints, work on economics of control tactics in a farm situation is frequently limited. In addition, the economics of a tactic may completely change almost overnight when grower practices, governmental regulations, or other crop economics change. With regard to constraints of the chemical industry, these are probably the least understood by outside researchers because of the proprietary secrecy of the companies and because economic strategies of companies differ. Because of these and other limitations on information, my arguments herein will be primarily based on my experiences, observations, and particular biases. Where citations are given it should be understood that with these topics there generally is no truth, just ideas that seem to mimic reality to some extent.

3 THE CONSTRAINTS IMPOSED BY IPM:

3.1 *Selectivity*

Insecticide selectivity is one of the cornerstones for use of insecticides within IPM systems. Selectivity is important because it will allow non-target species such as predators, parasites and pollinators to survive an insecticide treatment that is toxic to pest species. At present, relatively little research has been done on selectivity, so there are few broad generalizations concerning its importance on particular IPM systems. Our understanding of the biochemical and biophysical mechanisms of selectivity are poor in most cases where such mechanisms have been studied, and selectivity has been studied in relatively few cases relevant to IPM. In addition, the complexity and variability inherent in most agricultural systems has prevented the formulation of many generalizations.

An example of generalizations concerning selectivity is given by Metcalf (1975). He established a pest management selectivity rating for insecticides based on several "indicator" organisms. On the surface this measure of selectivity seems logical. However, no one has verified that the use of specific indicator species in agricultural situations in general or in specific cases is valid. The futility of such indicators is demonstrated by Hollingworth (1975) who has derived quantitative vertebrate selectivity ratios (VSRs) as indicators of selective toxicity. In this case, the researcher has documented the fact that the value for the VSR changed radically depending on the species of organisms within the broad taxonomic group (mammal or arthropod) being compared. Generalizations regarding selectivity do not hold at even the genera level.

Another criticism of any measurement of selectivity is the variable nature of bioassay results. Not only can bioassay results vary greatly depending on the small differences in methods, but bioassay results may differ from experiment to experiment where no variables were intended (Hollingworth 1975). Savin et al. (1977) have documented variability over the generations of a colony of insects. Based on this work, generalizations from toxicology data even within a single species or strain is questioned.

If broad generalizations concerning a chemical's IPM selectivity are not available or reliable, how can we hope to find chemicals which are selective? One way to find selective chemicals is to consider only specific agricultural systems where knowledge of the importance of specific target and non-target species is known. A number of IPM researchers have used this narrower approach for specific IPM problems (for example, Croft 1976, Hoy and Roush 1978, Granett et al. 1976, Jones et al. 1983). It should be noted that sole reliance on laboratory data is inappropriate because of the many field variables not assessed in the laboratory. It may be that the generalizations concerning agriculturally relevant selectivity will become obvious if enough specific systems are studied.

3.2 *Mechanisms of benzoylphenyl urea selectivity*

Differential toxicity of a chemical to two species can be based on a number of factors. Mechanisms of benzoylphenyl urea selectivity appear to be based on toxicological, ecological or behavioral differences between species.

There are several toxicological mechanisms of selectivity. First, a chemical may be inherently more active to one species than to another because of the site of action of the chemical. For benzoylphenyl ureas, selectivity between vertebrates and arthropods is based on this mechanism; since vertebrates do not synthesize chitin, and the action of these materials apparently is restricted to chitin synthesis, these materials should be exceedingly safe to vertebrates. This reasoning is a simplification based on the symptoms caused by these materials. The enzymatic site of action of the benzoylphenyl ureas in not known: when researchers have tested activity of these materials on cell free preparations of chitin synthetase little activity is seen (Leighton et al. 1981, Cohen and Casida 1980, others). One explanation is that activity may be on the proteinase activation thought necessary for chitin synthetase activity (Leighton et al. 1981). In addition, the benzoylphenyl ureas have been shown to be cytostatic in certain systems (Meola and Mayer 1980). These actions suggest additional sites at which these chemicals function and these sites may induce toxicity to vertebrates as well as invertebrates. The fact that invertebrates are apparently most notably affected may not be true for all benzoylphenyl urea analogues, and at least implies lack of an absolute safety for vertebrates.

In addition to observations of cross-phylum selectivity, there are probably differences in the susceptibilities of the active site(s) within groups of insect species. Large variations in susceptibilities have been noted between insect species (Hammann and Sirrenberg 1980, Retnakaran et al. 1985, others). Much of this variation is probably due to differences in chemical absorption, transportation and metabolism (pharmacological phenomena).

Pharmacological differences can occur within groups of species. For benzoylphenyl ureas, differences in rates of absorption of the materials through the gut or cuticle (Ascher and Nemny 1976, Ascher and Eliyahu 1981, Granett et al. 1983) have been documented. It is likely that sizable differences in rates of absorption will be found for insect species within specific IPM systems. With the benzoylphenyl urea, diflubenzuron, most species appear to be considerably more susceptible to ingested material compared to material applied topically. As a result of this difference, Mulder and Gijswijt (1973) hypothesized that parasites, predators, and pollinators which do not ingest treated leaf surfaces will be found to be unaffected in the field. In field experiments, researchers have indeed found that many such organisms survive field treatments, but topical insensitivity has been confirmed with relatively few species and benzoylphenyl urea analogues (Mulder and Gijswijt 1973, Jones et al. 1983).

Another important pharmacological variable is metabolic rate difference within several species that may contribute to selectivity. Differences in routes and rates of metabolism with the benzoylphenyl ureas (Granett et al. 1980, Hammock and Quistad 1981, Ivie and Wright 1978, Pimprikar and Georghiou 1979) have been noted with various experimental animals. Work relating to such metabolic differences to species complexes within specific agricultural systems has not been done.

Also relating to insect-insect selectivity are the effects on reproduction. Some species are severely affected by ovicidal action of benzoylphenyl ureas and others are not. Conversely, some analogues seem to induce reproductive effects whereas others do not. At present we do not understand why the reproductive effects are observed in some species but not in others. The phenomenon of reduced reproduction could be important in a number of agricultural situations. The reproductive effects of diflubenzuron on the boll weevil is such a case. If we understood the relationship between reproductive effects and benzoylphenyl urea structures, as well as the physiological differences between species which did and did not exhibit the reproductive effects, we might utilize reproductive effects more practically in agricultural systems.

A non-pharmacological mechanism for selectivity relates to the response of the animal to intoxication. Some species might withstand poisoning of the active site to a greater extent than other species. Retnakaran et al. (1980) presented evidence that spruce budworms were more sensitive to destruction of their ability to synthesize chitin than were tussock moths.

Some species are not affected greatly by benzoylphenyl ureas, but the mechanism of the selectivity is not known. Some insects such as honeybees, are apparently not greatly affected in field studies with benzoylphenyl ureas (Barker and Taber 1977, Schroeder al. 1980). Although poor penetration of the cuticle is probably important, even when these arthropods are fed the material, they tend not to be greatly affected. In the case of bees, the workers digest food fed to brood; the workers' metabolism may degrade the benzoylphenyl ureas and thus protect the brood from intoxication.

For the most part, the above mechanisms have not been explored in specific agricultural systems. Although the mechanisms have been demonstrated in the laboratory with the convenient species scientists tend to use for such work, and the field trials have demonstrated that selectivity does indeed exist, researchers have generally avoided the difficult task of correlating the field and laboratory work. Identification of the mechanisms to the field observations of selectivity with the same species has not been done.

Selectivity is not solely limited to chemical-organism interactions; the interaction between organisms and the ecosystem may cause differential toxicity between target and non-target species. An organism which does not come into contact with residues of the toxicant obviously will not be poisoned. For example, mining or boring insects will generally not be susceptible to non-systemic residues of insecticides on the plant surface. Non-systemic insecticides distributed on the soil surface will generally not be toxic to aerial insect species because of lack of contact with the chemical. Insecticides distributed in bait formulations will generally not effect organisms not attracted to the bait. Such ecological or behavioral mechanisms of selectivity can be just as effective as physiologically based selectivity in IPM programs for protecting beneficial species.

With the benzoylphenyl ureas, ecological mechanisms can be important. The benzoylphenyl ureas are largely non-systemic in plants and hence will not be effective against insects while they bore into plant materials. Boring insects such as the codling moth, however, are susceptible as early first instars as they enter the fruit (Hoying and Riedl 1980). Soil treatments with benzoylphenyl ureas have seldom been tried because of chemical instability in all but sterilized soils (Verloop and Ferrell 1977).

Such instability may be considered an important mechanism for the protection of non-pest soil borne arthropods.

Selectivity may also be based on temporal differences in insect prevalence or susceptibility. Conventional insecticides have been used as winter dormant sprays and as night sprays (to protect bees) to achieve temporal selectivity. Temporal selectivity is also frequently seen when materials with short residuals are used; such materials may be shown to be less detrimental to parasite populations because chances of exposure to lethal residues are reduced.

There are several examples of temporal selectivity with benzoylphenyl ureas. With ladybeetles in temporary row crop IPM systems (Jones et al. 1983), laboratory work has shown that the adults were not killed by two benzoylphenyl ureas, although no data were collected on reproductive effects. It was suggested that selectivity based on the migratory behavior of the beetles would occur. Because the beetles would migrate out of the fields for diapause prior to reproducing, the reproductive effects would be irrelevant to the celery program being studied. An additional example of temporal selectivity has been shown with the gypsy moth parasite, *Apanteles melanoscelus*. This wasp was apparently protected from the effects of benzoylphenyl urea treatments by its living host when treatments were made before the wasp reached the mature larval stage. After the third gypsy moth instar toxic effects of the benzoylphenyl urea treatment were seen on the wasp. Early field treatments, therefore, were critical to the parasites survival during a forest treatment (Granett et al. 1976).

3.3 *Use strategies for benzoylphenyl ureas*

Knowledge of the instances and mechanisms of selectivity may be useful for developing optimal usage strategies for benzoylphenyl ureas in IPM systems. Understanding potential selectivity is important early in the development of a candidate insecticide since such information will help determine which agricultural ecosystems are appropriate for utilization of the new material. With benzoylphenyl ureas, selectivity in agricultural IPM situations appears to be based on (1) action in arthropods but not in vertebrates, (2) action primarily in immature insects as opposed to mature insects, and (3) general nature of the materials such that they are more toxic by ingestion rather than topically. As discussed previously, these generalizations should not be considered to be absolute and therefore should be subject to testing within each system.

With regard to the first mechanism of selectivity, the strategy for use of these materials may include situations where contact with vertebrates is common. In fact, use of the materials as "feed-throughs" for controlling flies in domestic animal wastes has been extensively tested (Wright 1974,

Wright and Smalley 1976). The fact that these applications do not yet have U. S. Environmental Protection Agency (E. P. A.) registrations emphasizes the fact that safety is not absolute. Even though the site of action relates primarily to chitin synthesis, vertebrate problems with other target sites as well as general chronic toxicity problems (such as carcinogenicity tests which were not conclusive) have slowed if not prevented diflubenzuron from being registered in the United States for use in relation to human foods and close to human habitations. Apparently, only in the past year has the E. P. A. become satisfied with the safety of diflubenzuron when used around humans and human foods.

The second mechanism of selectivity, selectivity based on developmental stage and slow action, suggests that benzoylphenyl ureas would be most useful against a pest complex that does damage in immature stages and has several generations a year. Preferably the crop that is treated will be able to withstand some damage since the benzoylphenyl ureas tend to be slower acting than the conventional nerve poisons. In forestry, the use of benzoylphenyl ureas against gypsy moth or spruce budworm fits well into this strategy. These insects are pests solely as larvae and the trees can withstand considerable leaf or needle loss before they are severely damaged. Parasites and other insects in the forest situation are active primarily as adults. So, in some tests the selectivity is favorable (Granett et al. 1976, Buckner et al. 1975). However, the selectivity is not absolute as indicated by damage to some gyspy moth parasites (Madrid and Stewart 1981).

In agricultural situations the strategy of developmental stage and slow action also seems to be valid. Immature cotton and soybean pests (Ables et al. 1977, Keever et al. 1977, Turnipseed et al. 1974, Wilkinson et al. 1978) are affected but the mature parasites and pollinators for the most part are not. I must add here that the temporal mechanism of selectivity may not fully explain observed selectivity in the field; little work on selectivity has been done in relation to field experiments.

The third mechanism of selectivity that has been demonstrated is the slow topical penetration in contrast to more rapid penetration after ingestion. This mechanism suggests strong effects after residual sprays on leaf feeding insects but relatively less effects on sucking insects such as hemipterans and homopterans, or on parasites and predators which crawl over leaves but do not eat the plant surfaces containing the residues. Populations of hemipteran predators such as big-eyed bugs should not be directly harmed by benzoylphenyl ureas. This contention is very difficult to show in the field: treatments which kill the food of the predators will cause the predators themselves to disperse even if there is no direct toxicity to the predators. In field and laboratory experiments, the predictions based on this mechanism appear to be confirmed (Granett et al. 1976, Jones et al. 1983, Ables et al. 1977). However, slow cuticular penetration appears to be both species and chemical dependent and may not pertain to all situations.

3.4 *Needed research*

The foregoing discussion demonstrates the need for more critical work on benzoylphenyl urea selectivity. The following is a summary of some important areas where further research is needed. First, several mechanisms, specifically selectivity relating to site of action in insects and selectivity based on differential metabolism, although shown to be significant in strain or species pairs, have not been shown to be of importance in species complexes within agricultural systems. A better understanding of the important metabolic paths for benzoylphenyl ureas in specific insects is necessary before such work can proceed. With a better understanding of metabolism, perhaps selectophores can be chosen which, when installed on benzoylphenyl urea analogues, will enhance existing pharmacological differences between species.

A second thrust of research might concentrate on the structure-activity relationships (S. A. R.) in relation to benzoylphenyl urea penetration and reproductive effects. At this point published S. A. R. studies are limited to those which use mortality of the insect as a criterion of effect. Since considerable efficacy as well as selectivity can be related to reproductive effects (McLaughlin 1977, Wright and Spates 1976, Maas et al. 1981, Retnakaran et al. 1985) studies using other criteria seem critical in choosing more efficacious compounds as well as more selective compounds.

Third, the bulk of the selectivity research has either been with laboratory bioassays (the reductionist approach, See Retnakaran et al. 1985) or with field tests (the holistic approach). Each of these research directions is valuable, but more research must be done in which the two approaches are combined. If selectivity is observed in field trials, laboratory tests must be done to explain the mechanism(s) involved so that the phenomenon can be generated in other situations. Without a combination of field and laboratory data this science will be restricted to observations rather than predictions and technology for insect control will not be able to include the most logical field manipulations.

From the foregoing discussion I feel justified in saying that selectivity in IPM systems is potentially a great advantage to use of benzoylphenyl ureas. In a number of situations, selectivity has been documented, while in others, it is merely hypothesized to exist. Based on the selectivity criterion alone, benzoylphenyl ureas should be well suited to use in agricultural IPM. However, the agricultural constraints and industrial constraints in some cases diametrically oppose chemical characteristics which favor the use of these chemicals in IPM.

4 AGRICULTURAL CONSTRAINTS

Insecticide use is severely limited by constraints imposed by agricultural practices and farmer biases. Considerations of these constraints in the development of new insecticides is therefore necessary. Agricultural constraints fall into at least three main categories: cost-effectiveness of the chemical usage, consistent effect of the chemical, and favorable characteristics of use.

4.1 Cost-Effectiveness

Cost-effectiveness refers to the relative size of the monetary and non-monetary return as compared with both the direct monetary costs and indirect costs from an application of pesticide. For a farmer to use an insect control tactic, the total cost of the procedure should be substantially less than the value of the benefits derived. In an IPM program, costs of gathering information for making the treat-don't treat decisions can be considerable and must be considered in the treatment costs. In addition, costs of the chemical, applications, and detrimental effects of the chemicals (i.e., crop damage by secondary pest outbreaks and costs of treatments to control such outbreaks, costs of a chemical induced delay in crop maturity and reduced yield, costs of environmental contamination, and cost of insecticide resistance) are some of the direct and indirect costs.

The economics of growing, harvesting and marketing a crop vary tremendously. A crop with a high profit margin can withstand considerably more costs for pest control (and proportionally less insect damage) than a crop of lower value. The economics of a crop, and hence the economic thresholds, can change rapidly even within season as a result of market prices, weather conditions in all the growing regions for that crop, and the predicted size and quality of the crop. Minor changes in application technology, grower practices, and the pest complex can also alter the cost-benefit balance. In development of a chemical, therefore, consideration of specific uses in time, place, and crop is critical to success. The more specific the consideration of economics is, the greater is the chance of success of a candidate insecticide.

With regard to the benzoylphenyl ureas, some general statements can be made. Costs of producing benzoylphenyl ureas tend to be greater than the costs of producing conventional carbamate and phosphate insecticides. Manufacturers in general seem to set a limit on the amount of benzoylphenyl urea which can economically (competetively) be sold; this limit apparently is 2–4 oz per acre (or 140–281 g/hectare) given the production costs and other costs of these materials. This limit means that, if more than that amount is needed per treatment, the farmer would probably find alternative insecticides cheaper to use. Registrations of diflubenzuron have been in forest and field crop situations with applications at or below these rates.

The indirect benefits from use of benzoylphenyl ureas, specifically lack of secondary pest outbreaks, has not been shown to greatly alter marketing decisions. That is, up till now, farmers apparently have not put a premium on selectivity of benzoylphenyl ureas. However, if diflubenzuron is registered for codling moths in a crop such as walnuts, prevention of outbreaks of mites and walnut aphids may make use of this chemical more economical than the currently used organophosphates which, when sprayed for codling moths, may exacerbate the mite or aphid problems (Riedl et al. 1979).

4.2 *Consistency*

Consistency in prevention of crop damage is critical to the farmer and is therefore a basic constraint imposed by agriculture. Farmers use insecticides primarily because these chemicals reduce the risk of losses from pest damage and crop failure. If use of an insecticide does not reduce risks or inconsistently reduces risks, farmers will see no advantage of that method over use of no control method.

With chemicals, failure can result from a number of factors relating to the chemical and formulation. With regard to the benzoylphenyl ureas, these constraints are relatively unimportant. As with the use of most conventional nerve-poison insecticides, use of benzoylphenyl ureas produces consistent results from one time to the next. Residual activity is relatively long (weeks to months on foliage) so that inconsistencies due to degredation of residues are uncommon. Although there have been some formulation problems with benzoylphenyl ureas, these problems have related to lower than optimal toxicity as opposed to inconsistent activity.

Resistance among pest species is increasingly becoming the major flaw in the farmers' chemical arsenal. Resistance to benzoylphenyl ureas has not yet been shown to be a major field problem because of the relatively sparce use. One of the major advantages of benzoylphenyl ureas is that the mode of resistance is likely to be metabolic (Pimprikar and Georghiou 1979) a mechanism which is likely not to have cross resistance from the synthetic pyrethroids.

4.3 *Ease of use*

The last broad category of agricultural constraints concerns characteristics of use and relate to chemical timing requirements, speed of action, and spectrum of species susceptible to poisoning.

Timing of insecticide applications is frequently geared to the most sensitive or available stage of the pest. With each pest species, the stage of greatest susceptibility may be different. Nerve poisons and metabolic poisons tend to have very broad temporal windows of species sensitivity in contrast to the juvenile hormone analogues where the window of activity is restricted (in the worst cases) to a few days during the last instar. The breadth of a chemical's activity window influences the farmer's treatment strategy. Considerably more survey work and care in applications is needed by the farmer for use of the narrow-windowed compounds. Since farmers as a rule wish to spend as little time as is possible worrying about their pest problems, and materials with narrow windows of activity need more attention, farmers will tend to avoid such materials when given the choice.

With benzoylphenyl ureas, timing criteria are somewhat more restrictive than with conventional nerve poison insecticides. Immature stages are most susceptible because these are the stages that produce chitin. Adults are generally not killed because they are primarily past the stages that produce chitin. Hence, timing of treatments has to coincide with the immature stages. Where the benzoylphenyl ureas are used to decrease reproduction or affect insects before they mine into the plant, timing is more critical; timing of the conventional insecticides, however, would be almost equally critical.

Speed of action of an insecticide is not a problem with most conventional nerve poisons, because they frequently kill the insects within a few hours of contact. With benzoylphenyl ureas, symptoms of larval mortality are not seen until the insects begin their next molt. With insects that have infrequent molts, a considerable amount of time can pass before symptoms are seen. With agricultural pests, where crop damage can be extremely rapid once an economic threshold of pests is reached, a delay in insect mortality (or at least cessation of feeding) may be unacceptable to farmers. In cases where the delay of symptoms is not economically critical, farmers may still object to the slow development of symptoms since they are accustomed to seeing insects literally fall out of the foliage after a conventional insecticide treatment. Education about the slow response of the benzoylphenyl ureas may be all that is needed to address this problem.

The narrow spectrum of species poisoned by the benzoylphenyl ureas is probably the most serious agricultural constraint with these materials. Although a narrowed spectrum of activity is important to IPM, the opposite tends to be true for use of chemicals by farmers. To be simplisitic, if a farmer can use a single compound for several pests at the same time, he will prefer such a material since less planning is needed and multiple treatments may be minimized. In situations where sucking types of insects are important components of the pest complex along with leaf chewing types, the farmer would have to use a material in addition to the benzoylphenyl ureas. The logical alternative for the farmer would be to use only a broad spectrum poison rather than the narrower spectrum benzoylphenyl urea.

This constraint causes a severe dilemma for the development of any insecticide specifically for IPM programs. If materials are selective enough for IPM, then they may be too selective for general uses by farmers. Without general uses, the chemical industry, as will be discussed below, may be unable to develop them. Although farmers do not automatically restrict their insecticide use to broad spectrum materials, such materials tend to have the greater flexibility in their eyes and therefore are favored.

4.4 *Research and other work needed*

From an agricultural point of view, research is needed to determine potential uses for the narrow spectrum insecticides such as the benzoylphenyl ureas. Although such research may not be scientifically exciting, such development-type of research is required if we are to see a greater use of the benzoylphenyl ureas. We have already had a number of situations where chemical companies have halted development of benzoylphenyl ureas because agricultural situations demanding the unique selectivity characteristics of these insecticides were not apparent.

Where agricultural uses are scientifically sound, farmers must be educated about the unique benefits and problems relating to the benzoylphenyl ureas. Although this effort is an educational (extension) problem rather than a research problem, it is extremely important to the success of applied research with the benzoylphenyl ureas as well as other innovative pest control measures.

5 INDUSTRIAL CONSTRAINTS

In western countries where most insecticides are developed, the companies involved have strong profit motives. As a result of this financial orientation, chemical companies have critical requirements and constraints upon the types of materials that they can develop. The major categories of constraints are (1) the availability of markets suitable for making profits with the products, (2) the likelihood of meeting the U. S. Environmental Protection Agency (E. P. A.) requirements and restrictions on registrations for pesticide uses, (3) patent protection long enough to recoup expenses of development, and (4) other problems related to selling a product profitably.

5.1 *Market definition*

In order for a company to develop a compound as an insecticide, a potential market must be available that will yield sufficient profits to pay for the high cost of research, development, production, and sales-support. Although it is obviously not their job to identify commercial markets, academic scientists can aid in such work. The scientist has a direct interest in the identification of markets: if no markets are found the discoveries will be lost to society. In the past, scientists have shied away from these problems and as a result have found, in some cases, that the compounds on which they have worked have not been marketed. An example of such a situation was seen with the juvenile hormone analogues. An immense scientific effort went into the laboratory and field testing of the many juvenile hormone analogues available during the 1960s and 1970s. Although scientists were extremely excited about the applied potential of these materials, only a few were ever registered as insect control agents. The relative failure of commercial development of this class of compounds appeared to relate to the industrial constraints. If we wish to avoid such dead-ends for the benzoylphenyl ureas possibly we, as scientists, should address these constraints in our ongoing research and discussions concerning these materials.

As an outside observer of the chemical industry, I can identify two strategies used by chemical companies for defining a market for a candidate insecticide. The first strategy is to find a single or a few existing major markets for a potential product which, if penetrated suitably, would realize profits that would totally pay for the enormous product-development costs. The second possible strategy is to find a number of small markets for the product, which if penetrated suitably would, in composite, be profitable enough to pay for the developmental costs.

There are several major markets which, on first inspection, seem to be favourable for benzoylphenyl urea use. The forestry market with gypsy moth, spruce budworm or other major lepidopteran pests has been the most obvious, based on the amount of research that has been done, and the initial registration of diflubenzuron for gypsy moth control.

This market is not without its problems for company planners. The main problem is that the markets are not consistent from year to year. First, amounts of insecticides needed will vary depending on population cycles of the insects. Because applications in forests are frequently sponsored by government agencies, the extent of the areas sprayed will be inversely proportional to the environmental-protectionist pressures on those government agencies, a factor well outside the realm of corporate control. Also, because relatively few governmental units are involved in treatments, if a single sales bid is lost, this loss may be disastrous for total product sales. Such inconsistencies in markets play havoc with sales projections and will likely be avoided by company planners.

Within the agricultural realm, it is difficult to find large markets where leaf chewing insects (i.e. those susceptible to the benzoylphenyl ureas) are the only pests. Although a number of large markets seem to have some attributes favorable to benzoylphenyl urea uses, they all seem to have some serious drawbacks. The cotton markets are probably the closest to ideal. Various caterpillars are found; however, some of them are bud-burrowers and therefore have less susceptibility to these materials. Diflubenzuron is used extensively on cotton in Egypt. There, a conventional nerve poison is used simultaneously with diflubenzuron, and the selectivity benefits as well as company profits are somewhat diluted. Soybeans are hosts of a complex of leaf feeding insects which may constitute a suitably large market. Another large market is codling moth on fruit trees, although the pest complex is quite large and benzoylphenyl ureas would not be effective against all of the pests.

Diflubenzuron has a number of registrations in a variety of small-crop markets in Europe, with a similar complex of crop registrations being sought in the United States. Existing or pending registrations include codling moths, vegetable lepidopterans, soybean lepidopterans, mushroom fly pests, cattle fly pests and others (Maas et al. 1981). Although this strategy appears to have been somewhat successful for diflubenzuron, to date companies developing other candidate benzoylphenyl urea analogues do not appear to be enthusiastically following suit, indicating possibly that the profitability is just not sufficient.

Because of the complexity of agricultural systems, identifying markets is not an easy process. After an initial listing of the crops and locations, local specialists on the crops must be consulted to determine in detail where knowledge is available and where additional research is needed. Such planning must be done in a systematic way since with such complex problems obscure facts tend to have a way of disrupting large programs.

5.2 E. P. A. toxicology standards

Surmounting the E. P. A. hurdles has been a major stumbling block for diflubenzuron and appears to have slowed attainment of registrations on food crops in the United States by at least 6 years. A portion of the problem has been described well by Retnakaran (1982) who lamented the fact that the E. P. A. standards apparently favor the development of compounds which have moderately high acute toxicity for vertebrates. Such compounds when tested for chronic toxicity problems are tested at relatively low doses in comparison with compounds which have lower acute vertebrate toxicity which are tested at higher continual doses. The high chronic doses for toxicity tests are often above realistic environmental exposure rates and may result in apparent problems which would never exist in the real world because the environmental exposure would never reach the levels tested. Benzoylphenyl ureas as materials with low acute vertebrate toxicity are not favored by such testing procedures because of their selectivity, a favorable characteristic by all other criteria.

5.3 *Patents and Profits*

A patent is meant to protect a company by giving it a certain monopoly over a product that the company spends time and money developing. The patent lasts for 17 years and then the monopoly over the product may legally dissipate. Djerassi et al. (1974) has described a basic dilemma relating to patents and development of chemicals for pest control purposes. If a company takes 10 years to develop a product and a market for that product, then there are only 7 years left on the patent for the company to recoup its developmental expenses. Unless the markets for insecticides are unusually large or profitable, the patent life may be too short for recouping expenses. This dilemma has been made worse by the delays, sometimes irrational from the chemical companies' viewpoint, caused by the E. P. A. in issuing registrations.

Benzoylphenyl ureas have been especially susceptible to delay problems because of the fact that they act in a different manner biologically from conventional insecticides. This difference has slowed the registration process because of the need to set precedents, a particularly difficult thing for most of us including the responsible bureaucrats.

5.4 *Promotion and sales problems*

Companies, as with the bureaucrats cited above and people in general, deal with new ideas more slowly than with the tried and true. From the point of view of the chemical company, IPM compatible insecticides are something new. How do you deal with a narrow spectrum insecticide if all your other products are broad spectrum materials? Education of sales personnel about the new properties, benefits, and difficulties of use must be emphasized. But even before the education of the sales people comes the problem of educating the company researchers and managers to appreciate the different properties and solve the new problems which will come about.

6 WHAT DOES THE FUTURE HOLD?

Will researchers be able to make a predictive science out of insecticide selectivity as a support for IPM in agricultural systems? Will farmers be able to make use of narrow spectrum insecticides and operate within IPM guidelines? Will it be possible for companies, which have benzoylphenyl ureas under development now, to define suitable markets, and solve registration, patent and promotion problems? The answers to these questions are critical to the future use of the benzoylphenyl ureas. If the answers are no, then the benzoylphenyl ureas may go the way of the juvenile hormone analogues: they will become no more than minor use insecticides.

If selective insecticides in general and the benzoylphenyl ureas in particular are to become useful chemical control agents with wide applications, certain attitudes and directions must change among IPM researchers, with farmers, and with chemical industry planners. The following ideas are not meant to be a list of requirements or a formal proposal for action; they are meant as a point of departure for discussions on the direction in which we want to see pest control, particularly where chemicals are used, move in the future.

First, if IPM is going to replace the chemical control "paradigm" in agriculture, IPM gurus must recognize that chemicals are important and that chemicals used in IPM must have different characteristics from chemicals which are used alone. Researchers must recognize that, when we make suggestions to growers on how to alter practices, the alterations must be feasible and plausible with respect to time available for the procedure, cost of the procedure, and availability of materials needed. With respect to the importance of chemicals within agriculture, IPM researchers must outgrow their fixation on total elimination of chemical use — instead, they must concentrate on rationalizing that use. Such rationalization would include minimal use to offset development of resistance and use of materials resulting in minimal disruption of the agricultural ecosystem, the human ecosystem and wildlife. Such conventional procedures as sampling and use of economic thresholds, use of selective insecticides, and development of compatible alternative strategies for pest control meet the criteria I have mentioned.

The characteristics of the insecticides which are of prime importance, of course, relate to the compatibility with beneficial arthropods and vertebrates. In insecticide screening work, selectivity within specific agricultural applications must be considered as important as the efficacy against the pest species. Selectivity screening must be done, and indicator species for specific agricultural systems must be used. In addition, as discussed in research suggestions above, the mechanistic basis for selectivity must be understood for manipulative and predictive aspects of the science to become feasible. These suggestions may be generally applied to all narrow spectrum insecticides and are therefore relevant to the benzoylphenyl ureas. If benzoylphenyl ureas are to become widely used chemical control agents, certain information must be gathered and protocols developed. We have to determine what sort of selectivity the benzoylphenyl ureas have in specific agricultural situations and as much as possible about the mechanisms for selectivity. Without this additional information, we cannot hope to use their insect-insect selectivity effectively.

The community of farmers, the ultimate users of IPM and insecticides, must also undergo certain attitude and direction changes if IPM in general and benzoylphenyl ureas in particular are going to succeed. Of prime importance is the understanding by the farmers that the days of insect control by chemicals alone are over, or, in crops where chemicals are still the control tactic, will be over soon. Nature's answer to sole use of insecticides is resistance of pests to insecticides. Since registration of insecticides to replace materials to which resistance has developed has not kept pace with the development of resistance, agricultural systems which depend solely on insecticides will find themselves without effective insecticides.

Once the logical imperative of IPM is understood, implementation of sound and feasible IPM practices should be easier for the farmers to accept. The obvious need of the agricultural community is for better education concerning the importance of integrated pest control tactics and development of sound specific programs.

The third component, the chemical industry, has the greatest impact on the future of selective insecticides such as the benzoylphenyl ureas. The companies which have as yet undeveloped analogues may choose to let these selective insecticides be relegated to minor uses, or alternatively, they may take the necessary steps to solve the problems associated with development of IPM markets.

Of greatest importance, potential use patterns for these materials need to be identified. It must be recognized that each analogue will have a different optimal use pattern based on its unique species spectrum of activity, chemical characteristics, and the capabilities of the company developing the compound. If companies decide on development they face a major decision at this point: they can search for markets in the conventional way by looking for existing insecticide uses for which their own product might competitively be replaced, or they may use an alternative market-defining strategy. For the benzoylphenyl ureas, the conventional strategy has not been highly fruitful to date. Since their discovery 12 years ago, only two benzoylphenyl ureas have been registered for use. Since 1972 a number of compounds has begun development but their development has been stalled or they have been discarded. It appears as if companies will have to be much more aggressive with this strategy if it is to succeed, or try an alternative strategy. Several companies are working on being more aggressive: "second-generation" benzoylphenyl ureas (i.e. CGA-119783 {IKI-7899}, XRD-473, and CME-134) being developed by Ciba-Geigy, Dow and Celamerk appear to be an order of magnitude more active than previous benzoylphenyl ureas. In addition, materials with acaricidal activity are being tested. Increasing toxicity of juvenile hormone analogues to insects did not make these materials more marketable, so alternative strategies may be necessary.

Developing an alternative strategy requires some creativity and a bit of luck. One alternative strategy would be to develop pest management programs for an entire agricultural system, including a sampling package, a mechanism to pinpoint an economic threshold, and insecticide treatment procedures. If a company (or academic unit) were to develop the IPM program from scratch, it would be feasible to build in use of the selective insecticides. Instead of selling a chemical, the private sector could be called upon to sell pest management and possibly even risk management. As part of the program, resistance management would be integral since loss of the selective insecticides would mean loss of the whole package. The chemical company would by this approach be out of the chemical business, but be involved with the pest management or pest risk management business.

Another innovative approach would be to develop the chemicals specifically as control tools rather than as single insecticides. For example, a chemical (with suitable selectophores) or mixture of chemicals could be developed as a product which would be toxic to lepidopterans, coleopterans but not hemipterans, homopterans and hymenopterans. Another chemical or mixture might be developed to be toxic to hemipterans and homopterans but not hymenopterans or coleopterans, and so on. Such selectivity could then be fit into the appropriate markets once the pest and beneficial complex of that market were defined over the production season. In order for such an approach to be feasible the science of selectivity would have to be greatly improved. However, even if the mechanisms of selectivity were not understood, I believe with appropriate directed screening such selectivity could be discovered. The only constraint is on the *will* (translated as dollars) and the imagination.

7 CONCLUSION

The benzoylhphenyl ureas as a class of insecticides stand at a crossroads. Their insecticidal activity has been well documented and their selectivity, although not firmly established or understood, is believed to be present by most researchers and agriculturalists. If the three groups in society which have an interest in such materials, the academically oriented researchers, farmers, and the chemical industry, see fit to cooperate to further IPM goals, and to deal with these materials in innovative and aggressive ways, the benzoylphenyl ureas can be developed and be profitable to each of the three groups mentioned. If these actions are not taken, the benzoylphenyl ureas will fail to meet their potential in insect pest management.

8 ACKNOWLEDGEMENTS

I wish to thank Dr. J. L. Robertson, Dr. T. J. Dennehy, Dr. L. T. Wilson, and Mr. R. Adamchak for critically reviewing the manuscript.

9 REFERENCES

Ables, J. R., Jones, S. L., and Bee, M. J. 1977. Effect beneficial arthropods associated with cotton. S. W. Ent. 2: 66–72.

Ascher, K. R. S. and Eliyahu, M. 1981. The residual contact toxicity of BAY SIR 8514 to *Spodoptera littoralis* larvae. Phytoparasitica 9: 133–138.

Ascher, K. R. S. and Nemny, N. E. 1976. Contact activity of diflubenzuron against *Spodoptera littoralis* larvae. Pestic. Sci. 7: 447–452.

Barker. R. J. and Taber, S., III 1977. Effects of diflubenzuron fed to caged honey bees. Environ. Ent. 6: 167–168.

Buckner, C. H., McLeod, B. B., and Kingsbury, P. D. 1975. The effect of an experimental application of Dimilin upon selected forest fauna. Chemical Control Research Institute, Ottawa, Canada, Rept. CC-X-97.

Cohen, E. and Casida, J. E. 1980. Inhibition of *Tribolium* gut synthetase. Pestic. Biochem. Physiol. 13: 129–136.

Croft, B. A. 1976. Establishing insecticide-resistant phytoseiid mite predators in deciduous tree fruit orchards. Entomophaga 21: 383–399.

Dennehy, T. J. Granett, J., and Leigh, T. F. 1983. The relevance of slide-dip and residual bioassay comparisons to detection of resistance in spider mites. J. Econ. Ent. 76: 1225–1230.

Djerassi, C., Shih-Coleman, C., and Diekman, J. 1974. Insect control of the future: Operational and policy aspects. Science 186: 596–607.

Granett, J., Dunbar, D. M., and Weseloh, R. M. 1976. Gypsy moth control with Dimilin sprays timed to minimize effects on the parasite *Apanteles melanoscelus*. J. Econ. Ent. 69: 403–404.

Granett, J., Bisabri-Ershadi, B., and Hejazi, M. J. 1983. Some parameters of benzoylphenyl urea toxicity to beet armyworms (Lepidoptera: Noctuidae). J. Econ. Ent. 76: 399–402.

Granett, J., Robertson, J., and Retnakaran, A. 1980. Metabolic basis of differential susceptibility of two forest lepidopterans to diflubenzuron. Ent. Exp. & Appl. 28: 295–300.

Hammann, I. and Sirrenberg, W. 1980. Laboratory evaluation of SIR 8514, a new chitin synthesis inhibitor of the benzoylated urea class. Pflanzenschutz-Nachrichten Bayer 33: 1–34.

Hammock, B. D. and Quistad, G. B. 1981. Metabolism and mode of action of juvenile hormone, juvenoids, and other insect growth regulators. in Progress in Pesticide Biochemistry, edited by D. H. Hutson and T. R. Roberts, Vol. 1, pages 1–83, John Wiley and Sons, New York.

Hollingworth, R. M. 1975. The biochemical and physiological basis of selective toxicity. in Insecticide Biochemistry and Physiology, edited by C. F. Wilkinson, pages 431–506, Plenum Press, New York.

Hoy, M. A. and Roush, R. T. 1978. Spider mite predator tested for pesticide resistance on Cal. Ag. 33: 11–13.

Hoying, S. A. and Riedl, H. 1980. Susceptibility of the codling moth to diflubenzuron. J. Econ. Ent. 73: 556–560.

Ivie, G. W. and Wright, J. E. 1978. Fate of diflubenzuron in the stable fly and house fly. J. Agric. Food Chem. 26: 90–94.

Jones, D., Snyder, M., and Granett, J. 1983. Can chemical control be integrated with biological control for *Trichoplusia ni* in celery. Ent. Exp. & Appl. 33: 290–296.

Keever, D. W., Bradley, J. R., Jr., and Ganyard, M. C. 1977. Effects of diflubenzuron (Dimilin) on selected beneficial arthropods in cotton fields. Environ. Ent. 6: 732–736.

302

Leighton, T., Marks, E., and Leighton, F. 1981. Pesticides: Insecticides and fungicides are chitin synthesis inhibitors. Science 213: 905–907.

Maas, W., Van Hes, R., Grosscurt, A. C., and Deul, D. H. 1981. Benzoylphenyl urea insecticides. in Chemie der Pflanzenschutz und Schadlingsbekampfungsmittel, Edited by R. Wegler, Vol 6 pages 423–470, Springer-Verlag, Heidelberg.

Madrid, F. J. and Stewart, R. K. 1981. Impact of diflubenzuron spray on gypsy moth parasitoids in the field. J. Econ. Ent. 74: 1–2.

McLaughlin, R. E. 1977. Dose-response of the boll weevil to topical formulations of TH-6040. J. Ga. Ent. Soc. 12: 369–373.

Meola, S. M. and Mayer, R. T. 1980. Inhibition of cellular proliferation of the imaginal epidermal cells by diflubenzuron in pupae of the stable fly (Stomoxys calcitrans L.). Science 207: 985–987.

Metcalf, R. L. 1975. Insecticides in pest management. in Introduction to Insect Pest Management. edited by R. L. Metcalf and W. H. Luckmann, Chapter 12, pages 235–274, John Wiley & Sons, New York.

Mulder, R. and Gijswijt, M. J. 1973. The laboratory evaluation of two promising new insecticides which interfere with cuticle deposition. Pestic. Sci. 4: 737–745.

Pimprikar, G. D. and Georghiou, G. P. 1979. Mechanisms of resistance to diflubenzuron in the housefly, Musca domestica L. Pestic. Biochem. Physiol 12: 10–22.

Retnakaran, A. 1982. Do regulatory agencies unwittingly favor toxic pesticides? Bull. Ent. Soc. Amer. 28: 146.

Retnakaran, A., Granett, J., and Robertson, J. 1980. Possible physiological mechanisms for the differential susceptibility of two forest lepidoptera to diflubenzuron. J. Insect. Physiol. 26: 385–390.

Retnakaran, A., Granett, J., and Ennis T. 1985. Insect growth regulators. In: Comprehensive Insect Physiology, Biochemistry and Pharmacology, edited by G. A. Kerkut and L. I. Gilbert, Vol. 12, Chapter 14. Pergamon Press, New York.

Riedl, H., Barnes, M. M., and Davis, C. S. 1979. Walnut pest management: Historical perspective and present status. In: Pest Management Programs for Deciduous Tree Fruits and Nuts, edited by D. J. Boethel and R. D. Eikenbary, pages 15–80, Plenum Press, New York.

Savin, N. E., Robertson, J. L., and Russell,. R. M. 1977. A critical evaluation of bioassay in insecticide research: Likelihood ratio tests of dose-mortality regression. Bull. Ent. Soc. Amer. 23: 257–266.

Schroeder, W. J., Sutton, R. A., and Beavers, J. B. 1980. Diaprepes abbreviatus: Fate of diflubenzuron and effect on nontarget pests and beneficial species after application to citrus for weevil control. J. Econ. Ent. 73: 637–638.

Turnipseed, S. G., Heinrichs, E. A., DaSilva, R. F. P., and Todd, J. W. 1974. Response of soybean insects to foliar applications of a chitin synthesis inhibitor TH-6040. J. Econ. Ent. 67: 760–762.

Verloop, A. and Ferrell, C. D. 1977. Benzoylphenyl ureas – a new group of larvicides interfering with chitin deposition. In: A. C. S. Symposium Series, No. 37 Pesticide Chemistry in the 20th Century, Edited by J. R. Plimmer, pages 237–270.

Wilkinson, J. D., Biever, K. D., Ignoffo, C. M., Pons, W. J., Morrison, R. K., and Seay, R. S. 1978. Evaluation of diflubenzuron formulations on selected insect parasitoids and predators. J. Ga. Ent. Soc. 13: 277–236.

Wright, J. E. 1974. Insect growth regulators: Laboratory and field evaluations of Thompson-Hayward TH-6040 against the housefly and the stable fly. J. Econ. Ent. 67: 746–747.

Wright, J. E. and Smalley, H. E. 1976. House fly, Musca domestica L., insect growth regulator, Dimilin, 1976. Insect Acaric. Tests 1, 119.

Wright, J. E. and Spates, G. E. 1976. Reproductive inhibition activity of the insect growth regulatory TH-6040 against the stable fly and the housefly: Effects on hatchability. J. Econ. Ent. 69: 365–368.

Index